COMPUTATIONAL INTELLIGENCE AND FEATURE SELECTION

IEEE Press
445 Hoes Lane
Piscataway, NJ 08854

Books in the IEEE Press Series on Computational Intelligence

Introduction to Evolvable Hardware: A Practical Guide for Designing Self-Adaptive Systems
Garrison W. Greenwood and Andrew M. Tyrrell
2007 978-0471-71977-9

Evolutionary Computation: Toward a New Philosophy of Machine Intelligence, Third Edition
David B. Fogel
2006 978-0471-66951-7

Emergent Information Technologies and Enabling Policies for Counter-Terrorism
Edited by Robert L. Popp and John Yen
2006 978-0471-77615-4

Computationally Intelligent Hybrid Systems
Edited by Seppo J. Ovaska
2005 0-471-47668-4

Handbook of Learning and Appropriate Dynamic Programming
Edited by Jennie Si, Andrew G. Barto, Warren B. Powell, Donald Wunsch II
2004 0-471-66054-X

Computational Intelligence: The Experts Speak
Edited by David B. Fogel and Charles J. Robinson
2003 0-471-27454-2

Computational Intelligence in Bioinformatics
Edited by Gary B. Fogel, David W. Corne, Yi Pan
2008 978-0470-10526-9

COMPUTATIONAL INTELLIGENCE AND FEATURE SELECTION

Rough and Fuzzy Approaches

RICHARD JENSEN
QIANG SHEN
Aberystwyth University

IEEE Computational Intelligence Society, *Sponsor*

IEEE Press Series on Computational Intelligence
David B. Fogel, *Series Editor*

IEEE PRESS

WILEY

A John Wiley & Sons, Inc., Publication

Published by John Wiley & Sons, Inc., Hoboken, New Jersey.
Published simultaneously in Canada.

Library of Congress Cataloging-in-Publication Data is available.

ISBN: 978-0-470-22975-0

10 9 8 7 6 5 4 3 2 1

CONTENTS

PREFACE

The main purpose of this book is to provide both the background and fundamental ideas behind feature selection and computational intelligence with an emphasis on those techniques based on rough and fuzzy sets, including their hybridizations. For those readers with little familiarity with set theory, fuzzy set theory, rough set theory, or fuzzy-rough set theory, an introduction to these topics is provided. Feature selection (FS) refers to the problem of selecting those attributes that are most predictive of a given problem, which is encountered in many areas such as machine learning, pattern recognition, systems control, and signal processing. FS intends to preserve the meaning of selected attributes; this forms a sharp contrast with those approaches that reduce problem complexity by transforming the representational forms of the attributes.

Feature selection techniques have been applied to small- and medium-sized datasets in order to locate the most informative features for later use. Many FS methods have been developed, and this book provides a critical review of these methods, with particular emphasis on their current limitations. To help the understanding of the readership, the book systematically presents the leading methods reviewed in a consistent algorithmic framework. The book also details those computational intelligence based methods (e.g., fuzzy rule induction and swarm optimization) that either benefit from joint use with feature selection or help improve the selection mechanism.

From this background the book introduces the original approach to feature selection using conventional rough set theory, exploiting the rough set ideology in that only the supplied data and no other information is used. Based on demonstrated applications, the book reviews the main limitation of this approach in the sense that all data must be discrete. The book then proposes and develops a fundamental approach based on fuzzy-rough sets. It also presents optimizations, extensions, and further new developments of this approach whose underlying ideas are generally applicable to other FS mechanisms.

Real-world applications, with worked examples, are provided that illustrate the power and efficacy of the feature selection approaches covered in the book. In particular, the algorithms discussed have proved to be successful in handling tasks that involve datasets containing huge numbers of features (in the order of tens of thousands), which would be extremely difficult to process further. Such applications include Web content classification, complex systems monitoring, and algae population estimation. The book shows the success of these applications by evaluating the algorithms statistically with respect to the existing leading approaches to the reduction of problem complexity.

Finally, this book concludes with initial supplementary investigations to the associated areas of feature selection, including rule induction and clustering methods using hybridizations of fuzzy and rough set theories. This research opens up many new frontiers for the continued development of the core technologies introduced in the field of computational intelligence.

This book is primarily intended for senior undergraduates, postgraduates, researchers, and professional engineers. However, it offers a straightforward presentation of the underlying concepts that anyone with a nonspecialist background should be able to understand and apply.

Acknowledgments

Thanks to those who helped at various stages in the development of the ideas presented in this book, particularly: Colin Aitken, Stuart Aitken, Malcolm Beynon, Chris Cornelis, Alexios Chouchoulas, Michelle Galea, Knox Haggie, Joe Halliwell, Zhiheng Huang, Jeroen Keppens, Pawan Lingras, Javier Marin-Blazquez, Neil Mac Parthalain, Khairul Rasmani, Dave Robertson, Changjing Shang, Andrew Tuson, Xiangyang Wang, and Greg Zadora. Many thanks to the University of Edinburgh and Aberystwyth University where this research was undertaken and compiled.

Thanks must also go to those friends and family who have contributed in some part to this work; particularly Elaine Jensen, Changjing Shang, Yuan Shen, Sarah Sholl, Mike Gordon, Andrew Herrick, Iain Langlands, Tossapon Boongoen, Xin Fu, and Ruiqing Zhao.

The editors and staff at IEEE Press were extremely helpful. We particularly thank David Fogel and Steve Welch for their support, enthusiasm, and encouragement. Thanks also to the anonymous referees for their comments and suggestions that have enhanced the work presented here, and to Elsevier, Springer, and World Scientific for allowing the reuse of materials previously published in their journals. Additional thanks go to those authors whose research is included in this book, for their contributions to this interesting and ever-developing area.

Richard Jensen and Qiang Shen

Aberystwyth University
17th June 2008

CHAPTER 1

THE IMPORTANCE OF FEATURE SELECTION

1.1 KNOWLEDGE DISCOVERY

It is estimated that every 20 months or so the amount of information in the world doubles. In the same way, tools for use in the various knowledge fields (acquisition, storage, retrieval, maintenance, etc.) must develop to combat this growth. Knowledge is only valuable when it can be used efficiently and effectively; therefore knowledge management is increasingly being recognized as a key element in extracting its value. This is true both within the research, development, and application of computational intelligence and beyond.

Central to this issue is the knowledge discovery process, particularly knowledge discovery in databases (KDD) [10,90,97,314]. KDD is the nontrivial process of identifying valid, novel, potentially useful, and ultimately understandable patterns in data. Traditionally data was turned into knowledge by means of manual analysis and interpretation. For many applications manual probing of data is slow, costly, and highly subjective. Indeed, as data volumes grow dramatically, manual data analysis is becoming completely impractical in many domains. This motivates the need for efficient, automated knowledge discovery. The KDD process can be decomposed into the following steps, as illustrated in Figure 1.1:

- *Data selection.* A target dataset is selected or created. Several existing datasets may be joined together to obtain an appropriate example set.

Computational Intelligence and Feature Selection: Rough and Fuzzy Approaches, by Richard Jensen and Qiang Shen

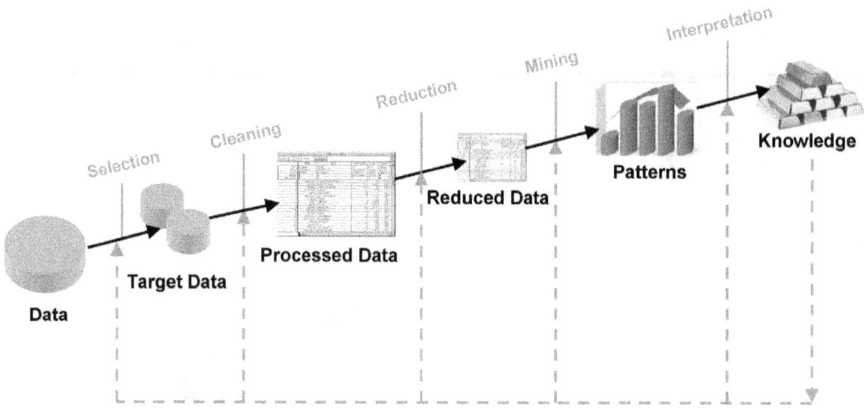

Figure 1.1 Knowledge discovery process (adapted from [97])

- *Data cleaning/preprocessing.* This phase includes, among other tasks, noise removal/reduction, missing value imputation, and attribute discretization. The goal is to improve the overall quality of any information that may be discovered.
- *Data reduction.* Most datasets will contain a certain amount of redundancy that will not aid knowledge discovery and may in fact mislead the process. The aim of this step is to find useful features to represent the data and remove nonrelevant features. Time is also saved during the data-mining step as a result.
- *Data mining.* A data-mining method (the extraction of hidden predictive information from large databases) is selected depending on the goals of the knowledge discovery task. The choice of algorithm used may be dependent on many factors, including the source of the dataset and the values it contains.
- *Interpretation/evaluation.* Once knowledge has been discovered, it is evaluated with respect to validity, usefulness, novelty, and simplicity. This may require repeating some of the previous steps.

The third step in the knowledge discovery process, namely data reduction, is often a source of significant data loss. It is this step that forms the focus of attention of this book. The high dimensionality of databases can be reduced using suitable techniques, depending on the requirements of the future KDD processes. These techniques fall into one of two categories: those that transform the underlying meaning of the data features and those that preserve the semantics. Feature selection (FS) methods belong to the latter category, where a smaller set of the original features is chosen based on a subset evaluation function. In knowledge discovery, feature selection methods are particularly desirable as these facilitate the interpretability of the resulting knowledge.

1.2 FEATURE SELECTION

There are often many features in KDD, and combinatorially large numbers of feature combinations, to select from. Note that the number of feature subset combinations with m features from a collection of N total features can be extremely large (with this number being $N!/[m!(N-m)!]$ mathematically). It might be expected that the inclusion of an increasing number of features would increase the likelihood of including enough information to distinguish between classes. Unfortunately, this is not true if the size of the training dataset does not also increase rapidly with each additional feature included. This is the so-called curse of dimensionality [26]. A high-dimensional dataset increases the chances that a data-mining algorithm will find spurious patterns that are not valid in general. Most techniques employ some degree of reduction in order to cope with large amounts of data, so an efficient and effective reduction method is required.

1.2.1 The Task

The task of feature selection is to select a subset of the original features present in a given dataset that provides most of the useful information. Hence, after selection has taken place, the dataset should still have most of the important information still present. In fact, good FS techniques should be able to detect and ignore noisy and misleading features. The result of this is that the dataset quality might even *increase* after selection.

There are two feature qualities that must be considered by FS methods: relevancy and redundancy. A feature is said to be relevant if it is predictive of the decision feature(s); otherwise, it is irrelevant. A feature is considered to be redundant if it is highly correlated with other features. An informative feature is one that is highly correlated with the decision concept(s) but is highly uncorrelated with other features (although low correlation does not mean absence of relationship). Similarly subsets of features should exhibit these properties of relevancy and nonredundancy if they are to be useful.

In [171] two notions of feature relevance, strong and weak relevance, were defined. If a feature is strongly relevant, this implies that it cannot be removed from the dataset without resulting in a loss of predictive accuracy. If it is weakly relevant, then the feature may sometimes contribute to accuracy, though this depends on which other features are considered. These definitions are independent of the specific learning algorithm used. However, this no guarantee that a relevant feature will be useful to such an algorithm.

It is quite possible for two features to be useless individually, and yet highly predictive if taken together. In FS terminology, they may be both redundant and irrelevant on their own, but their combination provides invaluable information. For example, in the exclusive-or problem, where the classes are not linearly separable, the two features on their own provide no information concerning this separability. It is also the case that they are uncorrelated with each other. However, when taken together, the two features are highly informative and can provide good class separation. Hence in FS the search is typically for high-quality feature subsets, and not merely a ranking of features.

1.2.2 The Benefits

There are several potential benefits of feature selection:

1. *Facilitating data visualization*. By reducing data to fewer dimensions, trends within the data can be more readily recognized. This can be very important where only a few features have an influence on data outcomes. Learning algorithms by themselves may not be able to distinguish these factors from the rest of the feature set, leading to the generation of overly complicated models. The interpretation of such models then becomes an unnecessarily tedious task.
2. *Reducing the measurement and storage requirements*. In domains where features correspond to particular measurements (e.g., in a water treatment plant [322]), fewer features are highly desirable due to the expense and time-costliness of taking these measurements. For domains where large datasets are encountered and manipulated (e.g., text categorization [162]), a reduction in data size is required to enable storage where space is an issue.
3. *Reducing training and utilization times*. With smaller datasets, the runtimes of learning algorithms can be significantly improved, both for training and classification phases. It can sometimes be the case that the computational complexity of learning algorithms even prohibits their application to large problems. This is remedied through FS, which can reduce the problem to a more manageable size.
4. *Improving prediction performance*. Classifier accuracy can be increased as a result of feature selection, through the removal of noisy or misleading features. Algorithms trained on a full set of features must be able to discern and ignore these attributes if they are to produce useful, accurate predictions for unseen data.

For those methods that extract knowledge from data (e.g., rule induction) the benefits of FS also include improving the readability of the discovered knowledge. When induction algorithms are applied to reduced data, the resulting rules are more compact. A good feature selection step will remove unnecessary attributes which can affect both rule comprehension and rule prediction performance.

1.3 ROUGH SETS

The use of rough set theory (RST) [261] to achieve data reduction is one approach that has proved successful. Over the past 20 years RST has become a topic of great interest to researchers and has been applied to many domains (e.g., classification [54,84,164], systems monitoring [322], clustering [131], and expert systems [354]; see *LNCS Transactions on Rough Sets* for more examples). This success is due in part to the following aspects of the theory:

- Only the facts hidden in data are analyzed.
- No additional information about the data is required such as thresholds or expert knowledge on a particular domain.
- It finds a minimal knowledge representation.

The work on RST offers an alternative, and formal, methodology that can be employed to reduce the dimensionality of datasets, as a preprocessing step to assist any chosen modeling method for learning from data. It helps select the most information-rich features in a dataset, without transforming the data, all the while attempting to minimize information loss during the selection process. Computationally, the approach is highly efficient, relying on simple set operations, which makes it suitable as a preprocessor for techniques that are much more complex. Unlike statistical correlation-reducing approaches [77], it requires no human input or intervention. Most importantly, it also retains the semantics of the data, which makes the resulting models more transparent to human scrutiny.

Combined with an automated intelligent modeler, say a fuzzy system or a neural network, the feature selection approach based on RST not only can retain the descriptive power of the learned models but also allow simpler system structures to reach the knowledge engineer and field operator. This helps enhance the interoperability and understandability of the resultant models and their reasoning.

As RST handles only one type of imperfection found in data, it is complementary to other concepts for the purpose, such as fuzzy set theory. The two fields may be considered analogous in the sense that both can tolerate inconsistency and uncertainty—the difference being the type of uncertainty and their approach to it. Fuzzy sets are concerned with vagueness; rough sets are concerned with indiscernibility. Many deep relationships have been established, and more so, most recent studies have concluded at this complementary nature of the two methodologies, especially in the context of granular computing. Therefore it is desirable to extend and hybridize the underlying concepts to deal with additional aspects of data imperfection. Such developments offer a high degree of flexibility and provide robust solutions and advanced tools for data analysis.

1.4 APPLICATIONS

As many systems in a variety of fields deal with datasets of large dimensionality, feature selection has found wide applicability. Some of the main areas of application are shown in Figure 1.2.

Feature selection algorithms are often applied to optimize the classification performance of image recognition systems [158,332]. This is motivated by a peaking phenomenon commonly observed when classifiers are trained with a limited

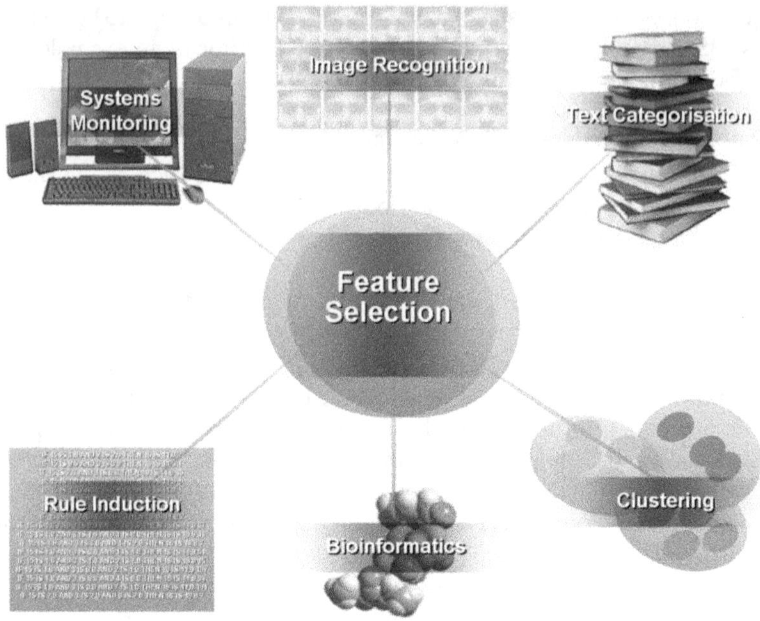

Figure 1.2 Typical feature selection application areas

set of training samples. If the number of features is increased, the classification rate of the classifier decreases after a peak. In melanoma diagnosis, for instance, the clinical accuracy of dermatologists in identifying malignant melanomas is only between 65% and 85% [124]. With the application of FS algorithms, automated skin tumor recognition systems can produce classification accuracies above 95%.

Structural and functional data from analysis of the human genome have increased many fold in recent years, presenting enormous opportunities and challenges for AI tasks. In particular, gene expression microarrays are a rapidly maturing technology that provide the opportunity to analyze the expression levels of thousands or tens of thousands of genes in a single experiment. A typical classification task is to distinguish between healthy and cancer patients based on their gene expression profile. Feature selectors are used to drastically reduce the size of these datasets, which would otherwise have been unsuitable for further processing [318,390,391]. Other applications within bioinformatics include QSAR [46], where the goal is to form hypotheses relating chemical features of molecules to their molecular activity, and splice site prediction [299], where junctions between coding and noncoding regions of DNA are detected.

The most common approach to developing expressive and human readable representations of knowledge is the use of *if-then* production rules. Yet real-life problem domains usually lack generic and systematic expert rules for mapping feature patterns onto their underlying classes. In order to speed up the rule induction process and reduce rule complexity, a selection step is required. This reduces the dimensionality of potentially very large feature sets while minimizing the loss of information needed for rule induction. It has an advantageous side effect in that it removes redundancy from the historical data. This also helps simplify the design and implementation of the actual pattern classifier itself, by determining what features should be made available to the system. In addition the reduced input dimensionality increases the processing speed of the classifier, leading to better response times [12,51].

Many inferential measurement systems are developed using data-based methodologies; the models used to infer the value of target features are developed with real-time plant data. This implies that inferential systems are heavily influenced by the quality of the data used to develop their internal models. Complex application problems, such as reliable monitoring and diagnosis of industrial plants, are likely to present large numbers of features, many of which will be redundant for the task at hand. Additionally there is an associated cost with the measurement of these features. In these situations it is very useful to have an intelligent system capable of selecting the most relevant features needed to build an accurate and reliable model for the process [170,284,322].

The task of text clustering is to group similar documents together, represented as a bag of words. This representation raises one severe problem: the high dimensionality of the feature space and the inherent data sparsity. This can significantly affect the performance of clustering algorithms, so it is highly desirable to reduce this feature space size. Dimensionality reduction techniques have been successfully applied to this area—both those that destroy data semantics and those that preserve them (feature selectors) [68,197].

Similar to clustering, text categorization views documents as a collection of words. Documents are examined, with their constituent keywords extracted and rated according to criteria such as their frequency of occurrence. As the number of keywords extracted is usually in the order of tens of thousands, dimensionality reduction must be performed. This can take the form of simplistic filtering methods such as word stemming or the use of stop-word lists. However, filtering methods do not provide enough reduction for use in automated categorizers, so a further feature selection process must take place. Recent applications of FS in this area include Web page and bookmark categorization [102,162].

1.5 STRUCTURE

The rest of this book is structured as follows (see Figure 1.3):

- *Chapter 2: Set Theory.* A brief introduction to the various set theories is presented in this chapter. Essential concepts from classical set theory, fuzzy set theory, rough set theory, and hybrid fuzzy-rough set theory are presented and illustrated where necessary.

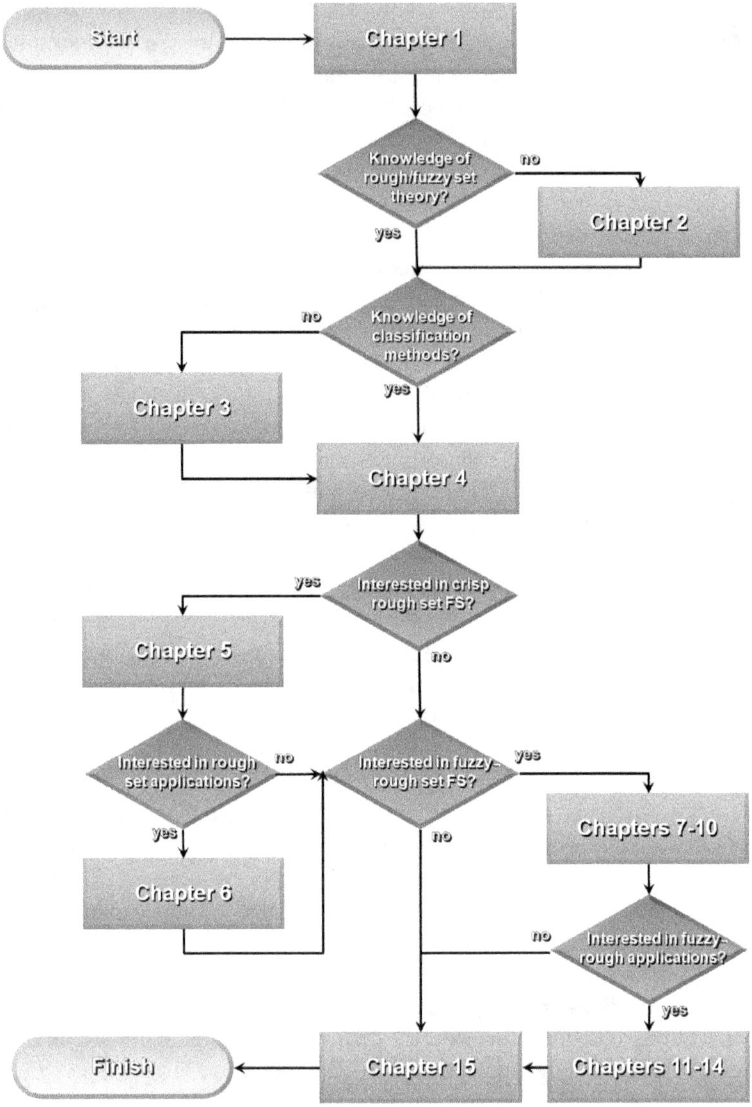

Figure 1.3 How to read this book

- *Chapter 3: Classification Methods.* This chapter discusses both crisp and fuzzy methods for the task of classification. Many of the methods presented here are used in systems later in the book.
- *Chapter 4: Dimensionality Reduction.* A systematic overview of current techniques for dimensionality reduction with a particular emphasis on feature selection is given in this chapter. It begins with a discussion of those

reduction methods that irreversibly transform data semantics. This is followed by a more detailed description and evaluation of the leading feature selectors presented in a unified algorithmic framework. A simple example illustrates their operation.

- *Chapter 5: Rough Set-based Approaches to Feature Selection.* This chapter presents an overview of the existing research regarding the application of rough set theory to feature selection. Rough set attribute reduction (RSAR), the precursor to the developments in this book, is described in detail. However, these methods are unsuited to the problems discussed in Section 5.11. In particular, they are unable to handle noisy or real-valued data effectively—a significant problem if they are to be employed within real-world applications.
- *Chapter 6: Applications I: Use of RSAR.* This chapter looks at the applications of RSAR in several challenging domains: medical image classification, text categorization, and algae population estimation. Details of each classification system are given with several comparative studies carried out that investigate RSAR's utility. Additionally a brief introduction to other applications that use a crisp rough set approach is provided for the interested reader.
- *Chapter 7: Rough and Fuzzy Hybridization.* There has been great interest in developing methodologies that are capable of dealing with imprecision and uncertainty. The large amount of research currently being carried out in fuzzy and rough sets is representative of this. Many deep relationships have been established, and recent studies have concluded at the complementary nature of the two methodologies. Therefore it is desirable to extend and hybridize the underlying concepts to deal with additional aspects of data imperfection. Such developments offer a high degree of flexibility and provide robust solutions and advanced tools for data analysis. A general survey of this research is presented in the chapter, with a focus on applications of the theory to disparate domains.
- *Chapter 8: Fuzzy-Rough Feature Selection.* In this chapter the theoretical developments behind this new feature selection method are presented together with a proof of generalization. This novel approach uses fuzzy-rough sets to handle many of the problems facing feature selectors outlined previously. A complexity analysis of the main selection algorithm is given. The operation of the approach and its benefits are shown through the use of two simple examples. To evaluate this new fuzzy-rough measure of feature significance, comparative investigations are carried out with the current leading significance measures.
- *Chapter 9: New Developments of FRFS.* Fuzzy-rough set-based feature selection has been shown to be highly useful at reducing data dimensionality, but possesses several problems that render it ineffective for datasets possessing tens of thousands of features. This chapter presents three new approaches to fuzzy-rough feature selection (FRFS) based on

fuzzy similarity relations. The first employs the new similarity-based fuzzy lower approximation to locate subsets. The second uses boundary region information to guide search. Finally, a fuzzy extension to crisp discernibility matrices is given in order to discover fuzzy-rough subsets. The methods are evaluated and compared using benchmark data.

- *Chapter 10: Further Advanced FS Methods.* This chapter introduces two promising areas in feature selection. The first, feature grouping, is developed from recent work in the literature where groups of features are selected simultaneously. By reasoning with fuzzy labels, the search process can be made more intelligent allowing various search strategies to be employed. The second, ant-based feature selection, seeks to address the nontrivial issue of finding the smallest optimal feature subsets. This approach to feature selection uses artificial ants and pheromone trails in the search for the best subsets. Both of these developments can be applied within feature selection, in general, but are applied to the specific problem of subset search within FRFS in this book.
- *Chapter 11: Applications II: Web Content Categorization.* With the explosive growth of information on the Web, there is an abundance of information that must be dealt with effectively and efficiently. This area, in particular, deserves the attention of feature selection due to the increasing demand for high-performance intelligent Internet applications. This motivates the application of FRFS to the automatic categorization of user bookmarks/favorites and Web pages. The results show that FRFS significantly reduces data dimensionality by several orders of magnitude with little resulting loss in classification accuracy.
- *Chapter 12: Applications III: Complex Systems Monitoring.* Complex application problems, such as reliable monitoring and diagnosis of industrial plants, are likely to present large numbers of features, many of which will be redundant for the task at hand. With the use of FRFS, these extraneous features can be removed. This not only makes resultant rulesets generated from such data much more concise and readable but can reduce the expense due to the monitoring of redundant features. The monitoring system is applied to water treatment plant data, producing better classification accuracies than those resulting from the full feature set and several other reduction methods.
- *Chapter 13: Applications IV: Algae Population Estimation.* Biologists need to identify and isolate the chemical parameters of rapid algae population fluctuations in order to limit their detrimental effect on the environment. This chapter describes an estimator of algae populations, a hybrid system involving FRFS that approximates, given certain water characteristics, the size of algae populations. The system significantly reduces computer time and space requirements through the use of feature selection. The results show that estimators using a fuzzy-rough feature selection step produce more accurate predictions of algae populations in general.
- *Chapter 14: Applications V: Forensic Glass Analysis.* The evaluation of glass evidence in forensic science is an important issue. Traditionally this

has depended on the comparison of the physical and chemical attributes of an unknown fragment with a control fragment. A high degree of discrimination between glass fragments is now achievable due to advances in analytical capabilities. A random effects model using two levels of hierarchical nesting is applied to the calculation of a likelihood ratio (LR) as a solution to the problem of comparison between two sets of replicated continuous observations where it is unknown whether the sets of measurements shared a common origin. This chapter presents the investigation into the use of feature evaluation for the purpose of selecting a single variable to model without the need for expert knowledge. Results are recorded for several selectors using normal, exponential, adaptive, and biweight kernel estimation techniques. Misclassification rates for the LR estimators are used to measure performance.

- *Chapter 15: Supplementary Developments and Investigations.* This chapter offers initial investigations and ideas for further work, which were developed concurrently with the ideas presented in the previous chapters. First, the utility of using the problem formulation and solution techniques from propositional satisfiability for finding rough set reducts is considered. This is presented with an initial experimental evaluation of such an approach, comparing the results with a standard rough set-based algorithm, RSAR. Second, the possibility of universal reducts is proposed as a way of generating more useful feature subsets. Third, fuzzy decision tree induction based on the fuzzy-rough metric developed in this book is proposed. Other proposed areas of interest include fuzzy-rough clustering and fuzzy-rough fuzzification optimization.

CHAPTER 2

SET THEORY

The problem of capturing the vagueness present in the real world is difficult to overcome using classical set theory alone. As a result several extensions to set theory have been proposed that deal with different aspects of uncertainty. Some of the most popular methods for this are fuzzy set theory, rough set theory, and their hybridization, fuzzy-rough set theory. This chapter starts with a quick introduction to classical set theory, using a simple example to illustrate the concept. Then an introduction to fuzzy sets is given, covering the essentials required for a basic understanding of their use. There are many useful introductory resources regarding fuzzy set theory, for example [66,267]. Following this, rough set theory is introduced with an example dataset to illustrate the fundamental concepts. Finally, fuzzy-rough set theory is detailed.

2.1 CLASSICAL SET THEORY

2.1.1 Definition

In classical set theory, elements can belong to a set or not at all. For example, in the set of old people, defined here as {*Rod, Jane, Freddy*}, the element *Rod* belongs to this set whereas the element *George* does not. No distinction is made within a set between elements; all set members belong fully. This may be considered to be a source of information loss for certain applications. Returning to the example, *Rod* may be older than *Freddy*, but by this formulation both

Computational Intelligence and Feature Selection: Rough and Fuzzy Approaches, by Richard Jensen and Qiang Shen

are considered to be equally old. The order of elements within a set is not considered.

More formally, let $\mathbb{U}$ be a space of objects, referred to as the universe of discourse, and x an element of $\mathbb{U}$. A classical (crisp) set A, $A \subseteq \mathbb{U}$, is defined as a collection of elements $x \in \mathbb{U}$, such that each element x can belong to the set or not belong. A classical set A can be represented by a set of ordered pairs $(x, 0)$ or $(x, 1)$ for each element, indicating $x \notin A$ or $x \in A$, respectively.

Two sets A and B are said to be equal if they contain exactly the same elements; every element of A is an element of B and every element of B is an element of A. The cardinality of a set A is a measure of the number of elements of the set, and is often denoted $|A|$. For example, the set $\{Rod, Jane, Freddy\}$ has a cardinality of 3 ($|\{Rod, Jane, Freddy\}| = 3$). A set with no members is called the empty set, usually denoted $\emptyset$ and has a cardinality of zero.

2.1.2 Subsets

If every element of a set A is a member of a set B, then A is said to be a *subset* of B, denoted $A \subseteq B$ (or equivalently $B \supseteq A$). If A is a subset of B but is not equal to B, then it is a *proper* subset of B, denoted $A \subset B$ (or equivalently $B \supset A$). For example, if $A = \{Jane, Rod\}$, $B = \{Rod, Jane, Geoffrey\}$ and $C = \{Rod, Jane, Freddy\}$:

$$
\begin{aligned}
&A \subseteq B(\{Jane, Rod\} \subseteq \{Rod, Jane, Geoffrey\})\\
&A \subseteq A(\{Jane, Rod\} \subseteq \{Jane, Rod\})\\
&A \subset B(\{Jane, Rod\} \subset \{Rod, Jane, Geoffrey\})\\
&A \subseteq C(\{Jane, Rod\} \subseteq \{Rod, Jane, Freddy\})\\
&B \nsubseteq C(\{Rod, Jane, Geoffrey\} \nsubseteq \{Rod, Jane, Freddy\})
\end{aligned}
$$

2.1.3 Operators

Several operations exist for the manipulation of sets. Only the fundamental operations are considered here.

2.1.3.1 Union The *union* of two sets A and B is a set that contains all the members of both sets and is denoted $A \cup B$. More formally, $A \cup B = \{x | (x \in A)$ or $(x \in B)\}$. Properties of the union operator include

- $A \cup B = B \cup A$
- $A \subseteq A \cup B$
- $A \cup A = A$
- $A \cup \emptyset = A$

2.1.3.2 Intersection The *intersection* of two sets A and B is a set that contains only those elements that A and B have in common. More formally, $A \cap B = \{x|(x \in A)$ and $(x \in B)\}$. If the intersection of A and B is the empty set, then sets A and B are said to be *disjoint*. Properties of the intersection operator include

- $A \cap B = B \cap A$
- $A \cap B \subseteq A$
- $A \cap A = A$
- $A \cap \emptyset = \emptyset$

2.1.3.3 Complement The *relative* complement (or set theoretic difference) of A in B is the set of all elements that are members of B but not of A. This is denoted $B - A$ (or $B \setminus A$). More formally, $B - A = \{x|(x \in B)$ and not $(x \in A)\}$.

In some situations, all sets may be considered to be subsets of a given universal set $\mathbb{U}$. In this case, $\mathbb{U} - A$ is the *absolute* complement (or simply *complement*) of A, and is denoted A' or A^c. More formally, $A^c = \{x|(x \in \mathbb{U})$ and not $(x \in A)\}$. Properties of complements include

- $A \cup A^c = \mathbb{U}$
- $A \cap A^c = \emptyset$
- $(A^c)^c = A$
- $A - A = \emptyset$
- $A - B = A \cap B^c$

2.2 FUZZY SET THEORY

A distinct approach to coping with reasoning under uncertain circumstances is to use the theory of fuzzy sets [408]. The main objective of this theory is to develop a methodology for the formulation and solution of problems that are too complex or ill-defined to be suitable for analysis by conventional Boolean techniques. It deals with subsets of a universe of discourse, where the transition between full membership of a subset and no membership is gradual rather than abrupt as in the classical set theory. Such subsets are called fuzzy sets.

Fuzzy sets arise, for instance, when mathematical descriptions of ambiguity and ambivalence are needed. In the real world the attributes of, say, a physical system often emerge from an elusive vagueness or fuzziness, a readjustment to context, or an effect of human imprecision. The use of the "soft" boundaries of fuzzy sets, namely the graded memberships, allows subjective knowledge to be utilized in defining these attributes. With the accumulation of knowledge the subjectively assigned memberships can, of course, be modified. Even in some cases where precise knowledge is available, fuzziness may be a concomitant of complexity involved in the reasoning process.

The adoption of fuzzy sets helps to ease the requirement for encoding uncertain domain knowledge. For example, labels like *small*, *medium*, and *large* have an intuitive appeal to represent values of some physical attributes. However, if they are defined with a semantic interpretation in terms of crisp intervals, such as

$$small = \{x|x > 0, x \ll 1\}$$

$$medium = \{x|x > 0, x \approx 1\}$$

the representation may lead to rather nonintuitive results. This is because, in practice, it is not realistic to draw an exact boundary between these intervals. When encoding a particular real number, it may be difficult to decide which of these the number should, or should not, definitely belong to. It may well be the case that what can be said is only that a given number belongs to the *small* set with a possibility of A and to the *medium* with a possibility of B. The avoidance of this problem requires gradual membership and hence the break of the laws of excluded-middle and contradiction in Boolean logic. This forms the fundamental motivation for the development of fuzzy logic.

2.2.1 Definition

A fuzzy set can be defined as a set of ordered pairs $A = \{x, \mu_A(x)|x \in \mathbb{U}\}$. The function $\mu_A(x)$ is called the membership function for A, mapping each element of the universe $\mathbb{U}$ to a membership degree in the range [0, 1]. The universe may be discrete or continuous. Any fuzzy set containing at least one element with a membership degree of 1 is called *normal*.

Returning to the example in Section 2.1.1, it may be better to represent the set of old people as a fuzzy set, *Old*. The membership function for this set is given in Figure 2.1, defined over a range of ages (the universe). Given that the age of *Rod* is 74, it can be determined that this element belongs to the set *Old* with a membership degree of 0.95. Similarly, if the age of *Freddy* is 38, the resulting

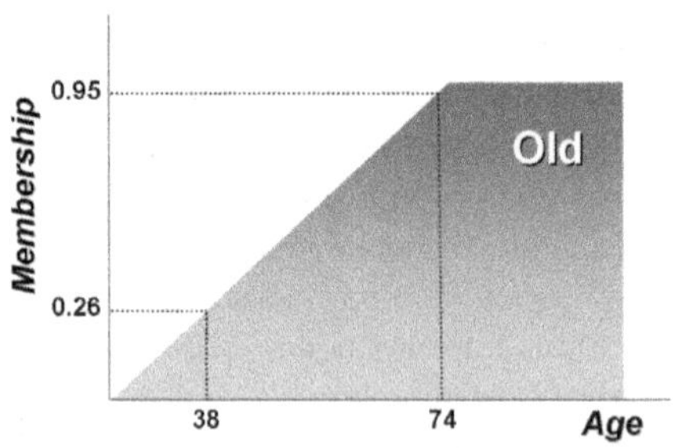

Figure 2.1 Fuzzy set representing the concept *Old*

degree of membership is 0.26. Here, both *Rod* and *Freddy* belong to the (fuzzy) set of old people, but *Rod* has a higher degree of membership to this set.

The specification of membership functions is typically subjective, as can be seen in this example. There are many justifiable definitions of the concept *Old*. Indeed people of different ages may define this concept quite differently. One way of constructing the membership function is by using a voting model, and this way the generated fuzzy sets can be rooted in reality with clear semantics. An alternative to this is the integration of the probability distribution of variables. For example, integrating the probability distribution of height gives a possible version of the tall fuzzy set. This does not help with hedges and sets such as very tall (see Section 2.2.6), and so although the integration of the distribution gives a voting model, it needs more to arrive at the hedges.

In fuzzy logic the truth value of a statement is linguistic (and no longer Boolean), of the form *very true, true, more or less true, not very false, false*. These logic values are themselves fuzzy sets; some may be compounded fuzzy sets from other atomic ones, by the use of certain operators. As with ordinary crisp sets, different operations can be defined over fuzzy sets.

2.2.2 Operators

The most basic operators on fuzzy sets are the union, intersection, and complement. These are fuzzy extensions of their crisp counterparts, ensuring that if they are applied to crisp sets, the results of their application will be identical to crisp union, intersection, and complement.

2.2.2.1 Fuzzy Intersection The intersection (t-norm) of two fuzzy sets, A and B, is specified by a binary operation on the unit interval; that is, a function of the form

$$t : [0, 1] \times [0, 1] \rightarrow [0, 1] \tag{2.1}$$

For each element x in the universe, this function takes as its arguments the memberships of x in the fuzzy sets A and B, and yields the membership grade of the element in the set constituting the intersection of A and B:

$$\mu_{A \cap B}(x) = t[\mu_A(x), \mu_B(x)] \tag{2.2}$$

The following axioms must hold for the operator t to be considered a t-norm, for all x, y, and z in the range [0,1]:

- $t(x, 1) = x$ (boundary condition)
- $y \leq z \rightarrow t(x, y) \leq t(x, z)$ (monotonicity)
- $t(x, y) = t(y, x)$ (commutativity)
- $t(x, t(y, z)) = t(t(x, y), z)$ (associativity)

The following are examples of t-norms that are often used as fuzzy intersections:

$$t(x, y) = \min(x, y) \quad \text{(standard intersection)}$$
$$t(x, y) = x.y \quad \text{(algebraic product)}$$
$$t(x, y) = \max(0, x + y - 1) \quad \text{(bounded difference)}$$

2.2.2.2 *Fuzzy Union* The fuzzy union (t-conorm or s-norm) of two fuzzy sets A and B is specified by a function

$$s : [0, 1] \times [0, 1] \rightarrow [0, 1] \tag{2.3}$$

$$\mu_{A \cup B}(x) = s[\mu_A(x), \mu_B(x)] \tag{2.4}$$

A fuzzy union s is a binary operation that satisfies at least the following axioms for all x, y, and z in [0,1]:

- $s(x, 0) = x$ (boundary condition)
- $y \leq z \rightarrow s(x, y) \leq s(x, z)$ (monotonicity)
- $s(x, y) = s(y, x)$ (commutativity)
- $s(x, s(y, z)) = s(s(x, y), z)$ (associativity)

The following are examples of t-conorms that are often used as fuzzy unions:

$$s(x, y) = \max(x, y) \quad \text{(standard union)}$$
$$s(x, y) = x + y - x.y \quad \text{(algebraic sum)}$$
$$s(x, y) = \min(1, x + y) \quad \text{(bounded sum)}$$

The most popular interpretation of fuzzy union and intersection is the max/min interpretation, primarily due to its ease of computation. This particular interpretation is used in the book.

2.2.2.3 *Fuzzy Complement* The complement of a fuzzy set A is specified by a function

$$c : [0, 1] \rightarrow [0, 1] \tag{2.5}$$

$$\mu_{cA}(x) = \mu_{\neg A}(x) = c[\mu_A(x)] \tag{2.6}$$

subject to the following:

- $c(0) = 1$ and $c(1) = 0$ (boundary condition)
- $\forall a, b \in [0, 1]$, if $a \leq b$ then $c(a) \geq c(b)$ (monotonicity)

- c is a continuous function (continuity)
- c is an involution. (i.e., $c(c(a)) = a$ for each $a \in [0, 1]$)

The complement of a fuzzy set can be denoted in a number of ways; $\neg A$ and $\overline{A}$ are also in common use. It may also be represented as A^c, the same way as in crisp set theory. A standard definition of fuzzy complement is

$$\mu_{cA}(x) = 1 - \mu_A(x) \tag{2.7}$$

It should not be difficult to see that these definitions cover the conventional operations on crisp sets as their specific cases. Taking the set intersection as an example, when both sets A and B are crisp, x is a member of $A \cap B$ if and only if (abbreviated to iff hereafter) it belongs to both A and B, and vice versa. This is covered by the definition of fuzzy set intersection because $\mu_{A\cap B}(x) = 1$ (or $\min(\mu_A(x), \mu_B(x)) = 1$), iff $\mu_A(x) = 1$ and $\mu_B(x) = 1$. Otherwise, $\mu_{A\cap B}(x) = 0$, given that $\mu_A(x)$ and $\mu_B(x)$ take values only from $\{0, 1\}$ in this case.

2.2.3 Simple Example

An example should help with the understanding of the basic concepts and operators introduced above. Suppose that the universe of discourse, X, is a class of students and that a group, A, of students within this class are said to be *tall* in height. Thus A is a fuzzy subset of X (but X itself is not fuzzy), since the boundary between *tall* and *not tall* cannot be naturally defined with a fixed real number. Rather, describing this vague concept using a gradual membership function as characterized in Figure 2.2. is much more appealing. Similarly the fuzzy term *very tall* can be represented by another fuzzy (sub-)set as also shown in this figure. Given such a definition of the fuzzy set $A = tall$, a proposition like "student x is *tall*" can be denoted by $\mu_A(x)$.

Assume that Andy is a student in the class and that he has a 80% possibility of being considered as a *tall* student. This means that $\mu_A(Andy) = 0.8$. Also

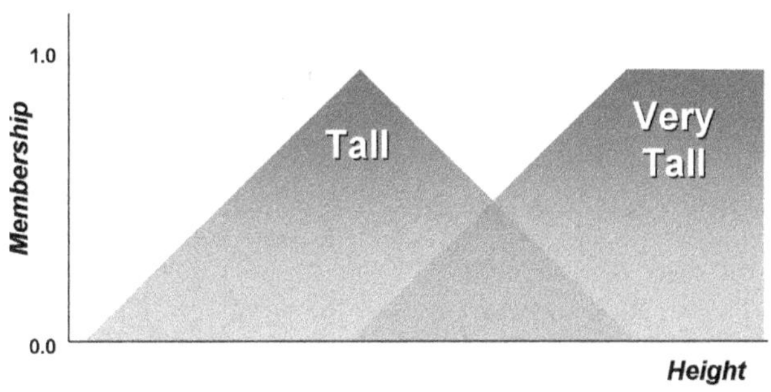

Figure 2.2 Representation of fuzzy sets "tall" and "very tall"

suppose that another fuzzy set B is defined on the same universe X, whose members are said to be *young* in age, and that $\mu_B(Andy) = 0.7$, meaning that Andy is thought to be *young* with a possibility of 70%. From this, using the operations defined above, the following can be derived that is justifiable with respect to common-sense intuitions:

- $\mu_{\neg A}(Andy) = 1 - 0.8 = 0.2$, indicating the possibility of Andy being *not tall*
- $\mu_{A \cup B}(Andy) = \max(\mu_A(Andy), \mu_B(Andy)) = \max(0.8, 0.7) = 0.8$, indicating the possibility of Andy being *tall or young*
- $\mu_{A \cap B}(Andy) = \min(\mu_A(Andy), \mu_B(Andy)) = \min(0.8, 0.7) = 0.7$, indicating the possibility of Andy being *both tall and young*

Not only having an intuitive justification, these operations are also computationally very simple. However, care should be taken not to explain the results from the conventional Boolean logic point of view. The laws of excluded-middle and contradiction do not hold in fuzzy logic anymore. To make this point clearer, let us attempt to compare the results of $A \cup \neg A$ and $A \cap \neg A$ obtained by fuzzy logic with those by Boolean probabilistic logic (although such a comparison is itself a common source for debate in the literature). Applying fuzzy logic to the case above of Andy gives that

- $\mu_{A \cup \neg A}(Andy) = \max(\mu_A(Andy), \mu_{\neg A}(Andy)) = \max(0.8, 0.2) = 0.8$
- $\mu_{A \cap \neg A}(Andy) = \min(\mu_A(Andy), \mu_{\neg A}(Andy)) = \min(0.8, 0.2) = 0.2$

However, using the theory of probability, different results would be expected such that

- p("Andy is *tall*" *or* "Andy is *not tall*") $= 1$
- p("Andy is *tall*" *and* "Andy is *not tall*") $= 0$

This important difference is caused by the deliberate avoidance of the excluded-middle and contradiction laws in fuzzy logic. This avoidance enables fuzzy logic to represent vague concepts that are difficult to capture otherwise. If the memberships of x belonging to A and $\neg A$ are neither 0 nor 1, then it should not be surprising that x is also of a nonzero membership belonging to A *and* $\neg A$ at the same time.

2.2.4 Fuzzy Relations and Composition

In addition to the three operators defined above, many conventional mathematical functions can be extended to be applied to fuzzy values. This is possible by the use of an *extension principle*, so called because it provides a general extension of classical mathematical concepts to fuzzy environments. This principle is stated as follows: If an n-ary function f maps the Cartesian product $X_1 \times X_2 \times \cdots \times X_n$

onto a universe Y such that $y = f(x_1, x_2, \ldots, x_n)$, and $A_1, A_2, \ldots, A_n$ are n fuzzy sets in $X_1, X_2, \ldots, X_n$, respectively, characterized by membership distributions $\mu_{A_i}(X_i), i = 1, 2, \ldots, n$, a fuzzy set on Y can then be induced as given below, where Ø is the empty set:

$$\mu_B(y) = \begin{cases} \max_{\{x_1,\ldots,x_n, y=f(x_1,\ldots,x_n)\}} \min(\mu_{A_1}(x_1), \ldots, \mu_{A_n}(x_n)) & \text{if } f^{-1}(Y) \neq \emptyset \\ 0 & \text{if} f^{-1}(Y) = \emptyset \end{cases}$$

A crucial concept in fuzzy set theory is that of *fuzzy relations*, which is a generalization of the conventional crisp relations. An n-ary fuzzy relation in $X_1 \times X_2 \times \cdots \times X_n$ is, in fact, a fuzzy set on $X_1 \times X_2 \times \cdots \times X_n$. Fuzzy relations can be composed (and this composition is closely related to the extension principle shown above). For instance, if U is a relation from X_1 to X_2 (or, equivalently, a relation in $X_1 \times X_2$), and V is a relation from X_2 to X_3, then the composition of U and V is a fuzzy relation from X_1 to X_3 which is denoted by $U \circ V$ and defined by

$$\mu_{U \circ V}(x_1, x_3) = \max_{x_2 \in X_2} \min(\mu_U(x_1, x_2), \mu_V(x_2, x_3)), \qquad x_1 \in X_1, x_3 \in X_3$$

A convenient way of representing a binary fuzzy relation is to use a matrix. For example, the following matrix, P, can be used to indicate that a computer-game addict is much more fond of multimedia games than conventional ones, be they workstation or PC-based:

$$P = \begin{matrix} \\ \textit{Multimedia} \\ \textit{Conventional} \end{matrix} \begin{matrix} \begin{matrix} \textit{Workstation} & \textit{PC} \end{matrix} \\ \begin{pmatrix} 0.9 & 0.8 \\ 0.3 & 0.2 \end{pmatrix} \end{matrix}$$

Suppose that the addict prefers keyboard-based games over mouse-based, another relation, Q, may then be used to describe this preference such that

$$Q = \begin{matrix} \\ \textit{Workstation} \\ \textit{PC} \end{matrix} \begin{matrix} \begin{matrix} \textit{Keyboard} & \textit{Mouse} \end{matrix} \\ \begin{pmatrix} 0.7 & 0.5 \\ 0.8 & 0.3 \end{pmatrix} \end{matrix}$$

Given these two relations, a composed relation, R, can be obtained as shown below:

$$\begin{aligned} R = P \circ Q &= \begin{pmatrix} 0.9 & 0.8 \\ 0.3 & 0.2 \end{pmatrix} \circ \begin{pmatrix} 0.7 & 0.5 \\ 0.8 & 0.3 \end{pmatrix} \\ &= \begin{pmatrix} \max(\min(0.9, 0.7), \min(0.8, 0.8)) & \max(\min(0.9, 0.5), \min(0.8, 0.3)) \\ \max(\min(0.3, 0.7), \min(0.2, 0.8)) & \max(\min(0.3, 0.5), \min(0.2, 0.3)) \end{pmatrix} \\ &= \begin{matrix} \\ \textit{Multimedia} \\ \textit{Conventional} \end{matrix} \begin{matrix} \begin{matrix} \textit{Keyboard} & \textit{Mouse} \end{matrix} \\ \begin{pmatrix} 0.8 & 0.5 \\ 0.3 & 0.3 \end{pmatrix} \end{matrix} \end{aligned}$$

This composition indicates that the addict enjoys multimedia games, especially keyboard-based multimedia games.

2.2.5 Approximate Reasoning

Fuzzy relations and fuzzy relation composition form the basis for approximate reasoning, sometimes termed fuzzy reasoning. Informally, approximate reasoning means a process by which a possibly inexact conclusion is inferred from a collection of inexact premises. Systems performing such reasoning are built upon a collection of fuzzy production (*if-then*) rules, which provide a formal means of representing domain knowledge acquired from empirical associations or experience. Such systems run upon a given set of facts that allows a (usually partial) instantiation of the premise attributes in some of the rules.

For example, a rule like

$$\text{if } x \text{ is } A_i \text{ and } y \text{ is } B_i, \text{ then } z \text{ is } C_i$$

which governs a particular relationship between the premise attributes x and y and the conclusion attribute z, can be translated into a fuzzy relation R_i:

$$\mu_{R_i}(x, y, z) = \min(\mu_{A_i}(x), \mu_{B_i}(y), \mu_{C_i}(z))$$

Here, A_i, B_i, and C_i are fuzzy sets defined on the universes X, Y, and Z of the attributes x, y, and z, respectively. Provided with this relation, if the premise attributes x and y actually take the fuzzy values A' and B', a new fuzzy value of the conclusion attribute z can then be obtained by applying the *compositional rule of inference*:

$$C' = (A' \times B') \circ R_i$$

Or,

$$\mu_{C'}(z) = \max_{x \in X, y \in Y} \min(\mu_{A'}(x), \mu_{B'}(y), \mu_{R_i}(x, y, z))$$

Of course, an approximate reasoning system would not normally function using only one specific rule, but a set. Given a set of production rules, two different approaches may be adopted to implement the reasoning in such a system; both rely on the use of the compositional rule of inference given above. The first is to apply the compositional rule after an overall relation associated with the entire rule set is found. The second is to utilize the compositional rule locally to each individual production rule first and then to aggregate the results of such applications to form an overall result for the consequent attribute.

Given a set of K *if-then* rules, basic algorithms for these two approaches are summarized below. For simplicity, it is assumed that each rule contains a fixed number of N premise attributes $x_1, x_2, \ldots, x_N$ and one conclusion attribute y. Also the logical connectives are to be applied strictly from left to right.

Method 1: Inference with Overall Interpreted Relation

Step 1: For each rule $k, k \in \{1, 2, \ldots, K\}$, compute its corresponding fuzzy relation. For instance, given the following two rules ($K = 2, N = 3$):

If x_1 is A_1 and x_2 is not B_1 or x_3 is C_1, then y is D_1

If x_1 is not A_2 or x_2 is B_2 and x_3 is not C_2, then y is not D_2

their associated fuzzy relations are respectively calculated by

$$R_1(x_1, x_2, x_3, y) = \min\{\max\{\min\{\mu_{A_1}(x_1), 1 - \mu_{B_1}(x_2)\}, \mu_{C_1}(x_3)\}, \mu_{D_1}(y)\}$$

$$R_2(x_1, x_2, x_3, y) = \min\{\min\{\max\{1 - \mu_{A_2}(x_1), \mu_{B_2}(x_2)\}, 1 - \mu_{C_2}(x_3)\}, \times 1 - \mu_{D_2}(y)\}$$

Step 2: Combine individual fuzzy relations, $R_k(x_1, x_2, \ldots, x_N, y)$, $k \in \{1, 2, \ldots, K\}$, to form the overall integrated relation, such that

$$R(x_1, x_2, \ldots, x_N, y) = \max_{k \in \{1,2,\ldots,K\}} R_k(x_1, x_2, \ldots, x_N, y)$$

Step 3: Apply the compositional rule of inference to obtain the inferred fuzzy value D of the conclusion attribute, given a fuzzy value for each antecedent attribute (e.g., A for x_1, B for x_2, …, and C for x_N). That is,

$$\mu_D(y) = \max_{x_1, x_2, \ldots, x_N} \min\{\mu_A(x_1), \mu_B(x_2), \ldots, \mu_C(x_N), \mu_R(x_1, x_2, \ldots, x_N, y)\}$$

Method 2: Inference Via Locally Interpreted Relations

Step 1: This is the same as the first step of method 1.

Step 2: For each rule k, given a fuzzy value for each premise attribute (e.g., again A for x_1, B for x_2, …, and C for x_N), compute the inferred fuzzy value D_k of the conclusion attribute by applying the compositional rule of inference, such that

$$\mu_{D_k}(y) = \max_{x_1, x_2, \ldots, x_N} \min\{\mu_A(x_1), \mu_B(x_2), \ldots, \mu_C(x_N), \times \mu_{R_k}(x_1, x_2, \ldots, x_N, y)\}$$

Step 3: Calculate the overall inferred fuzzy value D of the conclusion attribute by aggregating the inferred values derived from individual rules. That is,

$$\mu_D(y) = \max_{k, k \in \{1,2,\ldots,K\}} \mu_{D_k}(y)$$

To use a fuzzy logic controller (FLC) in real-world situations, the observations obtained from the physical system must be mapped onto a fuzzy set representation, and the outputs of the FLC must be mapped back to real-valued control signals. Hence two supplementary processes are necessary to enable the use of an FLC, which are respectively called the fuzzification and the defuzzification. Fuzzification is, however, largely a task of domain-specific data interpretation and thus is left out from the present discussion.

2.2.5.1 Defuzzification There exist a number of methods for defuzzification [297], though there is no clear winner. Two widely used methods are briefly summarized below.

- The *center of gravity* method finds the geometrical center $\hat{y}$ in the inferred fuzzy value D of the conclusion attribute y as the defuzzified value, such that

$$\hat{y} = \frac{\sum y\mu_D(y)}{\sum \mu_D(y)}$$

- The *mean of maxima* method finds the value whose membership degree is the largest in D as the defuzzified value; if there are more than one value that has the maximum degree, then the average of them is taken as the defuzzified value.

2.2.6 Linguistic Hedges

In fuzzy systems, *hedges* can be used to modify fuzzy values while retaining the meaning of fuzzy rules. A simple example of this is the transformation of the statement "Rod is old" to "Rod is very old." The application of a linguistic hedge modifies the shape of the membership function of a fuzzy set [66], transforming one fuzzy set into another. The meaning of the transformed set can easily be interpreted from the meaning of the original set and that embedded in the hedge applied. The definition of hedges has more to do with commonsense knowledge in a domain than with mathematical theory. Although a simple linguistic hedge may be used to express different modifications in different applications, the general type of the underlying operation remains the same, varying only in terms of detailed parameter settings.

For example, concentration/dilation hedges are often implemented by applying a power function to the original set's membership values [66,156]. That is, given the original membership function $\mu_S(x)$ of a fuzzy set S and hedge H, the membership function of $H \cdot S$ is $\mu_{H \cdot S}(x) = \mu_S^e(x)$, where the exponent e is greater than 1 for concentration and less than 1 for dilation. Different values can be assigned to the exponent e; for hedge EXTREMELY, for instance, $e = 2$ is used in [66], whereas $e = 8$ is employed in [156].

2.2.7 Fuzzy Sets and Probability

Before closing this section, it is worth emphasizing that the concept of fuzzy sets is not merely a disguised form of subjective probability. The theory of fuzzy sets and the theory of probability are two approaches to attaining realistic solutions to problems in reasoning with uncertainty. In essence, fuzzy set theory is aimed at dealing with sources of uncertainty or imprecision that are inherently vague and nonstatistical in nature. For example, the proposition "X is a *large number*," in which *large number* is a label of a fuzzy subset of nonnegative integers, defines a *possibility distribution* rather than the probability distribution of X. This implies that if the degree to which an integer I fits the subjective perception of *large number* is $\mu(I)$, then the possibility that X may take I as its value is numerically equal to $\mu(I)$. Consequently such a proposition conveys no information about the probability distribution of the values of X, as argued by Zadeh [411].

A simple example to show the difference between the two theories is throwing a dice: probability distribution indicates the random number achieved after each throw, while fuzzy membership functions describe whether the achieved number is small, medium, or large. Thus probability theory does not always provide an adequate tool for the analysis of problems in which the available information is ambiguous, imprecise, or even unreliable. An interesting combination of fuzziness and probability would be the chance for the number achieved to be small. Attempts to combine probability theory and fuzzy logic exist, aiming at providing powerful inference mechanisms for problems involving such ill-defined random variables.

2.3 ROUGH SET THEORY

Dealing with incomplete or imperfect knowledge is the core of much research in computational intelligence and cognitive sciences. Being able to understand and manipulate such knowledge is of fundamental significance to many theoretical developments and practical applications of automation and computing, especially in the areas of decision analysis, machine learning and data-mining, intelligent control, and pattern recognition. Rough set theory [89,260,261,314,336,337] offers one of the most distinct and recent approaches for modeling imperfect knowledge.

Indeed, since its invention, rough set theory has been successfully utilized to devise mathematically sound and often computationally efficient techniques for addressing problems such as hidden pattern discovery from data, data reduction, data significance evaluation, decision rule generation, and data-driven inference interpretation [263]. Owing to the recognition of the existing and potential important impact of this theory, it has attracted worldwide attention of further research and development, resulting in various extensions to the original theory and increasingly widening fields of application. This section attempts to offer a concise overview of the basic ideas of rough set theory, and its major extensions with sample applications. Further details can be found in the literature (e.g., *LNCS Transactions on Rough Sets*) and on the Internet (e.g., *www.roughsets.org*).

Rough set theory is an extension of conventional set theory that supports approximations in decision making. It possesses many features in common (to a certain extent) with the Dempster-Shafer theory of evidence [334] and fuzzy set theory [262,389]. A rough set is itself the approximation of a vague concept (set) by a pair of precise concepts, called lower and upper approximations, that are a classification of the domain of interest into disjoint categories. The lower approximation is a description of the domain objects that are known with certainty to belong to the subset of interest, whereas the upper approximation is a description of the objects that possibly belong to the subset.

It works by exploring and exploiting the granularity structure of the data only. This is a major difference when compared with Dempster–Shafer theory [76,315] and fuzzy set theory [408], which require probability assignments and membership values, respectively. However, this does not mean that *no* model assumptions are made. In fact, by using only the given information, the theory assumes that the data are a true and accurate reflection of the real world (which may not be the case). The numerical and other contextual aspects of the data are ignored, which may seem to be a significant omission but keeps model assumptions to a minimum.

2.3.1 Information and Decision Systems

An information system can be viewed as a table of data, consisting of objects (rows in the table) and attributes (columns). In medical datasets, for example, patients might be represented as objects and measurements such as blood pressure, form attributes. The attribute values for a particular patient is their specific reading for that measurement. Throughout this book, the terms attribute, feature, and variable are used interchangeably.

An information system may be extended by the inclusion of decision attributes. Such a system is termed a decision system. For example, the medical information system mentioned previously could be extended to include patient classification information, such as whether a patient is ill or healthy. A more abstract example of a decision system can be found in Table 2.1. The table consists of four conditional features (a, b, c, d), a decision feature (e), and eight objects. A decision system is *consistent* if for every set of objects whose attribute values are the same, the corresponding decision attributes are identical.

More formally, $I = (\mathbb{U}, \mathbb{A})$ is an information system, where $\mathbb{U}$ is a nonempty set of finite objects (the universe of discourse) and $\mathbb{A}$ is a nonempty finite set of attributes such that $a : \mathbb{U} \rightarrow V_a$ for every $a \in \mathbb{A}$. V_a is the set of values that attribute a may take. For decision systems, $\mathbb{A} = \{\mathbb{C} \cup \mathbb{D}\}$, where $\mathbb{C}$ is the set of input features and $\mathbb{D}$ is the set of class indexes. Here, a class index $d \in \mathbb{D}$ is itself a variable $d : \mathbb{U} \rightarrow \{0, 1\}$ such that for $a \in \mathbb{U}$, $d(a) = 1$ if a has class d and $d(a) = 0$ otherwise.

TABLE 2.1 An example dataset

$x \in \mathbb{U}$	a	b	c	d	$\Rightarrow$	e
0	S	R	T	T		R
1	R	S	S	S		T
2	T	R	R	S		S
3	S	S	R	T		T
4	S	R	T	R		S
5	T	T	R	S		S
6	T	S	S	S		T
7	R	S	S	R		S

2.3.2 Indiscernibility

With any $P \subseteq \mathbb{A}$ there is an associated equivalence relation $IND(P)$:

$$IND(P) = \{(x, y) \in \mathbb{U}^2 | \forall a \in P, a(x) = a(y)\} \tag{2.8}$$

Note that this relation corresponds to the equivalence relation for which two objects are equivalent if and only if they have the same vectors of attribute values for the attributes in P. The partition of $\mathbb{U}$, determined by *IND(P)*, is denoted $\mathbb{U}/IND(P)$ or $\mathbb{U}/P$, which is simply the set of equivalence classes generated by $IND(P)$:

$$\mathbb{U}/IND(P) = \otimes\{\mathbb{U}/IND(\{a\}) | a \in P\} \tag{2.9}$$

where

$$A \otimes B = \{X \cap Y | X \in A, Y \in B, X \cap Y \neq \emptyset\} \tag{2.10}$$

If $(x, y) \in IND(P)$, then x and y are indiscernible by attributes from P. The equivalence classes of the indiscernibility relation with respect to P are denoted $[x]_P$, $x \in \mathbb{U}$. For the illustrative example, if $P = \{b, c\}$, then objects 1, 6, and 7 are indiscernible, as are objects 0 and 4. *IND(P)* creates the following partition of $\mathbb{U}$:

$$\begin{aligned} \mathbb{U}/IND(P) &= \mathbb{U}/IND(\{b\}) \otimes \mathbb{U}/IND(\{c\}) \\ &= \{\{0, 2, 4\}, \{1, 3, 6, 7\}, \{5\}\} \\ &\quad \otimes\{\{2, 3, 5\}, \{1, 6, 7\}, \{0, 4\}\} \\ &= \{\{2\}, \{0, 4\}, \{3\}, \{1, 6, 7\}, \{5\}\} \end{aligned}$$

2.3.3 Lower and Upper Approximations

Let $X \subseteq \mathbb{U}$. X can be approximated using only the information contained within P by constructing the P-*lower* and P-*upper* approximations of the classical crisp set X:

$$\underline{P}X = \{x | [x]_P \subseteq X\} \tag{2.11}$$

$$\overline{P}X = \{x | [x]_P \cap X \neq \emptyset\} \tag{2.12}$$

It is such a tuple $\langle \underline{P}X, \overline{P}X \rangle$ that is called a rough set. Consider the approximation of concept X in figure 2.3. Each square in the diagram represents an equivalence class, generated by indiscernibility between object values. Using the features in set P, via these equivalence classes, the lower and upper approximations of X can be constructed. Equivalence classes contained within X belong to the lower approximation. Objects lying within this region can be said to belong definitely to concept X. Equivalence classes within X and along its border form the upper approximation. Those objects in this region can only be said to possibly belong to the concept.

2.3.4 Positive, Negative, and Boundary Regions

Let P and Q be sets of attributes inducing equivalence relations over $\mathbb{U}$, then the positive, negative, and boundary regions are defined as

$$POS_P(Q) = \bigcup_{X \in \mathbb{U}/Q} \underline{P}X \tag{2.13}$$

$$NEG_P(Q) = \mathbb{U} - \bigcup_{X \in \mathbb{U}/Q} \overline{P}X \tag{2.14}$$

$$BND_P(Q) = \bigcup_{X \in \mathbb{U}/Q} \overline{P}X - \bigcup_{X \in \mathbb{U}/Q} \underline{P}X \tag{2.15}$$

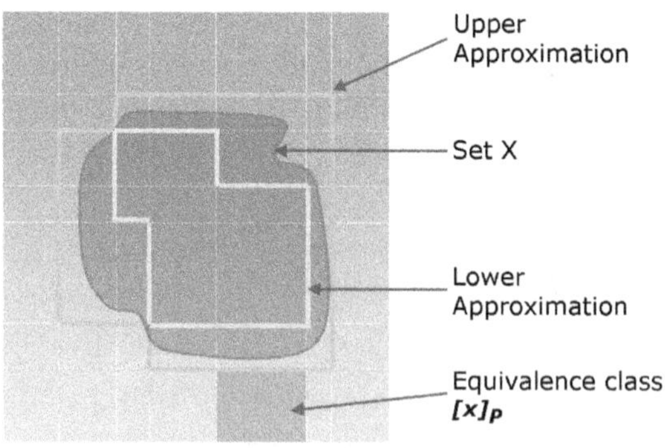

Figure 2.3 Rough set

The positive region comprises all objects of $\mathbb{U}$ that can be classified to classes of $\mathbb{U}/Q$ using the information contained within attributes P. The boundary region, $BND_P(Q)$, is the set of objects that can possibly, but not certainly, be classified in this way. The negative region, $NEG_P(Q)$, is the set of objects that cannot be classified to classes of $\mathbb{U}/Q$. Returning to the example in table 2.1, let $P = \{b,c\}$ and $Q = \{e\}$, then

$$\begin{aligned}
POS_P(Q) &= \cup\{\emptyset, \{2, 5\}, \{3\}\} = \{2, 3, 5\} \\
NEG_P(Q) &= \mathbb{U} - \cup\{\{0, 4\}, \{2, 0, 4, 1, 6, 7, 5\}, \{3, 1, 6, 7\}\} \\
&= \emptyset \\
BND_P(Q) &= \cup\{\{0, 4\}, \{2, 0, 4, 1, 6, 7, 5\}, \{3, 1, 6, 7\}\} - \{2, 3, 5\} \\
&= \{0, 1, 4, 6, 7\}
\end{aligned}$$

This means that objects 2, 3, and 5 can certainly be classified as belonging to a class in attribute e, when considering attributes b and c. The rest of the objects cannot be classified, as the information that would make them discernible is absent.

2.3.5 Feature Dependency and Significance

An important issue in data analysis is discovering dependencies between attributes. Intuitively, a set of attributes Q depends totally on a set of attributes P, denoted $P \Rightarrow Q$, if all attribute values from Q are uniquely determined by values of attributes from P. If there exists a functional dependency between values of Q and P, then Q depends totally on P. In rough set theory, dependency is defined in the following way:

For $P, Q \subset \mathbb{A}$, it is said that Q depends on P in a degree k $(0 \leq k \leq 1)$, denoted $P \Rightarrow_k Q$, if

$$k = \gamma_P(Q) = \frac{|POS_P(Q)|}{|\mathbb{U}|} \tag{2.16}$$

where $|S|$ stands for the cardinality of set S.

If $k = 1$, Q depends totally on P, if $0 < k < 1$, Q depends partially (in a degree k) on P, and if $k = 0$, then Q does not depend on P. In the example the degree of dependency of attribute $\{e\}$ on the attributes $\{b,c\}$ is

$$\begin{aligned}
\gamma_{\{b,c\}}(\{e\}) &= \frac{|POS_{\{b,c\}}(\{e\})|}{|\mathbb{U}|} \\
&= \frac{|\{2, 3, 5\}|}{|\{0, 1, 2, 3, 4, 5, 6, 7\}|} = \frac{3}{8}
\end{aligned}$$

By calculating the change in dependency when a feature is removed from the set of considered possible features, an estimate of the significance of that feature can be obtained. The higher the change in dependency, the more significant is the feature. If the significance is 0, then the feature is dispensible. More formally, given P,Q and a feature $x \in P$, the significance of feature x upon Q is defined by

$$\sigma_P(Q, a) = \gamma_P(Q) - \gamma_{P-\{a\}}(Q) \tag{2.17}$$

For example, if $P = \{a,b,c\}$ and $Q = e$, then

$$\begin{aligned}
\gamma_{\{a,b,c\}}(\{e\}) &= |\{2, 3, 5, 6\}|/8 = 4/8 \\
\gamma_{\{a,b\}}(\{e\}) &= |\{2, 3, 5, 6\}|/8 = 4/8 \\
\gamma_{\{b,c\}}(\{e\}) &= |\{2, 3, 5\}|/8 = 3/8 \\
\gamma_{\{a,c\}}(\{e\}) &= |\{2, 3, 5, 6\}|/8 = 4/8
\end{aligned}$$

And calculating the significance of the three attributes gives

$$\begin{aligned}
\sigma_P(Q, a) &= \gamma_{\{a,b,c\}}(\{e\}) - \gamma_{\{b,c\}}(\{e\}) = 1/8 \\
\sigma_P(Q, b) &= \gamma_{\{a,b,c\}}(\{e\}) - \gamma_{\{a,c\}}(\{e\}) = 0 \\
\sigma_P(Q, c) &= \gamma_{\{a,b,c\}}(\{e\}) - \gamma_{\{a,b\}}(\{e\}) = 0
\end{aligned}$$

From this it follows that attribute a is indispensable, but attributes b and c can be dispensed with when considering the dependency between the decision attribute and the given individual conditional attributes.

2.3.6 Reducts

For many application problems, it is often necessary to maintain a concise form of the information system. One way to implement this is to search for a minimal representation of the original dataset. For this, the concept of a *reduct* is introduced and defined as a minimal subset R of the initial attribute set $\mathbb{C}$ such that for a given set of attributes D, $\gamma_R(\mathbb{D}) = \gamma_{\mathbb{C}}(\mathbb{D})$. From the literature, R is a minimal subset if $\gamma_{R-\{a\}}(\mathbb{D}) \neq \gamma_R(\mathbb{D})$ for all $a \in R$. This means that no attributes can be removed from the subset without affecting the dependency degree. Hence a minimal subset by this definition may not be the *global* minimum (a reduct of smallest cardinality). A given dataset may have many reduct sets, and the collection of all reducts is denoted by

$$R_{all} = \{X | X \subseteq \mathbb{C}, \gamma_X(\mathbb{D}) = \gamma_{\mathbb{C}}(\mathbb{D}); \gamma_{X-\{a\}}(\mathbb{D}) \neq \gamma_X(\mathbb{D}), \forall a \in X\} \tag{2.18}$$

The intersection of all the sets in R_{all} is called the *core*, the elements of which are those attributes that cannot be eliminated without introducing more contradictions to the representation of the dataset. For many tasks (e.g., feature selection [67]), a reduct of minimal cardinality is ideally searched for. That is, an attempt is to be made to locate a single element of the reduct set $R_{min} \subseteq R_{all}$:

$$R_{min} = \{X | X \in R_{all}, \forall Y \in R_{all}, |X| \leq |Y|\} \tag{2.19}$$

The problem of finding a reduct of an information system has been the subject of much research [5,355] (see Chapter 5).

2.3.7 Discernibility Matrix

Many applications of rough sets make use of discernibility matrices for finding rules or reducts. A discernibility matrix [181,333] of a decision table $(\mathbb{U}, \mathbb{C} \cup \mathbb{D})$ is a symmetric $|\mathbb{U}| \times |\mathbb{U}|$ matrix with entries defined by

$$c_{ij} = \{a \in \mathbb{C} | a(x_i) \neq a(x_j)\}, \qquad i, j = 1, \ldots, |\mathbb{U}| \tag{2.20}$$

Each c_{ij} contains those attributes that differ between objects i and j.

For finding reducts, the so-called decision-relative discernibility matrix is of more interest. This matrix only considers those object discernibilities that occur when the corresponding decision attributes differ. Returning to the example dataset, the decision-relative discernibility matrix is produced as found in Table 2.2. For example, it can be seen from the table that objects 0 and 1 differ in each attribute. Although some attributes in objects 1 and 3 differ, their corresponding decisions are the same, so no entry appears in the decision-relative matrix. Grouping all entries containing single attributes forms the core of the dataset (those attributes appearing in *every* reduct). Here, the core of the dataset is $\{d\}$.

From this matrix, the concept of discernibility functions can be introduced. This is a concise notation of how each object within the dataset may be distinguished from the others. A discernibility function f_D is a Boolean function of

TABLE 2.2 Decision-Relative Discernibility Matrix

$x \in \mathbb{U}$	0	1	2	3	4	5	6	7
0								
1	a, b, c, d							
2	a, c, d	a, b, c						
3	b, c		a, b, d					
4	d	a, b, c, d		b, c, d				
5	a, b, c, d	a, b, c		a, b, d				
6	a, b, c, d		b, c		a, b, c, d	b, c		
7	a, b, c, d	d		a, c, d			a, d	

m Boolean variables $a_1^*, \ldots, a_m^*$ (corresponding to the membership of attributes $a_1, \ldots, a_m$ to a given entry of the discernibility matrix), defined as below:

$$f_D(a_1^*, \ldots, a_m^*) = \wedge\{\vee c_{ij}^* | 1 \leq j \leq i \leq |\mathbb{U}|, c_{ij} \neq \emptyset\} \tag{2.21}$$

where $c_{ij}^* = \{a^* | a \in c_{ij}\}$. The notation $\vee\{a, b, c, d\}$ and $\wedge\{a, b, c, d\}$ denotes $a \vee b \vee c \vee d$ and $a \wedge b \wedge c \wedge d$, respectively. By finding the set of all prime implicants of the discernibility function, all the minimal reducts of a system may be determined. From Table 2.2 the decision-relative discernibility function is (with duplicates removed)

$$\begin{aligned} f_D(a^*, b^*, c^*, d^*) = {} & (a^* \vee b^* \vee c^* \vee d^*) \wedge (a^* \vee c^* \vee d^*) \\ & \wedge (b^* \vee c^*) \wedge (d^*) \wedge (a^* \vee b^* \vee c^*) \\ & \wedge (a^* \vee b^* \vee d^*) \wedge (b^* \vee c^* \vee d^*) \\ & \wedge (a^* \vee d^*) \end{aligned}$$

Further simplification can be performed by removing those clauses that are subsumed by others:

$$f_D(a^*, b^*, c^*, d^*) = (b^* \vee c^*) \wedge (d^*)$$

The reducts of the dataset may be obtained by converting the expression above from conjunctive normal form to disjunctive normal form (without negations). Hence the minimal reducts are $\{b, d\}$ and $\{c, d\}$. Although this operation is guaranteed to discover all minimal subsets, it is a costly operation rendering the method impractical for even medium-sized datasets. As all that is required is the discovery of a single reduct for many applications, efficient heuristic methods may be applied.

2.4 FUZZY-ROUGH SET THEORY

This section covers the fundamental concepts behind fuzzy extensions to rough set theory. For those readers interested only in the basic ideas of rough set-based feature selection, Chapters 3 through 6 are more directly of interest.

The reliance on discrete data for the successful operation of rough set theory can be seen as a significant drawback of the approach. Indeed this requirement of rough set theory implies an objectivity in the data that is simply not present [178]. For example, in a medical dataset, values such as *Yes* or *No* cannot be considered objective for a *Headache* attribute as it may not be straightforward to decide whether a person has a headache to a high degree of accuracy. Again, consider an attribute *Blood Pressure*. In the real world this is a real-valued measurement, but for the purposes of RST must be discretized into a small set of labels *Normal*,

High, and so forth. Subjective judgments are required for establishing boundaries for objective measurements.

A better way of handling this problem is the use of *fuzzy-rough* sets. Subjective judgments are not entirely removed as fuzzy set membership functions still need to be defined. However, the method offers a high degree of flexibility when dealing with real-valued data, enabling the vagueness and imprecision present to be modeled effectively.

2.4.1 Fuzzy Equivalence Classes

In the same way that crisp equivalence classes are central to rough sets, *fuzzy* equivalence classes are central to the fuzzy-rough set approach [86,359,401]. For typical applications, this means that the decision values and the conditional values may all be fuzzy. The concept of crisp equivalence classes can be extended by the inclusion of a fuzzy similarity relation S on the universe, which determines the extent to which two elements are similar in S. For example, if $\mu_S(x, y) = 0.9$, then objects x and y are considered to be quite similar. The usual properties of reflexivity ($\mu_S(x, x) = 1$), symmetry ($\mu_S(x, y) = \mu_S(y, x)$), and transitivity ($\mu_S(x, z) \geq \mu_S(x, y) \wedge \mu_S(y, z)$) hold. The transitivity requirement may not actually be useful when modeling real world situations. For example, let x, y, and z be a large box, a medium-sized box, and a small box. In this case x and y are somewhat similar (e.g., their similarity is 0.5). It may also be the case that y and z are somewhat similar to each other (e.g., their similarity is 0.5). But, in this example, x and z are not similar (e.g., their similarity is 0.1), and hence transitivity does not hold. However, the properties of transitive fuzzy relations are often desirable from a mathematical viewpoint and are used here.

Using the fuzzy similarity relation, the fuzzy equivalence class $[x]_S$ for objects close to x can be defined:

$$\mu_{[x]_S}(y) = \mu_S(x, y) \tag{2.22}$$

The following axioms should hold for a fuzzy equivalence class F [132]:

- $\exists x$, $\mu_F(x) = 1$ (μ_F is normalized)
- $\mu_F(x) \wedge \mu_S(x, y) \leq \mu_F(y)$
- $\mu_F(x) \wedge \mu_F(y) \leq \mu_S(x, y)$

The first axiom corresponds to the requirement that an equivalence class is nonempty. The second axiom states that elements in y's neighborhood are in the equivalence class of y. The final axiom states that any two elements in F are related via S. Obviously this definition degenerates to the normal definition of equivalence classes when S is nonfuzzy.

The family of normal fuzzy sets produced by a fuzzy partitioning of the universe of discourse can play the role of fuzzy equivalence classes [86]. Consider the crisp partitioning of a universe of discourse, $\mathbb{U}$, by the attributes in Q: $\mathbb{U}/Q = \{\{1, 3, 6\}, \{2, 4, 5\}\}$. This contains two equivalence classes ($\{1, 3, 6\}$ and $\{2, 4, 5\}$) that can be thought of as degenerated fuzzy sets, with those elements belonging to the class possessing a membership of one, and zero otherwise. For the first class, for instance, the objects 2, 4, and 5 have a membership of zero. Extending this to the case of fuzzy equivalence classes is straightforward: objects can be allowed to assume membership values, with respect to any given class, in the interval [0, 1]. $\mathbb{U}/Q$ is not restricted to crisp partitions only; fuzzy partitions are equally acceptable [224].

2.4.2 Fuzzy-Rough Sets

From the literature, the fuzzy P-lower and P-upper approximations are defined as [86]

$$\mu_{\underline{P}X}(F_i) = \inf_x \max\{1 - \mu_{F_i}(x), \mu_X(x)\} \qquad \forall i \tag{2.23}$$

$$\mu_{\overline{P}X}(F_i) = \sup_x \min\{\mu_{F_i}(x), \mu_X(x)\} \qquad \forall i \tag{2.24}$$

where F_i denotes a fuzzy equivalence class belonging to $\mathbb{U}/P$. Note that although the universe of discourse in feature selection is finite, this is not the case in general, hence the use of sup and inf. These definitions diverge a little from the crisp upper and lower approximations, as the memberships of individual objects to the approximations are not explicitly available. As a result the fuzzy lower and upper approximations are redefined here as

$$\mu_{\underline{P}X}(x) = \sup_{F \in \mathbb{U}/P} \min(\mu_F(x), \inf_{y \in \mathbb{U}} \max\{1 - \mu_F(y), \mu_X(y)\}) \tag{2.25}$$

$$\mu_{\overline{P}X}(x) = \sup_{F \in \mathbb{U}/P} \min(\mu_F(x), \sup_{y \in \mathbb{U}} \min\{\mu_F(y), \mu_X(y)\}) \tag{2.26}$$

In implementation, not all $y \in \mathbb{U}$ need to be considered—only those where $\mu_F(y)$ is nonzero, meaning where object y is a fuzzy member of (fuzzy) equivalence class F. The tuple $\langle \underline{P}X, \overline{P}X \rangle$ is called a *fuzzy-rough set*.

It can be seen that these definitions degenerate to traditional rough sets when all equivalence classes are crisp. It is useful to think of the crisp lower approximation as characterized by the following membership function:

$$\mu_{\underline{P}X}(x) = \begin{cases} 1, & x \in F, F \subseteq X \\ 0, & \text{otherwise} \end{cases} \tag{2.27}$$

This states that an object x belongs to the P-lower approximation of X if it belongs to an equivalence class that is a subset of X. Obviously the behavior of the fuzzy lower approximation must be exactly that of the crisp definition for crisp situations. This is indeed the case as the fuzzy lower approximation can be rewritten as

$$\mu_{\underline{P}X}(x) = \sup_{F \in \mathbb{U}/P} \min(\mu_F(x), \inf_{y \in \mathbb{U}} \{\mu_F(y) \rightarrow \mu_X(y)\}) \tag{2.28}$$

where "→" stands for fuzzy implication (using the conventional min-max interpretation). In the crisp case, $\mu_F(x)$ and $\mu_X(x)$ will take values from $\{0, 1\}$. Hence it is clear that the only time $\mu_{\underline{P}X}(x)$ will be zero is when at least one object in its equivalence class F fully belongs to F but not to X. This is exactly the same as the definition for the crisp lower approximation. Similarly the definition for the P-upper approximation can be established.

2.4.3 Rough-Fuzzy Sets

Also defined in the literature are rough-fuzzy sets [349], which can be seen to be a particular case of fuzzy-rough sets. A rough-fuzzy set is a generalization of a rough set, derived from the approximation of a fuzzy set in a crisp approximation space. This corresponds to the case where only the decision attribute values are fuzzy; the conditional values are crisp. The lower and upper approximations incorporate the extent to which objects belong to these sets, and are defined as

$$\mu_{\underline{R}X}([x]_R) = \inf\{\mu_X(x) | x \in [x]_R\} \tag{2.29}$$

$$\mu_{\overline{R}X}([x]_R) = \sup\{\mu_X(x) | x \in [x]_R\} \tag{2.30}$$

where $\mu_X(x)$ is the degree to which x belongs to fuzzy equivalence class X, and each $[x]_R$ is crisp. The tuple $\langle \underline{R}X, \overline{R}X \rangle$ is called a rough-fuzzy set. It can be seen that in the crisp case (where $\mu_X(x)$ is 1 or 0) the definitions above become identical to that of the traditional lower and upper approximations.

Rough-fuzzy sets can be generalized to fuzzy-rough sets [86], where *all* equivalence classes may be fuzzy. When applied to dataset analysis, this means that both the decision values and the conditional values may be fuzzy or crisp.

2.4.4 Fuzzy-Rough Hybrids

In addition to the fuzzy-rough definitions given previously, other generalizations are possible [266]. In [25] the concepts of information theoretic measures are related to rough sets, comparing these to established rough set models of uncertainty. This work has been applied to the rough and fuzzy-rough relational database models, where an alternative definition of fuzzy-rough sets that originates from the rough membership function is chosen [261].

Rough sets may be expressed by a fuzzy membership function to represent the negative, boundary, and positive regions [389]. All objects in the positive region have a membership of one, and those belonging to the boundary region have a membership of 0.5. Those that are contained in the negative region (and therefore do not belong to the rough set) have zero membership. In so doing, a rough set can be expressed as a fuzzy set, with suitable modifications to the rough union and intersection operators.

The reason for integrating fuzziness into rough sets is to quantify the levels of roughness in the boundary region by using fuzzy membership values. It is necessary therefore to allow elements in the boundary region to have membership values in the range of 0 to 1, not just the value 0.5. Hence a fuzzy rough set Y is defined (by this approach) as a membership function $\mu_Y(x)$ that associates a grade of membership from the interval [0, 1] with every element of $\mathbb{U}$. For a rough set X and a crisp equivalence relation R:

$$\mu_Y(\underline{R}X) = 1$$

$$\mu_Y(\mathbb{U} - \overline{R}X) = 0$$

$$0 < \mu_Y(\overline{R}X - \underline{R}X) < 1$$

However, this relation is not a true hybridization of the two approaches, it merely assigns a degree of membership to the elements depending on the crisp positive, boundary, or negative region they belong to. Fuzzy equivalence classes are not used, and so this approach does not offer a particularly useful means for fuzzy-rough attribute reduction.

Another approach that blurs the distinction between rough and fuzzy sets has been proposed in [266]. The research was fueled by the concern that a purely numeric fuzzy set representation may be too precise; a concept is described exactly once its membership function has been defined. By this approach, excessive precision is required in order to describe imprecise concepts.

The solution proposed is termed a *shadowed set*, which itself does not use exact membership values but instead employs basic truth values and a zone of uncertainty (the unit interval). A shadowed set could be thought of as an approximation of a fuzzy set or family of fuzzy sets where elements may belong to the set with certainty (membership of 1), possibility (unit interval), or not at all (membership of 0). This can be seen to be analogous to the definitions of the rough set regions: the positive region (certainty), the boundary region (possibility), and the negative region (no membership).

Given a fuzzy set, a shadowed set can be induced by elevating those membership values around 1 and reducing membership values around 0 until a certain threshold level is achieved. Any elements that do not belong to the set with a membership of 1 or 0 are assigned a unit interval, [0, 1], considered to be a non-numeric model of membership grade. These regions of uncertainty are referred to as "shadows" (see Figure 2.4.). In fuzzy set theory, vagueness is distributed across the entire universe of discourse, but in shadowed sets this vagueness is localized in the shadow regions. As with fuzzy sets, the basic set operations (union, intersection, and complement) can be defined for shadowed sets, as well as shadowed relations.

Shadowed sets have been applied to domains such as fuzzy clustering and image processing with some success [266]. They are particularly useful

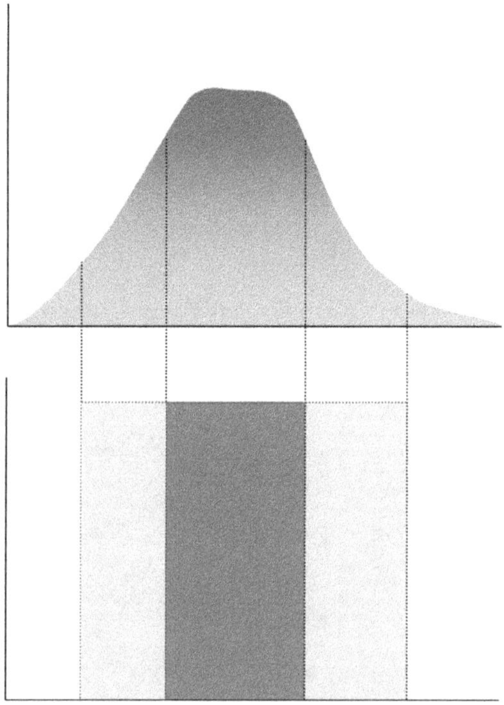

Figure 2.4 Fuzzy set and corresponding shadowed set

in situations where there is a trade-off between numerical precision and computational effort as they reduce the amount of processing involved compared to fuzzy sets. However, there is still a need for a method that uses object membership values when dealing with equivalence classes.

For a more comprehensive discussion on the hybridization of rough and fuzzy sets, including applications, see Chapter 7.

2.5 SUMMARY

This chapter has provided an introduction to the various set theories that are required to understand the concepts presented later in this book. Traditional set-theoretic concepts and operations were initially presented. From these definitions fuzzy sets, rough sets, and fuzzy-rough sets are all constructed. Each of these extensions of set theory attempts to address different uncertainties encountered in real-world data.

CHAPTER 3

CLASSIFICATION METHODS

This chapter presents a brief overview of several classification methods, with a particular focus on fuzzy approaches, including those used in this book. Learning comes in several general forms: supervised learning, unsupervised learning, and reinforcement learning.

Supervised learning aims to devise a method or construct a model for assigning instances to one of a finite set of classes on the basis of a vector of variables (or attributes) measured on the instances. The information on which the rule is to be based is called a training set of instances with known vectors of measurements and known classifications. A typical supervised classification problem has a database of the form: $D = (\mathbf{x}_1, y_1), (\mathbf{x}_2, y_2), \ldots, (\mathbf{x}_n, y_n)$.

Here $\mathbf{x}$ values are typically vectors of the form: $\mathbf{x} = \langle x_1, \ldots, x_n \rangle$ whose components can be discrete or real valued. These components are the attributes (or features) of the database. This approach is essentially the same as that of the decision systems described in Section 2.3.1. The objective is to infer the unknown function (or relation) $y = f(\mathbf{x})$, where the y value is drawn from a discrete set of classes $C = \{C_1, \ldots, C_k\}$ that characterize the given data. Models learned from training data are then evaluated with a different test set in order to determine if they can be generalized to new cases. Supervised learning techniques are the main focus of this chapter.

Unsupervised learning concerns problems where there are no explicit target outputs. Techniques in this field only have the observed input patterns $\mathbf{x}_i$ to work with, often assumed to be independent samples from an underlying unknown

Computational Intelligence and Feature Selection: Rough and Fuzzy Approaches, by Richard Jensen and Qiang Shen

probability distribution. In addition there may be some explicit or implicit a priori information as to the importance of aspects of the data.

Reinforcement learning is the mechanism used by agents that must learn behavior through trial-and-error interactions with a dynamic environment. Two main strategies exist. The first way is to examine the space of behaviors, with the purpose of locating a suitable behavior that performs well. The second is to employ dynamic programming and statistical methods to estimate the utility of taking actions in states of the world.

3.1 CRISP APPROACHES

3.1.1 Rule Inducers

The task of learning rules from data has a long history in machine learning, with the first rule learning algorithms traceable back to its early days. One of the main attractions of rule induction is that the rules are much more transparent and easier to interpret than, say, a regression model or a trained neural network. Rules are arguably the most comprehensive concept representation formalism.

The task itself is to find a set of classification rules from data that can be used for the prediction or classification of new unlabeled examples. Restrictions imposed by the language used to describe the data and the language used to describe the induced ruleset must be taken into account. The data description language imposes a bias on the form of data, and the hypothesis restriction language imposes a bias on the form of the induced rules. Induction takes place using a set of classified objects described in the data description language, with the aim of finding a set of rules in the hypothesis language that is both consistent (does not cover any negative examples) and complete (covers all positive examples).

3.1.1.1 AQ15 AQ15 [229] associates conjunctions in a generated rule with weights; those with the least weights are removed to avoid overfitting the data. First, a complete matching is considered where all attribute-value pairs of a rule must match all values of the corresponding attributes for the example. If this is not possible, a partial matching is done, where some attribute-value pairs of a rule match the values of corresponding attributes. The best rule is chosen based on probability estimates. Each rule is weighted by the percentage of positive examples in the set of examples covered by it. The weights of rules of the same class are combined to a weight for the entire class and the class with the highest weight will be returned.

3.1.1.2 CN2 CN2 induces propositional classification rules from data [58]. The algorithm consists of two main procedures: the search procedure and the control procedure. The search procedure performs beam search to locate a single rule, employing the classification accuracy of the rule as a guiding heuristic. For this, conditional probabilities need to be estimated (the probability of a class given the rule consequents) allowing different probability estimates to be used, for example, the Laplace [57] or the m-estimate [91]. Additionally CN2 can apply a significance test to the induced rule. The rule is considered to be significant if it locates regularity unlikely to have occurred by chance. The likelihood ratio statistic is typically used for this purpose. This ratio measures the difference between the distribution of class probabilities in the set of objects covered by the rule and the class probability distribution in the set of all training objects.

There are two different control procedures in CN2: one for the induction of ordered rulesets (where rules have a specific order) and the second for the unordered case. For ordered ruleset induction, the search mechanism looks for the best rule determined by the heuristic and removes all training objects covered by it. A new search is initiated by the search procedure, repeating this process until all objects are covered. For unordered rulesets, rules are induced for each class. Only objects covered by the rule and belonging to the class are removed, leaving the negative examples in the training data. This prevents CN2 from inducing exactly the same rule.

3.1.1.3 PRISM PRISM [47] is a method that generates rulesets for each class present in training data. Rules in each ruleset predict the same class. As a result the sets of rulesets produced are not order dependent. The induction process for PRISM is as follows: For each class C, the set of training objects in the dataset is examined. Rules are created while there are objects remaining in the dataset with this class. This rule generation process begins by creating a rule with an empty antecedent part that predicts the class C. Attribute-value pairs are iteratively added to the consequent part in order to maximize the classification accuracy. Instances are then removed that are covered by the finished rule, and the algorithm starts to construct another rule to cover the same class C. This is repeated for all classes. As a result of the rule construction method it is quite possible for rules to overlap.

3.1.1.4 LERS LERS (learning from examples based on rough sets) [119] induces a set of rules from examples and also handles inconsistent data through the use of rough sets. Lower and upper approximations of concepts are used for this, generating *certain* and *possible* rules. The process operates by examining attribute-value pairs and how well they approximate the decision concepts. Through the definition of minimal complexes and local coverings [120,351], minimal sets of attribute-value pairs can be discovered from data that adequately describe decision values. These sets are then directly translated into decision rules, forming the antecedent part of the rule with the associated decision value forming the consequent.

For unseen data, the decision to which concept the data belongs to is based on three factors: strength, specificity, and support. Strength is the total number of objects correctly classified by the rule during training. Specificity is the number of attribute-value pairs in the antecedent part of the rule. Rules containing a larger number of such pairs that match a test object are considered to be more specific.

Finally, support is defined as the sum of scores of all matching rules from the concept.

3.1.1.5 JRip JRip [59] learns propositional rules by repeatedly growing rules and pruning them. During the growth phase antecedents are added greedily until a termination condition is satisfied. Antecedents are then pruned in the next phase subject to a pruning metric. Once the ruleset is generated, a further optimization is performed where rules are evaluated and deleted based on their performance on randomized data.

3.1.1.6 PART PART [381] generates rules by means of repeatedly creating partial decision trees from data. The algorithm adopts a separate-and-conquer strategy in that it removes instances covered by the current ruleset during processing. Essentially a rule is created by building a pruned tree for the current set of instances; the leaf with the highest coverage is made into a rule.

3.1.2 Decision Trees

Decision trees are a popular hypothesis language as they are easy to comprehend and give an explicit model of the decision-making process. An internal node of an induced decision tree specifies a test on an attribute of the data set (though more complex trees may be built by specifying tests at nodes based on more than one attribute). Each outgoing branch of the node corresponds to a possible result of the test and leaf nodes represent the class label to be assigned to an instance. To classify an instance, a path from the root node of the decision tree is traced to a leaf node; each internal node reached specifies which outgoing branch should be taken, depending on the value of the relevant attribute in the instance. The instance is assigned the class label of the leaf node reached at the end of the path. Each individual path of the tree may also be equated with an *if-then* rule.

In traditional machine learning, decision trees are induced by following a divide-and-conquer strategy whereby, depending on the splitting criterion used, the training data are partitioned into disjoint subsets and the algorithm is applied recursively to each subset. Examples of such algorithms that induce decision trees are ID3 [279], C4.5 [281], and CART: classification and regression trees [44]. Such learning algorithms generally use an attribute-value representation as the example language; that is, each instance in the training data is described by specific values of a given set of attributes. The domain of each attribute may be finite or nominal, or numerical.

3.1.3 Clustering

Clustering is an unsupervised learning problem—it deals with finding structure in a collection of unlabeled data. The clustering method itself can be thought of as the process of organizing objects into groups whose members share similar properties. A cluster, then, is a collection of objects that are similar but are dissimilar to objects belonging to other clusters. This has been seen, to some extent, in Section 2.3 where the concept of equivalence classes is similar to the clustering concept. Objects belong to an equivalence class if they contain the same attribute values for a given set of attributes. Each class can be thought of as a cluster, with the similarity criterion being the strict requirement that objects are identical. Two of the main approaches to clustering are partitional and hierarchical.

Partitional clustering seeks to construct partitions of a dataset of n objects into k clusters. The goal of these methods is to optimize a chosen partitioning criterion for a given k. In order to find the global optimal, an exhaustive enumeration of all partitions would have to be carried out. This is impractical, so heuristic methods are typically employed to find local optima.

The k-means algorithm [219] assigns each object to a cluster whose centroid (center) is closest. The center is the average of all the points in the cluster—the arithmetic mean of each attribute for all objects in the cluster. The algorithm begins by randomly generating k clusters and determining cluster centers or generating k seed points as cluster centers directly. Every object is assigned to its nearest cluster center. With these newly formed clusters, the new cluster centers are calculated. The process is continued until some convergence criterion is met, usually when assignments have remained unchanged. The algorithm is relatively efficient: $O(tkn)$, where n is the number of objects, k the number of clusters, and t the number of iterations. However, the same result does not get systematically produced with each run of the algorithm as the resulting clusters depend on the initial assignments. There is no guarantee that the solution given is a global optimum and not just a local optimum.

A related approach is the k-mediods clustering method. Here the task is to find representative objects (mediods) in clusters. To achieve this, only the definition of distance between objects is required. Initially k objects are chosen as the initial mediods. Each remaining object is assigned to the cluster with nearest mediods. Mediods are iteratively replaced by nonmediods if they improve the total distance of the resulting clustering. Typical methods for this approach include PAM [174] and CLARANS [247].

One of the problems of the k-means algorithm is that it gives a hard partitioning of the data, meaning each point is attributed to one and only one cluster. But points on the edge of the cluster, or near another cluster, may not be as much in the cluster as points in the center of cluster. Fuzzy c-means [31,88] is an approach that extends k-means where cluster membership is relaxed so that objects have a degree of belonging to one or more clusters.

Hierarchical clustering is another approach to the clustering problem, which builds (agglomerative) or breaks up (divisive) a hierarchy of clusters. In the agglomerative method, clusters are merged iteratively. Each object is placed in its own cluster at the start, and these clusters successively merged into larger clusters until all clusters are merged. A suitable distance measure between clusters must be defined. Each agglomeration occurs at a greater distance between clusters than the previous agglomeration. Clustering may be terminated either when clusters are too far apart to be merged or when a sufficiently small number of clusters have been obtained. In the divisive method, clusters are split iteratively. Initially all objects are contained within one cluster which is then subdivided into smaller clusters. Standard hierarchical clustering methods are AGNES [174] and DIANA [174].

3.1.4 Naive Bayes

Naive Bayes classifiers, although simple, often exhibit high classification accuracies, comparable in performance with the best decision trees and neural network based approaches. Based on probabilities determined from data, new objects may be determined to belong to classes with varying degrees of probability.

Consider a fruit dataset with fruits described by their shape and color. Given a new fruit that requires classification, a Bayesian classifier would compare its attributes with those objects present in the dataset and determine which fruit it is most likely to be. Probabilities are estimated based on the frequencies of occurrence of attribute values within the dataset, so a reasonable number of objects are required in order to achieve suitably high accuracy. This can present problems when a large number of attributes and classes are considered.

Naive Bayes classification avoids this problem by assuming that attributes are independent, and hence attribute combinations do not need to be considered. In other words, the effect of an attribute value on a particular class is independent of the values of other variables. For instance, the probability of a green, round, firm fruit being an apple can be determined from the independent probabilities that the fruit is green, that it is round, and that it is firm. Although the independence assumption is a strong one, and is often not applicable, it does not affect performance greatly as it is the order of the probabilities, not their exact values, that determine the classifications.

The starting point for the naive Bayes classification is the Bayes theorem:

$$P(h|D) = \frac{P(D|h)P(h)}{P(D)} \tag{3.1}$$

where $P(h)$ is the prior probability of hypothesis h, $P(D)$ is the prior probability of training data D, $P(h|D)$ is the probability of h given D, and $P(D|h)$ is the probability of D given h. In general, the most probable hypothesis given the training data is required, the *maximum a posteriori* hypothesis h_{MAP}:

$$\begin{aligned} h_{MAP} &= \arg\max_{h\in H} P(h|D) \\ &= \arg\max_{h\in H} \frac{P(D|h)P(h)}{P(D)} \\ &= \arg\max_{h\in H} P(D|h)P(h) \end{aligned} \tag{3.2}$$

If it is assumed that $P(h_i) = P(h_j)$, then further simplification can be carried out, choosing the *maximum likelihood* (ML) hypothesis:

$$h_{ML} = \arg\max_{h_i \in H} P(D|h_i) \tag{3.3}$$

For classification, the naive Bayes probability model is combined with a decision rule. A common rule for this purpose is to choose the most probable hypothesis given an unlabeled vector, either via MAP or ML above.

3.1.5 Inductive Logic Programming

The particular subfield of machine learning that is concerned with learning more complex knowledge from training data than can be represented by propositional logic is that of inductive logic programming (ILP) [241]. This type of learning is often termed relational learning as the learned description may specify relations among parts of an object. ILP is distinguished by the use of background knowledge (in addition to the use of concrete examples) that acts as a model or template for the solution that must be learned by the algorithm. This background knowledge therefore restricts the solution search space and makes the computation more tractable.

ILP was originally focused on program synthesis in logic languages such as Prolog, but such programs may be considered rules. A consequent shift has occurred in research that also emphasizes the use of ILP methods for knowledge discovery in databases. Here a distinction is often made between predictive ILP, where the aim is the learning of classification and prediction rules, and descriptive ILP, where the aim is the learning of clausal theories and association rules. A more detailed discussion of ILP status, issues, and future trends can be found in [196].

For first-order classification rule induction the examples in the training data are commonly described by ground facts, namely logical formulas that contain no variables and have exactly one predicate that is positive. The hypothesis language is frequently a restriction of Horn clauses, namely clauses containing at most one literal. Well-known nonfuzzy algorithms include FOIL [280], GOLEM [242], and PROGOL [240], which are mainly concerned with the learning of single predicates from positive and negative examples and background knowledge

3.2 FUZZY APPROACHES

There are several different approaches for reasoning with imperfect or imprecise knowledge, including fuzzy rule-based systems that are based on fuzzy set theory and fuzzy logic [408]. At the core of a such a system are a knowledge base and an inference procedure:

1. The knowledge base is composed of fuzzy production *if-then* rules that conceptualize domain knowledge, and membership functions defining the fuzzy sets associated with the linguistic terms employed in the fuzzy rules.
2. The inference procedure uses this stored knowledge to formulate a mapping from a given input (e.g., in classification, conditions denoted by attribute values) to an output (e.g., in classification, a conclusion denoted by a class label).

Fuzzy rule-based systems capture and reason with imprecise or inexact knowledge (in fuzzy logic everything is a measure of degree [412]), and since many real-world problems contain a measure of imprecision and noise, the application of such approximate reasoning systems in these situations is a viable approach. This is supported by many successful applications in industry and commerce that deal with automated classification, diagnosis, monitoring, and control [20,130,265]. Approaches in the field of fuzzy rule-based systems can be categorized as fuzzy grid-based [149], fuzzy clustering-based [313], fuzzy decision tree-based [157], and neuro-fuzzy methods [245,246]. Evolutionary computation has often been used in the design of fuzzy rule-based classification systems [150,151]. This section provides an overview of several such fuzzy rule induction methods used in this book.

3.2.1 Lozowski's Method

The rule induction algorithm presented in [216] extracts fuzzy rules from real-valued examples. Although this data-driven RIA was proposed to be used in conjunction with neural network based classifiers, it is independent of the type of classifier used [321]. Provided with training data, the RIA induces approximate relationships between the characteristics of the conditional attributes and those of the decision attributes. The conditional attributes of the induced rules are represented by fuzzy variables, facilitating the modeling of the inherent uncertainty of application domains.

The algorithm generates a hyperplane of candidate fuzzy rules ($p_1 \wedge p_2 \wedge \cdots \wedge p_n \Rightarrow c$) by fuzzifying the entire dataset using all combinations of rule conditions. Thus a domain with n conditional attributes, each of which is a fuzzy region fuzzified by f_x fuzzy sets ($1 \leq x \leq n$), is fuzzified into $\prod_{i=1}^{n} f_i$ n-dimensional clusters, each representing one vector of rule conditions. Each cluster $\mathbf{p} = \langle \mu_1, \mu_2, \ldots, \mu_n \rangle$ may lead to a fuzzy rule provided that training examples support it. To obtain a measure of what classification applies to a cluster, fuzzy min-max composition is used. The conditional attribute values of each training example are fuzzified according to the fuzzy conditions $\langle \mu_1, \mu_2, \ldots, \mu_n \rangle$ that make up cluster $\mathbf{p}$. For each example $\mathbf{x} = \langle x_1, x_2, \ldots, x_n \rangle$, $S_c^{\mathbf{p}}\mathbf{x} = \min\left(\mu_1(x_1), \mu_2(x_2), \ldots, \mu_n(x_n)\right)$ is calculated. This procedure is the s-norm of example $\mathbf{x}$ with respect to cluster $\mathbf{p}$ and classification c. To give a measure of the applicability of a classification to cluster $\mathbf{p}$, the maximum of all s-norms with respect to $\mathbf{p}$ and c is calculated (this is dubbed a t-norm): $T_c^{\mathbf{p}} = \max\left\{S_c^{\mathbf{p}}\mathbf{x} \ \forall \ \mathbf{x} \in D_c\right\}$, where D_c is the set of all dataset examples that can be classified as c. This procedure is iterated over all possible classifications c to provide a full indication of how well each cluster applies to each classification.

A cluster generates at most one rule. The rule's conditions are the cluster's n coordinate fuzzy sets. The conclusion is the classification attached to the cluster. Since there may be t-norms for more than one classification, it is necessary to decide on one classification for each of the clusters. Such contradictions are resolved by using the *uncertainty margin*, ε ($0 \leq \varepsilon < 1$). This means that a t-norm assigns its classification on its cluster if and only if it is greater by at least ε than all other t-norms for that cluster. If this is not the case, the cluster is considered undecidable and no rule is generated. The uncertainty margin introduces a trade-off to the rule generation process. In general, the higher ε is, the fewer rules are generated, but the classification error may increase.

Lozowski's RIA is NP-hard, and it may become intractable when the rules induced for the datasets have many conditional attributes [321]. The most important problem, in terms of both memory and runtime is dealing with the large numbers of combinations of fuzzy values. This is not so important when only a few attributes are involved. Applied to a more complex problem without some means of attribute reduction, the algorithm's intractable nature becomes evident, both in terms of time and space.

It is thus convenient and helpful to treat the creation of fuzzy-set vectors as the creation of a tree. In this context a leaf node is one combination of membership functions, and each arc represents one evaluation of a membership function. The minimum membership is retained when creating the t-norms. Any membership function that evaluates to zero means that all leaf nodes in the subtree will eventually evaluate to zero, too, because of the use of the $\min(\cdot)$ function. A subtree is therefore useless and can be pruned iff its root node evaluates to zero.

In an application domain where a reasonable degree of resolution is required, it is not unusual to see quantities partitioned into five or seven fuzzy sets. Assuming an average of six fuzzy sets per attribute and 40 attributes, the data may be seen as a 40-dimensional ($\mathbb{R}^{40}$) hyperplane, each dimension of which is a fuzzy region covered by six fuzzy sets. The RIA would attempt to produce rules to cover the entire space by using each fuzzy set of each dimension. Thus it would need to generate at most 6^{40} possible rules.

In most applications of fuzzy logic, any given value x in a fuzzy region will belong to *at most* two fuzzy sets A and B with membership $\mu_A(x) \geq \mu_B(x) > 0$. Thus, for any other fuzzy set F_i, it may be assumed that $\mu_{F_i}(x) = 0$.

The RIA pruning algorithm will detect this pattern at an early stage and not consider fuzzy sets F_i as containing candidate rules. Therefore, for each of the 40 fuzzy regions (dimensions of the hyperplane), two of the six fuzzy sets will be allowed to generate candidate rules. This reduces the number of combinations to at *worst* 2^{40}. If some values in a fuzzy region only belong to *one* fuzzy set with nonzero membership, this number becomes smaller.

TABLE 3.1 Example dataset: SBA

Object	*a*			*b*			*c*		*Plan*		
	*A*1	*A*2	*A*3	*B*1	*B*2	*B*3	*C*1	*C*2	X	Y	Z
1	0.3	0.7	0.0	0.2	0.7	0.1	0.3	0.7	0.1	0.9	0.0
2	1.0	0.0	0.0	1.0	0.0	0.0	0.7	0.3	0.8	0.2	0.0
3	0.0	0.3	0.7	0.0	0.7	0.3	0.6	0.4	0.0	0.2	0.8
4	0.8	0.2	0.0	0.0	0.7	0.3	0.2	0.8	0.6	0.3	0.1
5	0.5	0.5	0.0	1.0	0.0	0.0	0.0	1.0	0.6	0.8	0.0
6	0.0	0.2	0.8	0.0	1.0	0.0	0.0	1.0	0.0	0.7	0.3
7	1.0	0.0	0.0	0.7	0.3	0.0	0.2	0.8	0.7	0.4	0.0
8	0.1	0.8	0.1	0.0	0.9	0.1	0.7	0.3	0.0	0.0	1.0
9	0.3	0.7	0.0	0.9	0.1	0.0	1.0	0.0	0.0	0.0	1.0

Even given the worst-case scenario, however, the time needed by the enhanced algorithm for this example is approximately 19 orders of magnitude less than that needed for the full tree traversal. The savings is significant, but the number of combinations is still far too large.

3.2.2 Subsethood-Based Methods

3.2.2.1 SBA The method proposed in [51] uses the notion of subsethood to derive fuzzy rules from data. For simplicity in outlining this induction procedure the dataset given in Table 3.1. is used. There are three features each with corresponding linguistic terms; for example, a has terms $A1$, $A2$, and $A3$. The decision feature $Plan$ is also fuzzy, separated into three linguistic decisions X, Y, and Z.

The algorithm begins by organizing the dataset objects into subgroups according to their highest decision value. Within each subgroup the fuzzy subsethood [185,406] is calculated between the decisions of the subgroup and each feature term. Fuzzy subsethood is defined as follows:

$$S(A, B) = \frac{M(A \cap B)}{M(A)} = \frac{\sum_{u \in \mathbb{U}} \min(\mu_A(u), \mu_B(u))}{\sum_{u \in \mathbb{U}} \mu_A(u)} \tag{3.4}$$

From this, the subsethood values listed in Table 3.2. can be obtained. For instance, $S(X, A1) = 1$ is obtained by examining the subgroup of objects that belong to the decision X in Table 3.3. The objects concerned are 2, 4, and 7:

Object	*A*1	*X*
2	1.0	0.8
4	0.8	0.6
7	1.0	0.7

TABLE 3.2 Subsethood Values between Conditional Feature Terms and the Decision Terms

Plan	Linguistic term							
	A1	A2	A3	B1	B2	B3	C1	C2
X	1	0.1	0	0.71	0.43	0.14	0.52	0.76
Y	0.33	0.58	0.29	0.42	0.58	0.04	0.13	0.92
Z	0.14	0.64	0.29	0.32	0.61	0.14	0.82	0.25

From these values the necessary components of equation (3.4) can be calculated as follows:

$$M(X) = 0.8 + 0.6 + 0.7 = 2.1$$

$$\begin{aligned} M(X \cap A1) &= \min(0.8, 1) + \min(0.6, 0.8) + \min(0.7, 1) \\ &= 0.8 + 0.6 + 0.7 = 2.1 \end{aligned}$$

These subsethood values are an indication of the relatedness of the individual terms of the conditional features (or values of the features) to the decisions. This measure is central to the fuzzy RIA [51]. A suitable level threshold, $\alpha \in [0,1]$, must be chosen beforehand in order to determine whether terms are close enough or not. At most, one term is selected per feature. For example, setting $\alpha = 0.9$ means that the term with the highest fuzzy subsethood value (or its negation) above this threshold will be chosen. Applying this process to the first two decision values X and Y generates the following rules:

Rule 1: IF a is $A1$, THEN $Plan$ is X
Rule 2: IF b is NOT $B3$ AND c is $C2$, THEN $Plan$ is Y

A problem is encountered here when there are no suitably representative terms for a decision (as is the case for decision Z). In this situation a rule is produced that classifies cases to the decision value if the other rules do not produce reasonable classifications, in order to entail full coverage of the learned rules over the entire problem domain. This requires another threshold value, $\beta \in [0, 1]$, that determines whether a classification is reasonable or not. For decision Z, the following rule is produced:

Rule 3: IF $MF(Rule\ 1) < \beta$ AND $MF(Rule\ 2) < \beta$, THEN $Plan$ is Z

where MF(Rule i) = MF(condition part of Rule i) and MF means the membership function value. The use of thresholds α and β pose a problem for the RIA, as it is not clear how to acquire such information. Typically these are estimated from repeated experimentation over the threshold range, but this procedure is not guaranteed to find the optimum values.

The classification results when using these rules on the example dataset can be found in Table 3.3. It shows the membership degrees of the cases to each

TABLE 3.3 Classified Plan with all Features and the Actual Plan

Object	Classified			Actual		
	X	*Y*	*Z*	*X*	*Y*	*Z*
1	0.3	0.7	0.0	0.1	0.9	0.0
2	1.0	0.3	0.0	0.8	0.2	0.0
3	0.0	0.4	1.0	0.0	0.2	0.8
4	0.8	0.7	0.0	0.6	0.3	0.1
5	0.5	1.0	0.0	0.6	0.8	0.0
6	0.0	1.0	0.0	0.0	0.7	0.3
7	1.0	0.8	0.0	0.7	0.4	0.0
8	0.1	0.3	1.0	0.0	0.0	1.0
9	0.3	0.0	1.0	0.0	0.0	1.0

classification for the classified plan and the underlying plan present in the training dataset. Clearly, the resulting classifications are the same when the min t-norm is used.

This technique has been shown to produce highly competitive results [51] in terms of both classification accuracy and number of rules generated. However, as is the case for most rule induction algorithms, the resultant rules may be unnecessarily complex due to the presence of redundant or misleading features. Fuzzy-rough feature selection (FRFS; see chapter 8) may be used to significantly reduce dataset dimensionality, removing redundant features that would otherwise increase rule complexity and reducing the time for the induction process itself. Applying this feature selection method to the example dataset removes feature b, and results in the set of rules given in Figure 3.1 upon application of SBA:

From this ruleset, it can be seen that rule 2 has been simplified because of the redundancy of feature b. Although the extent of simplification is small in this case, with larger datasets the effect can be expected to be greater.

The results using the FRFS-reduced dataset are provided in Table 3.4. The differences between the classifications of the reduced and unreduced approaches have been highlighted (objects 4 and 7). In object 4 only the membership degree for Y has changed. This value has increased from 0.7 to 0.8, resulting in an ambiguous classification. Again, for case 7 the membership degree for Y is the only value to have changed; this time it more closely resembles the classification present in the training dataset.

Rule 1: IF *a* is *A1*, THEN *Plan* is *X*
Rule 2: IF *c* is *C2*, THEN *Plan* is *Y*
Rule 3: IF *MF(Rule1)* < β AND *MF(Rule2)* < β, THEN *Plan* is *Z*

Figure 3.1 Generated rules using the reduced dataset

TABLE 3.4 Classified Plan with Reduced Features and the Actual Plan

Object	Classified			Actual		
	X	Y	Z	X	Y	Z
1	0.3	0.7	0.0	0.1	0.9	0.0
2	1.0	0.3	0.0	0.8	0.2	0.0
3	0.0	0.4	1.0	0.0	0.2	0.8
4	0.8	0.8	0.0	0.6	0.3	0.1
5	0.5	1.0	0.0	0.6	0.8	0.0
6	0.0	1.0	0.0	0.0	0.7	0.3
7	1.0	0.3	0.0	0.7	0.4	0.0
8	0.1	0.3	1.0	0.0	0.0	1.0
9	0.3	0.0	1.0	0.0	0.0	1.0

3.2.2.2 WSBA

In an attempt to tackle the limitations of the SBA method, the weighted subsethood-based approach was developed [286]. For this method *no* threshold values are required in the operation of the induction or classification algorithm. Additionally default rules are avoided.

WSBA uses the subsethood values calculated in SBA to attach weights to the constituent parts of fuzzy complete rules. Fuzzy complete rules take the form:

Rule 1: IF A is (A_1 OR A_2 OR ... OR A_i) AND B is (B_1 OR B_2 OR ... OR B_j) AND ... AND H IS (H_1 OR H_2 OR ... OR H_k), THEN decision is D_1

...

Rule n: IF A is (A_1 OR A_2 OR ... OR A_i) AND B is (B_1 OR B_2 OR ... OR B_j) AND ... AND H IS (H_1 OR H_2 OR ... OR H_k), THEN decision is D_n

All linguistic terms of each attribute are used to describe the antecedent of each rule initially. This may look tedious, but the reason for keeping this complete form is that every linguistic term may contain important information that should be taken into account.

WSBA employs the subsethood values for each fuzzy set and corresponding decision fuzzy set (representing a decision concept) to modify these rules as follows:

Rule 1: IF A is $(S(D_1, A_1).A_1$ OR $S(D_1, A_2).A_2$ OR ... OR $S(D_1, A_i) \cdot A_i)$ AND B is $(S(D_1, B_1).B_1$ OR $S(D_1, B_2).B_2$ OR ... OR $S(D_1, B_j).B_j)$ AND ... AND H IS $(S(D_1, H_1).H_1$ OR $S(D_1, H_2).H_2$ OR ... OR $S(D_1, H_k).H_k)$, THEN decision is D_1

...

Rule n: IF A is $(S(D_n, A_1).A_1$ OR $S(D_n, A_2).A_2$ OR ... OR $S(D_n, A_i).A_i)$ AND B is $(S(D_n, B_1).B_1$ OR $S(D_n, B_2).B_2$ OR ... OR $S(D_n, B_j).B_j)$ AND ... AND H IS $(S(D_n, H_1).H_1$ OR $S(D_n, H_2).H_2$ OR ... OR $S(D_n, H_k).H_k)$, THEN decision is D_n

These rules can be interpreted as a combination of fuzzy general rules and fuzzy quantifier rules. The weights for each linguistic term are considered as the quantifiers "some" or "all." If the weight is 1, the quantifier is regarded as "all;" otherwise, it is considered to represent "some." The extent to which "some" is interpreted depends on the value of the weights of the respective linguistic terms.

3.2.2.3 *QSBA* The weights used in WSBA outlined above are not fuzzy, and hence limit the interpretability of the induced rulesets. Quantified subsethood-based approach (QSBA) [287] is an attempt to remedy this situation, by replacing weights with *fuzzy quantifiers*.

Fuzzy quantifiers are one way to attach semantic labels to fuzzy sets as with the use of hedges (see Section 2.2.6). They can be seen as flexible tools for the representation of natural language, making existing fuzzy models more readable. Fuzzy quantification techniques can be based on the generalization of first-order logic quantifiers, where the quantification mechanism involves the definition of the existential quantifier, ∃ (exists at least one) and of the universal quantifier, ∀ (for all). However, the two-valued quantification technique seems too strict as it will return two extreme values, thus ignoring the existence of other quantifications that are readily available in fuzzy terms and natural language, such as "almost half," "nearly all," "few," and "most." Extending this representation language to fuzzy sets, the truth-value of the existential relative quantifier and the universal relative quantifier can be defined as

$$T_{\exists,A/D} = \wedge_{k=1}^{N}(\mu(d_k) \vee \mu(a_k)) \tag{3.5}$$

$$T_{\forall,A/D} = \vee_{k=1}^{N}(\mu(d_k) \rightarrow \mu(a_k)) \tag{3.6}$$

where $\mu(d_k)$ and $\mu(a_k)$ are the membership functions of fuzzy sets A and D respectively, and $\rightarrow$ denotes fuzzy implication.

It is obvious that this definition covers, as its specific cases, classical existential and universal quantifiers. The multi-valued fuzzy quantification can be defined using any available functions such as nondecreasing, nonincreasing, or unimodal within the above definition. The multi-valued quantifiers can be expanded further because the number of quantifiers that can exist may not only be limited to a few specific ones. However, the problems in expanding this kind of quantifier lie in the need to pre-define each of the quantifiers.

For the purposes of QSBA the continuous fuzzy quantifiers proposed in [367] are used. This employs linear interpolation between the two extreme cases (existential and universal):

$$T_{Q,A/D}(\lambda_Q) = (1 - \lambda_Q)T_{\forall,A/D} + \lambda_Q T_{\exists,A/D} \tag{3.7}$$

where Q is the quantifier for fuzzy set A relative to fuzzy set D and λ is the degree of neighborhood of the two extreme quantifiers.

3.2.3 Fuzzy Decision Trees

One significant problem that has faced crisp tree induction is how to effectively handle continuous features. A standard approach that aims to address this is the C4.5 tree construction algorithm [281], which considers intervals of values during its construction. However, some interpretability of the tree is lost as the intervals, although useful for classification purposes, may not have any direct physical relevance or meaning to the problem at hand.

Fuzzy decision trees (FDTs) are able to handle continuous features through the use of fuzzy sets. Unlike crisp decision tree induction, FDTs do not use the original numerical feature values directly in the tree. They instead use fuzzy sets generated either from a fuzzification process beforehand or expert-defined partitions to construct comprehensible trees. As a result there are several key differences between FDT induction and the original crisp approaches:

- *Membership of objects.* Traditionally objects/examples belonged to nodes with a membership of {0, 1}; now these memberships may take values from the interval [0, 1]. In each node an example has a different membership degree to the current example set, and this degree is calculated from the conjunctive combination of the membership degrees of the example to the fuzzy sets along the path to the node and its degrees of membership to the classes.
- *Measures of feature significance.* As fuzzy sets are used, the measures of significance should incorporate this membership information to decide which features form nodes within the tree. This is particularly important as the quality of the tree can be greatly reduced by a poor measure of feature significance.
- *Fuzzy tests.* Within nodes, fuzzy tests are carried out to determine the membership degree of a feature value to a fuzzy set.
- *Stopping criteria.* Learning is usually terminated if all features are used on the current path, or if all objects in the current node belong to the same class. With fuzzy trees, objects can belong to any node with any degree of membership. As a result fuzzy trees tend to be larger in size, which can lead to poorer generalization performance. An additional threshold can be introduced, based on the feature significance measure, to terminate construction earlier in induction. For classification the decision tree is converted to an equivalent ruleset.

In [16,157,406] fuzzy decision tree-building algorithms based on ID3 are presented. After all attributes' values have been fuzzified, this algorithm recursively partitions the training data set until a termination criterion is satisfied. Note that the training examples within a node are not partitioned further if they all belong to the same class, or all attributes have been used in the path leading to the node, or the information content of the node reaches a certain threshold level. Nodes are then processed according to an ordering measure; the node containing the greatest number of examples is processed first, and the partitioning of the training examples is then dependent on selecting the attribute that maximizes the information gain measure based on fuzzy entropy.

Let $I = (\mathbb{U}, \mathbb{A})$ be an information system, where $\mathbb{U}$ is a nonempty set of N finite objects (the universe) and $\mathbb{A}$ is a nonempty finite set of n attributes, $\{A^1, A^2, \ldots, A^n\}$. An attribute A^k takes m_k values of fuzzy subsets $\{A_1^k, A_2^k, \ldots, A_{m_k}^k\}$. Based on the attributes, an object is classified into C fuzzy subsets $\omega_1, \omega_2, \ldots, \omega_C$.

The fuzzy entropy for a subset can be defined as

$$H_i^k = \sum_{j=1}^{C} -p_i^k(j) \log p_i^k(j) \tag{3.8}$$

where, $p_i^k(j)$ is the relative frequency of the ith subset of attribute k with respect to ω_j $(1 \leq j \leq C)$ and defined as

$$p_i^k(j) = \frac{|A_i^k \cap \omega_j|}{|A_i^k|} \tag{3.9}$$

The cardinality of a fuzzy set is denoted by $|\cdot|$. An attribute s is chosen to split the instances at a given node according to

$$s = \arg \min_{1 \leq k \leq n} E^k \tag{3.10}$$

where

$$E^k = \sum_{i=1}^{m_k} \frac{|A_i^k|}{\sum_{j=1}^{m_k} |A_j^k|} H_i^k \tag{3.11}$$

3.2.4 Evolutionary Approaches

Evolutionary computation (EC) is the application of methods inspired by Darwinian principles of evolution to computationally difficult problems. Evolutionary algorithms (EAs) re-iteratively apply genetic-inspired operators to a population of solutions, modifying or replacing members of the population so that, on average, each new generation tends to be better than the previous one according to some predefined fitness criteria.

EAs have been extensively and successfully applied to combinatorial and search problems. Reasons for their popularity include their broad range of application (e.g., robotics, control, and natural language processing), their relative simplicity of implementation that requires little domain knowledge, and their development of multiple solutions that search different parts of the solution space simultaneously. More detailed discussion of EC strengths, issues, and theoretical foundations and aspects such as computational complexity and algorithm convergence may be found in [13,101,110,173,380,399].

Evolutionary techniques have been used to generate the rulebase of a fuzzy rule-based system, or fine-tune the membership functions, or both [61,62,127]. Other applications of evolutionary techniques are related to the preprocessing and postprocessing stages of the knowledge discovery process. These include feature construction or selection [347], training example subset selection [93], and rulebase optimization [152,243].

3.2.4.1 General Framework From some of the early work on classifier systems two terms have emerged that are still in common usage today: Michigan-style [134] and Pittsburgh-style [348], the names being in recognition of the institutions of the originators of the two approaches. In the first approach one rule is encoded as one individual of the population, while in the second, later, approach a rulebase is encoded as one individual of the population.

If Michigan-style encoding is adopted then the EA generally evolves either a rule antecedent or an entire rule. In the first case, at each iteration a separate deterministic procedure is used to determine the rule consequent before evaluating the rule for fitness (e.g., [243]). Alternatively, the rule consequent (i.e., class) may already be specified for each rule antecedent during the entire run of the EA. This is where an iterative rule learning approach is followed and an EA is run several times in succession, with each run concentrating on evolving rule antecedents pertaining to a specific class (e.g., [116,290]).

If a complete rule is encoded in an individual then the rule consequent may also be subject to evolution and change by the genetic operators. A restriction may be placed on the crossover operator to ensure that only parents belonging to the same rule consequent are combined to produce offspring [371,407].

With a Pittsburgh-style approach, generally, both rule antecedents and associated consequents are encoded. The genetic operators may then act on the individual rules within a RB, or on the composition of the RB itself. An alternative is proposed in [397], in which each individual represents a fixed number of rule antecedents. The consequents of the rules are dependent on the number of positive and negative examples they match in the training set and are determined prior to evaluation of the rule set.

3.2.4.2 Ant-Based Rule Induction FRANTIC (Fuzzy Rules from ANT-Inspired Computation) [108] implements a class-dependent iterative rule learning strategy, based on ant colony optimization. More detail concerning ACO in general and its application to feature selection can be found in Section 10.2. For each class in the dataset one or more ACOs are run, and from each the best rule constructed is determined and added to the final ruleset. However, before the next ACO is run to find another rule describing the same class, the cases belonging to that class that are covered by the previous best rule are removed from the training set. Before ACOs are run to find rules describing the next class, the full training set is reinstated.

A simplified version of the algorithm is to run just one ACO algorithm for each class, the assumption being that one rule is sufficient to describe a class. Further developments allow FRANTIC to induce rules sequentially. In the simplified form one ACO algorithm is also run for each class. However, instead of running the ACOs in succession, they are run in parallel (in principle; i.e., this is not as yet a true parallel implementation running on multiple processors), with each maintaining its own problem graph, pheromone levels, and heuristic values.

After each class has had its rules created for a particular iteration, all possible combinations of rules (one from each class) are formed into a ruleset, and this is tested on the training set. The rules in the best-performing ruleset are used to update the pheromone levels, with the rule describing a specific class being used to update the pheromone levels of the associated ACO.

When creating a rule antecedent, an ant traverses a problem graph where each node represents a term that may be added. In the case of constructing rules with negated terms, the graph has double the number of nodes—one extra for each original linguistic term. The choice of the next node to visit depends on both a heuristic value and the pheromone level associated with the node. It is made probabilistically but is biased toward terms that have relatively higher heuristic and pheromone values. However, after selection and before a term is added to a rule antecedent, a check is made; this ensures that the resultant rule antecedent covers a minimum number of the appropriate class instances from the training set, and is a way of avoiding overfitting to the training data. Here a fuzzy rule describing a specific class is said to cover or match a fuzzy instance if the rule and instance belong to the same class, and if the degree of match between the condition parts of rule and instance is equal to or greater than a pre-defined threshold value.

For simple propositional rules, or rules with negated terms, if an ant adds a term to its rule antecedent, then it will not consider other linguistic terms belonging to the same linguistic variable. If this restriction is removed, then it is possible for ants to add more than one linguistic term from each variable, with the interpretation being of a disjunctive operator between the terms added. The heuristic used to guide ants when selecting terms is based on fuzzy subsethood values (see Section 3.2.2.1), giving a degree to which one fuzzy set is a subset of another fuzzy set.

At the start of an ACO run, all nodes in the graph have an equal amount of pheromone, which is set to the inverse of the number of nodes. The pheromone level of individual nodes, however, changes between iterations. Toward the end of each iteration, rules created by all ants are evaluated. This is done by assessing how accurate it is in classifying the training instances. However, in standard FRANTIC-IRL each rule is evaluated individually, without taking into account how it may interact with other rules describing other classes, while in FRANTIC-SRL [109] a rule forms part of a ruleset that is evaluated as a whole. For rule evaluation the fitness function combines a measure of the sensitivity of a rule (its accuracy among instances of the same class as the rule) with a measure of the specificity of the rule (its accuracy among instances of different classes).

3.3 RULEBASE OPTIMIZATION

There are several ways in which a set of rules may be optimized. It can be the case that when a ruleset is generated automatically, rules are induced that are of little significance. The removal of such rules can improve not only the readability of the ruleset but also the classification performance. In the rules there may be unnecessary antecedent parts that may also be removed. In addition the fuzzifications may be adjusted to increase classification power while maintaining readability. This section provides information for the interested reader on two optimization techniques: fuzzy interpolation and fuzzy rule optimization.

3.3.1 Fuzzy Interpolation

The problem of interpolation can be informally illustrated in Figure 3.2. Here it is known that if an apple is green, then it is good and should be tasty. If an apple is brown-colored, then it is likely to be bad and not suitable for consumption. These rules work well when dealing with green and brown apples only, but it is not clear what should happen if a yellow apple is encountered (even if it is known that yellow goes in between green and brown). However, an estimate classification based on the existing rules may be obtained by means of reasoning with these rules. This process of inferring a result through neighboring rules is termed interpolation.

Fuzzy rule interpolation helps reduce the complexity of fuzzy models and supports inference in systems that employ sparse rulesets [179]. With interpolation, fuzzy rules that may be approximated from their neighboring rules can be omitted from the rule base. This leads to the complexity reduction of fuzzy models. When the given observations have no overlap with the antecedent values of rules, classical fuzzy inference methods have no rule to fire, but interpolative reasoning methods can still obtain certain conclusions. Suppose that two fuzzy rules are given:

If X is A_1, then Y is B_1
If X is A_2, then Y is B_2

Figure 3.2 Interpolation example

Also suppose that these two rules are adjacent, meaning there is no rule existing that the antecedent value A of that rule is between the region of A_1 and A_2. If an observation $A*$ is encountered, traditional fuzzy methods may fail to generate a reasonable outcome with the given ruleset. Interpolative methods consider the region between the antecedent values of these two rules, meaning they can determine a new conclusion $B*$ when an observation $A*$ located between fuzzy sets A_1 and A_2 is given.

Despite these significant advantages earlier work in fuzzy interpolative reasoning does not guarantee the convexity of the derived fuzzy sets [327,394]; some may even result in an inferred value that cannot be defined with a membership function. This complexity will make it very difficult, if not impossible, to attain more easily interpretable practical results, especially for applications where defuzzification of interpolated values is required.

In order to eliminate the nonconvexity drawback, there has been considerable work reported in the literature. For instance, an algorithm was proposed [366] that reduces the problem of nonconvex conclusions. An improved method has been published in [277] that uses similarity transfer reasoning to guarantee the convex results. Another interpolative method that exploits the slopes of the fuzzy sets to obtain convex conclusions was introduced in [139]. General fuzzy interpolation and extrapolation techniques [18] and a modified α-cut based method [19] have also been proposed. In addition an interpolative method based on graduality is reported in [41].

Nevertheless, some of the existing methods include complex computation. It becomes more complicated when they are extended to multiple variables interpolation. Others may only apply to simple fuzzy membership functions limited to triangular or trapezoidal shapes. These shortcomings are avoided by a novel method reported in [144], with further improvements and extensions to extrapolation covered in [145]. This method works first by constructing a new inference rule via manipulating two given adjacent rules, and then by using scale and move transformations to convert the intermediate inference results into the final derived conclusions. The approach can handle interpolation, and also extrapolation, of multiple antecedent variables with simple computation, while guaranteeing the reasoning results to be unique as well as convex and normal.

3.3.2 Fuzzy Rule Optimization

Fuzzy rule bases are typically assumed to be given by domain experts. However, the acquisition of knowledge on rule structures often forms the bottleneck to advance the success of fuzzy systems in practice [288], though the linguistic labels or fuzzy sets that are used within the rules may be subjectively defined. For many applications there exists a considerable volume of historical data obtained by observing the behavior of the system concerned. It is therefore desirable to be able to automatically generate rules from given data. Many techniques exist for this, most of which follow the so-called approximative[1] approach, which works by creating and tuning the fuzzy rule bases to best fit the data. The rules generated are not encoded to keep the meaning of the linguistic labels of the fuzzy sets used. Such an approach is, under minor restrictions, functionally equivalent to neural networks [155], and the resulting systems offer little explanatory power over their inferences.

Opposite approximative modeling stands the descriptive approach in which model transparency is as important as accuracy. Prescribed fuzzy sets are either not allowed to be modified or, at most, are permitted to have very slight modifications. The descriptive sets used (i.e., the fuzzy sets defined by humans with a preconceived linguistic label attached) induce a fuzzy grid in the product space of the domain variables. As little modification is permitted, the grid and the hyperboxes delimited by these sets are almost fixed. Often these hyperboxes may contain examples of different output states, and as they are fixed, there is no way to separate these outputs directly.

It is possible to implicitly modify fuzzy rule bases, without disrupting the definition of the underlying fuzzy sets, by the use of linguistic hedges [409], which allow more freedom in manipulating the hyperboxes. Not all pure descriptive methods support the addition of hedges though (e.g., the well-established work as reported in [373]). When no hedges can be applied, an increase in the number of fuzzy sets, the addition of confidence factors or prioritizing some data as critical, may increase the performance. However, these methods typically also give rise to a loss in the interpretability.

An alternative approach to gaining accurate, descriptive fuzzy rules is given in [224,225] via a two-step mechanism. The first step is to use an approximative method to create accurate rules, and the second is to convert the resulting approximative rules to descriptive ones. The conversion is, in general, one to many, implemented by a heuristic method that derives potentially useful translations and then by performing a fine-tuning of these translations via evolutionary computation. Both steps are computationally efficient. The resultant descriptive system is ready to be directly applied for inference; no approximative rules are needed in runtime.

Figure 3.3 shows the basic idea. Instead of using a direct descriptive rule induction technique, which is generally slow and inaccurate, a fast and accurate approximative rule induction algorithm is used initially. The approximative fuzzy model induced can be tuned to improve its modeling accuracy. This model is then converted into a fuzzy linguistic model that utilizes predefined descriptive sets. A GA is used to perform the optimization in the search for the best corresponding descriptive ruleset.

A relatively new trend in this area is the multi-objective design of fuzzy rule-based systems [63,153] where conflicting objectives (accuracy maximization and complexity minimization) are considered simultaneously. Several fuzzy

[1]The word approximative is used here instead of approximate to mirror the word descriptive in descriptive modeling, which is in fact an approximate approach.

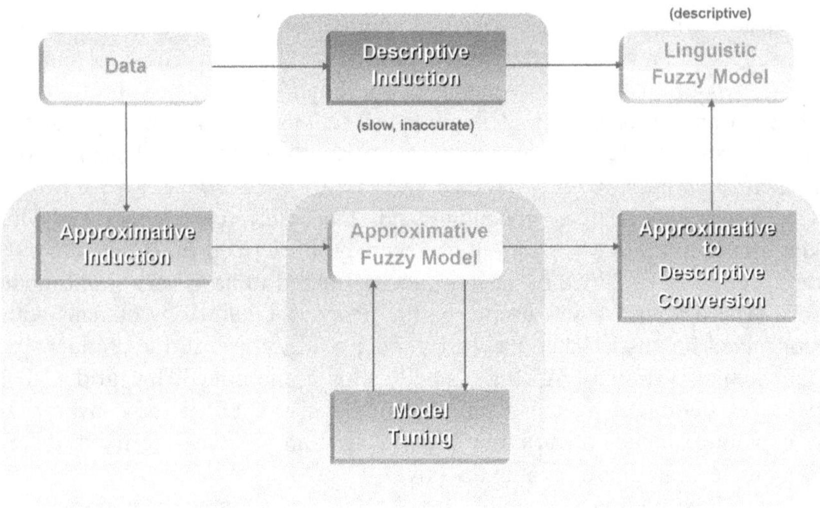

Figure 3.3 Descriptive rule generation

rule-based systems are generated based on the trade-off between these objectives, with the final compromise system selected by the user. By analyzing the obtained fuzzy rule-based systems, the user can develop a better understanding of this trade-off.

3.4 SUMMARY

This chapter has reviewed several crisp and fuzzy classification methods. Crisp rule and decision tree induction, in particular, have been successful but are limited in their utility when dealing with continuous data. For these methods to be able to operate effectively, a discretization step must be carried out beforehand (or during induction). Fuzzy methods are better equipped to handle this type of data. Several fuzzy rule induction methods were outlined. Many of these will be used later in this book for challenging classification tasks.

There are various ways to optimize fuzzy rulebases. Fuzzy interpolation is one technique that can remove redundant rules in rulebases and allow reasoning given observations that are unexpected. Rule accuracy can be improved through fuzzification optimization. The fuzzy sets used in fuzzy rules can be tuned to improve classification performance, at the same time maintaining their original meaning. Linguistic hedges are useful for this purpose as these can bring additional clarity to the interpretation of the discovered rules.

CHAPTER 4

DIMENSIONALITY REDUCTION

There are many factors that motivate the inclusion of a dimensionality reduction (DR) step in a variety of problem-solving systems [45]. Many application problems process data in the form of a collection of real-valued vectors (e.g., text classification [395] or bookmark categorization [164]). If these vectors exhibit a high dimensionality, then processing becomes infeasible. For this reason it is often useful, and sometimes necessary, to reduce the data dimensionality to a more manageable size with as little information loss as possible. The process is summarized in Figure 4.1 (adapted from [45]), where the dimensionality reduction step is a preprocessing stage in the whole system.

It can be the case that high-dimensional phenomena are governed by significantly fewer, simple variables [95]. Here the process of dimensionality reduction will act as a tool for modeling these phenomena, improving their clarity. There is often a significant amount of redundant or misleading information present; such information will need to be removed before any further processing can be carried out. For example, the problem of deriving classification rules from large datasets often benefits from a data reduction preprocessing step [321]. Not only does this reduce the time required to perform induction, but it makes the resulting rules more comprehensible and can increase the resulting classification accuracy.

However, many dimensionality reduction techniques destroy the underlying meaning behind the features present in a dataset (the semantics)—an undesirable property for many applications. This is particularly the case where the understanding of the data processing method and that of the resulting processed data is as important as the accuracy of the resultant lower dimensional dataset

Computational Intelligence and Feature Selection: Rough and Fuzzy Approaches, by Richard Jensen and Qiang Shen

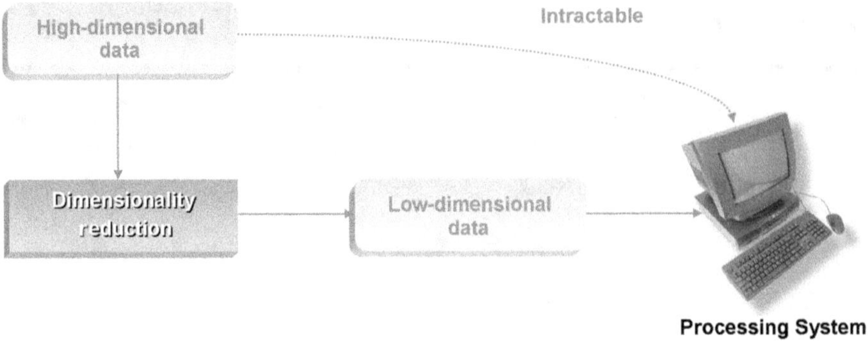

Figure 4.1 The dimension reduction problem

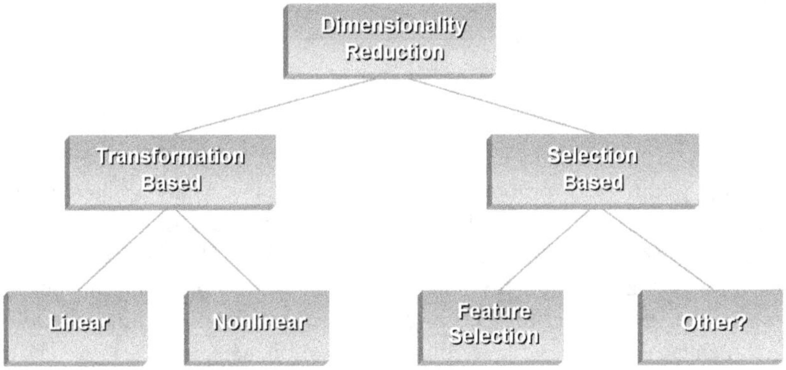

Figure 4.2 Taxonomy of dimension reduction approaches

in use. The primary focus of this chapter therefore is on those techniques that perform dimensionality reduction while preserving the meaning of the original dataset.

A taxonomy of dimensionality reduction techniques is presented in Figure 4.2. The key distinction made within the taxonomy is whether a DR technique transforms or preserves the dataset semantics in the process of reduction. The choice of DR technique is often guided by the particular application that it will be a part of. For example, if an application needs to use the meaning of the original feature set, then a DR technique must be chosen that will ensure this preservation. If, on the other hand, an application requires a visualization of relationships within the dataset, then a DR approach that transforms the data into two or three dimensions while emphasizing those relationships may be more beneficial.

Included in this taxonomy is the possibility of a semantics-preserving dimensionality reduction technique other than feature selection. Techniques that perform a task where semantics-preserving dimensionality reduction is a side effect may be classified here. For example, the machine learning algorithm C4.5 [281] constructs decision trees from data, selecting features and partitioning the dataset in the process. The resulting decision trees often involve fewer features than the original training data, so a degree of dimensionality reduction has been performed.

This chapter will first examine several representative techniques that irreversibly destroy feature semantics in the process of dataset dimensionality reduction, separating these into linear and nonlinear techniques. Next, semantics-preserving (feature selection) techniques are examined and their advantages and disadvantages discussed. For those methods that require it, an algorithmic outline is given.

4.1 TRANSFORMATION-BASED REDUCTION

Common throughout the DR literature are approaches that reduce dimensionality but in the process irreversibly transform the descriptive dataset features. These methods are employed in situations where the semantics of the original dataset are not needed by any future process. This section briefly discusses several such popular techniques. These dimensionality reduction techniques are separated into two categories: those methods that handle linearity and those that attempt to handle nonlinearity.

4.1.1 Linear Methods

Linear methods of dimensionality reduction have been well developed over the years and include techniques such as principal component analysis [172] and multidimensional scaling [360]. These techniques are used to determine the Euclidean structure of a dataset's internal relationships. However, when such relationships are of a higher dimensionality, these methods generally fail to detect this. This

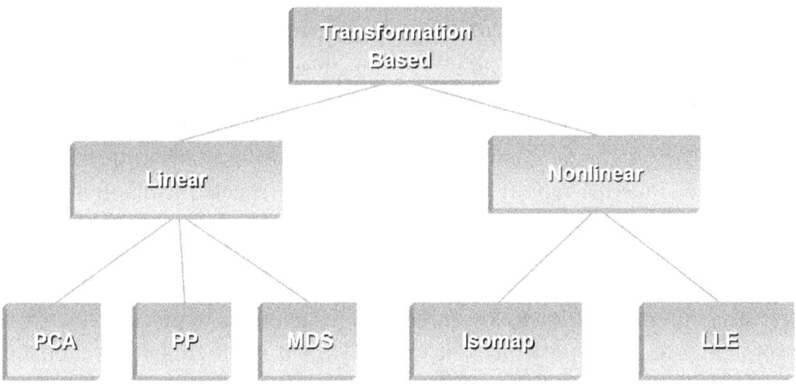

Figure 4.3 Classification of representative semantics-destroying DR techniques

problem is not too restrictive as for many applications linear DR is all that is needed.

Principal component analysis (PCA) is a dimensionality reduction tool in common use, perhaps due to its conceptual simplicity and the existence of relatively efficient algorithms for its computation. PCA transforms the original features of a dataset to a (typically) reduced number of uncorrelated ones, termed principal components. The method works on the assumption that a large feature variance corresponds to useful information, with small variance equating to information that is less useful. Data are transformed in such a way as to allow the removal of those transformed features with small variance. This is achieved by finding the eigenvectors of the covariance matrix of data points (objects), constructing a transformation matrix from the ordered eigenvectors, and transforming the original data by matrix multiplication. PCA is employed in Section 12.2.3 to perform dimensionality reduction within a monitoring system. Linear discriminant analysis (LDA) [99] maximizes the ratio of between-class variance to the within-class variance in any particular dataset, thereby guaranteeing maximal separability. The main difference between this approach and PCA is that LDA does not change the location of the original data but attempts to provide more class separability.

Projection pursuit (PP) [103,104] attempts to optimize a quality metric in the search for a lower dimensional projection of data. Potentially interesting projections are selected by the local optimization over projection directions of a certain index of "interestingness." This notion is motivated by the observation that for high-dimensional data, most low-dimensional projections are approximately normal [78]. Hence those projections that produce single-dimensional projected distributions far from the normal distribution are determined to be interesting. Different projection indexes arise from alternative definitions of normality deviation.

Typically linear projections are used because of their simplicity and interpretability. Often there will be several projections determined to be interesting that correspond to projection index local optima. Each such projection may highlight a different (but equally interesting) aspect of the structure of the high-dimensional data. This can be a problem for some projection indexes as the existence of many local maxima makes locating global maxima difficult. Also the method suffers from many of the disadvantages of PCA. In fact it can be shown that PCA is a special case of PP [45]. As PP works with linear projections, it is not suited to deal with highly nonlinear structure. In addition PP methods are computationally intensive. For instance, the projection index and derivatives must be rapidly computable due to repeated evaluations.

Multidimensional scaling (MDS) refers to a class of techniques that use proximities of objects as input and display its structure as a geometrical picture. Proximities are a measure of the similarities (or dissimilarities) of data points. The resulting transformation to lower dimensions attempts to preserve these original proximities as far as possible. Classical MDS has its origins in psychometrics, where it was proposed to help understand people's judgments of the similarity of members of a set of objects [289].

The first developments of MDS were metric [360,227]; the values used had to be quantitative, complete, and symmetric. The inability of this class of MDS to handle asymmetric and incomplete data proves too restrictive for most real datasets. This led to the development of nonmetric MDS to allow for these possibilities and additionally enabling the use of ordinal data [188,326]. Other extensions to the approach include replicated MDS (RMDS) [404] where the simultaneous analysis of several matrices of similarity data is allowed, and weighted MDS (WMDS) [404].

WMDS generalizes the distance model so that several similarity matrices can be assumed to differ from each other in systematically nonlinear or nonmonotonic ways. Whereas RMDS only accounts for individual differences in the ways subjects use the response scale (in psychological terms), WMDS incorporates a model to account for individual differences in the fundamental perceptual or cognitive processes that generate the responses. For this reason WMDS is often called individual differences scaling.

4.1.2 Nonlinear Methods

As useful as the previous methods are for reducing dimensionality, their utility fails for data exhibiting nonlinearity. Given a dataset containing nonlinear relationships, these methods detect only the Euclidean structure. This brought about the need for methods that can effectively handle nonlinearity. The first attempts were extensions to the original PCA process, either by clustering data initially and performing PCA within clusters [43] or by greedy optimization processes [187]. Both suffer from problems brought on as a result of simply attempting to extend linear PCA [111]. This motivates the development of techniques designed to suitably and successfully handle nonlinearity.

Isomap [358] is an extension of MDS in which embeddings are optimized to preserve geodesic distances between pairs of data points, estimated by calculating the shortest paths through large sublattices of data. The algorithm can discover nonlinear degrees of freedom as these geodesic distances represent the true low-dimensional geometry of the manifold.

The success of Isomap depends on being able to choose a neighborhood size (either ϵ or K) that is neither so large that it introduces "short-circuit" edges into the neighborhood graph nor so small that the graph becomes too sparse to approximate geodesic paths accurately. Short-circuit edges occur where there are links between data points that are not near each other geodesically and can lead to low-dimensional embeddings that do not preserve a manifold's true topology.

Locally linear embedding (LLE) [293] is an eigenvector method for the problem of nonlinear DR. It calculates low-dimensional neighborhood-preserving reconstructions (embeddings) of data of high dimensionality. LLE achieves this by exploiting the local symmetries of linear reconstructions. To conceptualize this, consider the following informal analogy. The initial data is three-dimensional and forms the topology of a two-dimensional rectangular manifold bent into a three-dimensional S-curve. Scissors are then used to cut this manifold into small squares representing locally linear patches of the nonlinear surface. These squares can then be arranged onto a flat tabletop while angular relationships between neighboring squares are maintained. This linear mapping is due to the fact that all transformations involve translation, scaling, or rotation only. This way the algorithm identifies the nonlinear structure through a series of linear steps.

LLE avoids the need to solve large dynamic programming problems. It also tends to accumulate very sparse matrices whose structure can be exploited to save time and space. However, in [293] there is no indication as to how a test data point may be mapped from the input space to the manifold space, or how a data point may be reconstructed from its low-dimensional representation. Additionally, LLE suffers from the problem of short-circuit edges as described previously for Isomap.

Multivariate adaptive regression splines (MARS) [105] is an implementation of techniques for solving regression-type problems, with the main purpose of predicting the values of a continuous decision feature from a set of conditional features. It is a nonparametric regression procedure that makes no assumptions about the underlying functional relationships. MARS instead constructs this relation from a set of coefficients and basis functions, defined by control points, that are selected based on the data only.

However, this is a relatively complex process, and if suffers from the curse of dimensionality. Each dimension of the hyperplane requires one dimension for the approximation model, and an increase in the time and space required to compute and store the splines. The time required to perform predictions increases exponentially with the number of dimensions. Noise can also distort the model by causing MARS to generate a much more complex model as it tries to incorporate the noisy data into its approximation.

Also worth mentioning are methods based on artificial neural networks (ANNs). ANNs are mathematical models that are inspired by the connections and the functioning of neurons in biological systems, based on the topology of nodes and connections between them, and transfer functions that relate the input and output of each node. ANNs are often used as a way of optimizing a classification (or pattern recognition) procedure. They also usually have more input than output nodes; they may thus also be viewed as performing a dimensionality reduction on input data, in a way more general than principal component analysis and multidimensional scaling [36,187,192].

4.2 SELECTION-BASED REDUCTION

Whereas semantics-destroying dimensionality reduction techniques irreversibly transform data, semantics-preserving DR techniques (referred to as feature selection) attempt to retain the meaning of the original feature set. The main aim of feature selection (FS) is to determine a minimal feature subset from a problem domain while retaining a suitably high accuracy in representing the original

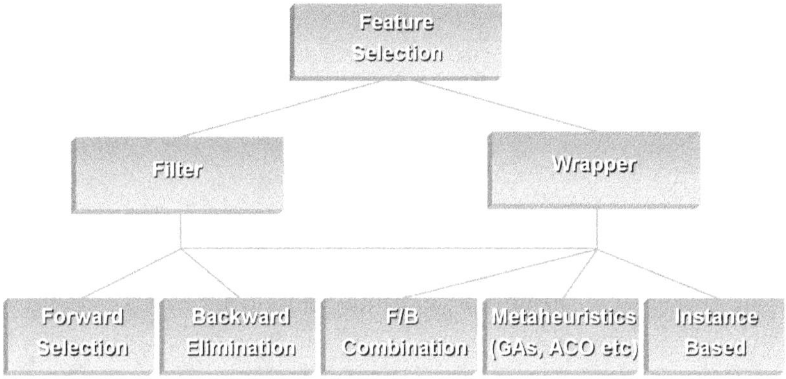

Figure 4.4 Aspects of feature selection

features. In many real-world problems FS is needed because of the abundance of noisy, irrelevant or misleading features. For instance, with these factors removed, learning from data techniques can benefit greatly. A detailed review of feature selection techniques devised for classification tasks can be found in [67].

The usefulness of a feature or feature subset is determined by both its *relevancy* and *redundancy*. A feature is said to be relevant if it is predictive of the decision feature(s); otherwise, it is irrelevant. A feature is considered to be redundant if it is highly correlated with other features. Hence the search for a good feature subset involves finding those features that are highly correlated with the decision feature(s) but are uncorrelated with each other.

A taxonomy of feature selection approaches can be seen in Figure 4.4. Given a feature set size n, the task of FS can be seen as a search for an "optimal" feature subset through the competing 2^n candidate subsets. The definition of what an optimal subset is may vary depending on the problem to be solved. An exhaustive method can be used for this purpose in theory but is quite impractical for most datasets. Usually FS algorithms involve heuristic or random search strategies in an attempt to avoid this prohibitive complexity. However, the degree of optimality of the final feature subset is often reduced. The overall procedure for any feature selection method is given in Figure 4.5.

The generation procedure implements a search method [194,329] that generates subsets of features for evaluation. It may start with no features, all features, a selected feature set, or some random feature subset. Those methods that start with an initial subset usually select these features heuristically beforehand. Features are added (*forward selection*) or removed (*backward elimination*) iteratively in the first two cases [67]. In the last case, features are either iteratively added or removed or produced randomly thereafter. An alternative selection strategy is to select instances and examine differences in their features. The evaluation function calculates the suitability of a feature subset produced by the generation procedure and compares this with the previous best candidate, replacing it if found to be better.

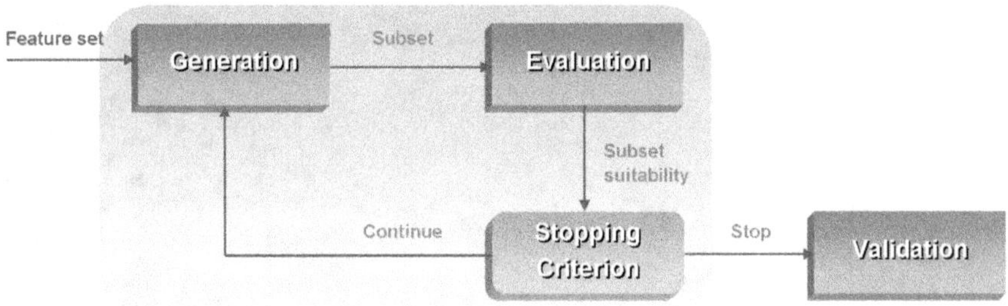

Figure 4.5 Feature selection [67]

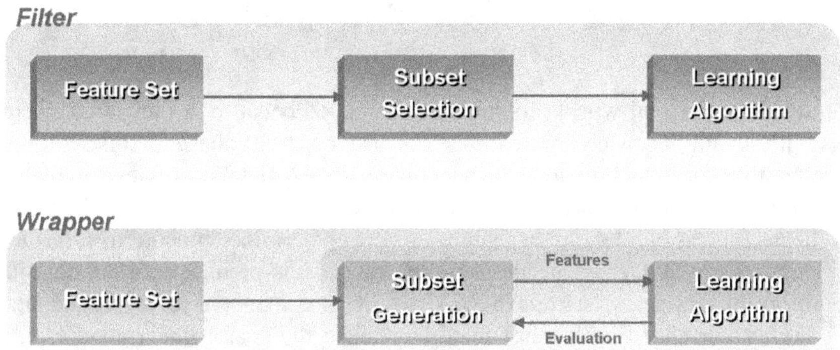

Figure 4.6 Filter and wrapper approaches to feature selection

A stopping criterion is tested every iteration to determine whether the FS process should continue or not. For example, such a criterion may be to halt the FS process when a certain number of features have been selected if based on the generation process. A typical stopping criterion centered on the evaluation procedure is to halt the process when an optimal subset is reached. Once the stopping criterion has been satisfied, the loop terminates. For use, the resulting subset of features may be validated.

Determining subset optimality is a challenging problem. There is always a trade-off in nonexhaustive techniques between subset minimality and subset suitability—the task is to decide which of these must suffer in order to benefit the other. For some domains (particularly where it is costly or impractical to monitor many features), it is much more desirable to have a smaller, less accurate feature subset. In other areas it may be the case that the modeling accuracy (e.g., the classification rate) using the selected features must be extremely high at the expense of a nonminimal set of features.

Feature selection algorithms may be classified into two categories based on their evaluation procedure. If an algorithm performs FS independently of any learning algorithm (i.e., it is a completely separate preprocessor), then it is a *filter* approach. In effect, irrelevant attributes are filtered out before induction. Filters tend to be applicable to most domains as they are not tied to any particular induction algorithm.

If the evaluation procedure is tied to the task (e.g., classification) of the learning algorithm, the FS algorithm employs the *wrapper* approach. This method searches through the feature subset space using the estimated accuracy from an induction algorithm as a measure of subset suitability. Although wrappers may produce better results, they are expensive to run and can break down with very large numbers of features. This is due to the use of learning algorithms in the evaluation of subsets, some of which can encounter problems when the datasets are large.

4.2.1 Filter Methods

To illustrate the operation of the algorithms outlined in this section, an example dataset will be used, as given in Table 4.1. The dataset is restricted to containing only binary values due to the different requirements of the algorithms. Throughout the text, $\mathbb{C}$ indicates the set of conditional features while $\mathbb{D}$ denotes the set of decision features.

4.2.1.1 Relief The first feature selection algorithms were based on the filter approach. In RELIEF [176] each feature is given a relevance weighting that reflects its ability to discern between decision class labels. An overview of this algorithm can be found in Figure 4.7. The first threshold, *its*, specifies the number of sampled objects used for constructing the weights. For each sampling an object x is randomly chosen, and its nearHit and nearMiss are calculated. These are x's nearest objects with the same class label and different class label, respectively.

TABLE 4.1 Example 2-Class Dataset

Object	a	b	c	d	e	f	⇒	g
0	0	1	1	1	0	0		1
1	1	0	1	1	0	0		1
2	1	1	0	0	0	0		1
3	1	1	0	1	0	1		1
4	1	1	1	0	1	0		1
5	0	0	1	1	0	0		1
6	0	0	0	1	1	1		0
7	0	0	1	0	0	1		0
8	1	0	0	0	1	1		0
9	1	0	1	0	0	1		0
10	0	1	0	0	0	1		0
11	0	1	0	1	1	1		0
12	0	1	1	0	1	0		0

```
RELIEF(O, ℂ, its, ε)
Input: O, the set of all objects; ℂ, the set of conditional
features; its, the number of iterations; ε, weight threshold
value.
Output: R, the feature subset
    R ← {}
    foreach W_a, W_a ← 0
    foreach i = 1...its
        choose an object x in O randomly
        calculate x's nearHit (nH) and nearMiss (nM)
        foreach j = 1...|ℂ|
            W_j ← W_j − d(x_j, nH_j)/its + d(x_j, nM_j)/its
    foreach j = 1...|ℂ|
        if W_j ≥ ε; R ← R ∪ {j}
    return R
```

Figure 4.7 RELIEF algorithm

The distance between object o and x is defined here as the sum of the number of features that differ in value between them (i.e., the Hamming distance between the two feature vectors):

$$dist(o, x) = \sum_{i=1}^{|\mathbb{C}|} d(o_i, x_i) \tag{4.1}$$

where

$$d(o_i, x_i) = \left\{ \begin{array}{ll} 1, & o_i \neq x_i \\ 0, & o_i = x_i \end{array} \right\} \tag{4.2}$$

The user must supply a threshold that determines the level of relevance that features must surpass in order to be finally chosen. This method is ineffective at removing redundant features as two predictive but highly correlated features are both likely to be given high relevance weightings. The method has been extended to enable it to handle inconsistency, noise, and multi-class datasets [182].

RELIEF is applied to the example dataset in the following way: An object is chosen randomly, say object 0, and its nearest neighbors are found, nearHit, and nearMiss. In this case object 5 is the nearHit and object 12 is the nearMiss. For each feature the weight is updated according to the difference of the feature and that of x's nearHit and nearMiss. This process continues until the desired number of iterations has elapsed. Features are then added to the final subset if their weights surpass the desired level, ϵ. Running RELIEF on the dataset with $its = 100$ and $\epsilon = 0$ produces the final subset $\{a, d, e, f\}$. Removing from the dataset all but these attributes will result in a smaller yet still consistent set of data. However, upon examination it can be seen that a further reduction can take place; no inconsistencies are introduced by eliminating feature e from this newly reduced dataset.

4.2.1.2 Focus FOCUS [4], another filter method, conducts a breadth-first search of all feature subsets to determine the minimal set of features that can provide a consistent labeling of the training data. The FOCUS algorithm, as summarized in Figure 4.8., generates all subsets of the current size (initially one) and checks each for at least one inconsistency. If an inconsistency is found, then that particular subset is removed. This continues until a consistent subset is found or all possible subsets have been evaluated. The consistency criterion makes FOCUS very sensitive to noise or inconsistencies in the training data. Moreover the exponential growth of the size of the power set of features makes this impractical for domains with medium to large dimensionality. By introducing a threshold value to line (5) in the algorithm, the sensitivity can be reduced, allowing a certain amount of inconsistency within the dataset.

Given the example dataset, FOCUS first evaluates the consistency of all subsets of size 1, namely $\{a\}$, $\{b\}$, $\{c\}$, $\{d\}$, $\{e\}$, $\{f\}$. It determines that none of these subsets produces a reduced dataset with zero inconsistency. For example, selecting the subset $\{f\}$ results in objects 0 and 12 conflicting. Again, all subsets of size 2 are examined ($\{a, b\}$, $\{a, c\}$, etc.), and no suitable subset is found. The algorithm continues until the subset $\{a, d, f\}$ is chosen—this will result in no inconsistencies. Hence the dataset can now be reduced to one only involving these attributes. This subset of features is the minimal subset for this dataset in terms of the consistency criterion.

```
FOCUS(O, c)
Input: O, the set of all objects; ℂ, the set of conditional
features
Output: R, the feature subset
(1)   R ← {}
(2)   foreach num = 1...|ℂ|
(3)      foreach subset L of size num
(4)         cons = determineConsistency(L, O)
(5)         if cons = = true
(6)            R ← L
(7)            return R
(8)         else
(9)            continue
```

Figure 4.8 FOCUS algorithm

```
LVF(O, ℂ, its, ϵ)
Input: O, the set of all objects; ℂ, the set of conditional
features; its, the number of iterations of the algorithm;
ϵ, consistency threshold
Output: R, the feature subset
    R ← ℂ
    foreach num = 1...its
        S ← randomFeatureSubset()
        if |S| ≤ |R|
            if inconsistency(S, O) ≤ ϵ
                if |S| < |R|
                    R ← S; output R
    return R
```

Figure 4.9 LVF algorithm

4.2.1.3 LVF The Las Vegas filter (LVF) algorithm employs an alternative generation procedure—that of choosing random feature subsets, accomplished by the use of a Las Vegas algorithm [212]. An outline of LVF is given in Figure 4.9. Initially the best feature subset is considered to be the entire conditional feature set. A feature subset is randomly chosen; if the subset has a smaller cardinality than the current best and its inconsistency rate is less than or equal to a threshold ϵ, it is considered to be the new best subset. Here the inconsistency count is defined as the sum of the number of inconsistent objects minus the number of inconsistent objects with the most frequent class label. For example, if there are n inconsistent objects for a 2-class dataset, c_1 such objects belong to class 1, c_2 to class 2, then $c_1 + c_2 = n$. If the largest of these is, say, c_1 then the inconsistency count will be $n - c_1$. The inconsistency rate is simply the sum of all inconsistency counts divided by the number of objects. Every time a better subset is encountered, it is outputted. Once the maximum number of iterations has been reached, the loop terminates and the algorithm returns the best subset found overall. A problem with this approach is that it will tend to take longer to locate an optimal subset than algorithms that employ heuristic generation procedures. Additionally, when datasets are huge, checking the consistency of the dataset takes a long time.

Returning to the example, LVF randomly chooses feature subsets from among the six features present, such as the subset $\{a, b, c\}$. For class 1 there are 3 inconsistent objects, for class 2 there are also 3, so the inconsistency count will be $(6 - 3)/12 = \frac{1}{4}$. If no inconsistency is allowed ($\epsilon = 0$), then this subset will not be kept. Given sufficient time, the algorithm should eventually reach the subset $\{a, d, f\}$, which is the smallest subset that produces the required inconsistency rate.

4.2.1.4 *Scrap* Selection construction ranking using attribute pattern (SCRAP) [285] is an instance-based filter that determines feature relevance by performing a sequential search within the instance space. SCRAP considers objects (instances) one at a time, instead of typical forward or backward search. The central idea is to identify those features that change at decision boundaries in the data table—the features assumed to be the most informative. An algorithmic overview may be seen in Figure 4.10. A sequential search is conducted, starting from a random object, which becomes the first point of class change (PoC). Its nearest object with a different class label becomes the next PoC. These two PoCs define a neighborhood; features that change between them define the dimensionality of decision boundary between the two classes. If there is only one such feature that changes, this is determined to be absolutely relevant and is included in the feature subset. If more than one feature changes, their associated relevance weights, initially zero, are incremented. If objects of the same class label are closer than this new PoC and differ in only one feature, then that feature's weight is decremented. Objects determined to belong to neighborhoods are then removed from processing. The process stops when all objects have been assigned to a neighborhood. Features that have a positive relevance weight and those that have been determined to be absolutely relevant are chosen as the final feature subset.

```
SCRAP(O)
Input: O, the set of all objects
Output: R, the feature subset
A ← {}; ∀Wi, Wi = 0;
T ← randomObject(); PoC ← T
while O ≠ {}
   O ← O − PoC; PoC_new ← NewPoC(PoC)
   n = dist(PoC,PoC_new)
   if n == 1
      i = diffFeature(PoC,X); A ← A ∪ {i}
   N ← getClosestNeighbours(PoC,n)
   foreach X ∈ N
      if classLabel(X) == classLabel(N)
         O ← O − X
         if dist(PoC,X) == 1
            i = diffFeature(PoC,X); Wi = Wi − 1
      else if dist(PoC,X) > 1
         incrementDifferingFeatures(X,W)
R ← A
foreach Wi
   if Wi > 0 R ← R ∪ { i}
return R
```

Figure 4.10 SCRAP algorithm

From the example dataset, SCRAP first chooses a random object, say object 1, and proceeds to find its nearest neighbor with a different class label. In this case object 12 is the PoC with two features that differ, d and e. These features are said to be weakly relevant, and their weights (initially zero) are incremented. The objects of a lesser distance away than object 12 with the same label as object 1 are assigned to this neighborhood. Only object 5 is closer as it differs in one feature, b. This results in b's weight being decremented. If more than one feature differed here, the weights would not have been affected—only those cases where one feature differs result in weights being reduced. Object 12 now becomes the new PoC, and the algorithm continues. Object 4 is the nearest neighbor with a different class label, with only one feature differing in value, feature a. This feature is determined to be absolutely relevant and is added to the final subset regardless of its final relevance weight. When the algorithm eventually terminates, the subset $\{a, b, c, d, e, f\}$ is returned: a is absolutely relevant, the rest have a positive final weight. No reduction of this dataset is achieved.

The example above serves to illustrate one of the main weaknesses of this instance-based approach—it regularly chooses too many features. This is due, in part, to the situation where weights are decremented. If more than one feature changes between a PoC and an object of the same class label, then the corresponding feature weights remain unaffected. This drawback could be tackled by reducing the weights of the affected features when the change occurs. With the modification in place and the algorithm run on the dataset, a smaller subset, $\{a, b, e, f\}$, is obtained. Another alteration may be to decrement each weight proportionally, such as for three irrelevant features whose weights will be reduced by one-third each. This modification combined with a similar one for incrementing weights may produce a more accurate reflection of a feature's importance within a dataset. Currently SCRAP will only handle nominal values, although it is relatively straightforward to extend this algorithm to continuously valued features.

4.2.1.5 EBR Another technique for filter-based feature selection is entropy-based reduction (EBR), which developed from work carried out in [164]. This approach is based on the entropy heuristic employed by machine learning techniques such as C4.5 [281]. A similar approach was adopted in [67] where an entropy measure is used for ranking features. EBR is concerned with examining a dataset and determining those attributes that provide the most gain in information. The entropy of attribute A (which can take values $a_1 \dots a_m$) with respect to the conclusion C (of possible values $c_1 \dots c_n$) is defined as

$$H(C|A) = -\sum_{j=1}^{m} p(a_j) \sum_{i=1}^{n} p(c_i|a_j) \log_2 p(c_i|a_j) \tag{4.3}$$

This expression can be extended to deal with *subsets* of attributes instead of individual attributes only. By this entropy measure the algorithm used in rough

```
EBR(ℂ,𝔻)
Input: ℂ, the set of all conditional features;
𝔻, the set of all decision features
Output: R, the feature subset
    R ← {}
    while H(𝔻|R) ≠ H(𝔻|ℂ)
      T ← R
      foreach x ∈ (ℂ − R)
         if H(R ∪ {x}) < H(T)
            T ← R ∪ {x}
      R ← T
    return R
```

Figure 4.11 Entropy-based algorithm

set-based attribute reduction [54] can be modified to that shown in Figure 4.11. This algorithm requires no thresholds in order to function—the search for the best feature subset is stopped when the resulting subset entropy is equal to that of the entire feature set. For consistent data the final entropy of the subset will be zero. Interestingly any subset with an entropy of 0 will also have a corresponding rough set dependency of 1.

Returning to the example dataset, EBR first evaluates the entropy of each individual attribute:

Subset	Entropy
$\{a\}$	0.8885861
$\{b\}$	0.9543363
$\{c\}$	0.9543363
$\{d\}$	0.8885860
$\{e\}$	0.8650087
$\{f\}$	0.6186034

The subset with the lowest entropy here is $\{f\}$, so this subset is added to the current feature subset. The next step is to calculate the entropy of all subsets containing f and one other attribute:

Subset	Entropy
$\{a, f\}$	0.42382884
$\{b, f\}$	0.46153846
$\{c, f\}$	0.55532930
$\{d, f\}$	0.42382884
$\{e, f\}$	0.40347020

Here the subset $\{e, f\}$ is chosen. This process continues until the lowest entropy for the dataset is achieved (for a consistent dataset, this is zero). The algorithm eventually reaches this lowest value when it encounters the feature subset $\{a, b, e, f\}$ $(H(\mathbb{D}|\{a, b, e, f\}) = 0)$. The dataset can now be reduced to these features only. As has been shown previously, this feature subset is close to the best feature subset for this dataset. The optimal subset was discovered by the FOCUS algorithm, which works well for small datasets such as this but cannot be applied to datasets of medium to large dimensionality. EBR does not suffer from this problem as its complexity is $O((n^2 + n)/2)$. It also does not require any user-defined thresholds in order to function, a drawback of RELIEF.

4.2.1.6 FDR Fractal dimension reduction (FDR) [362] is a novel approach to feature selection based on the concept of fractals—the self-similarity exhibited by data on different scales. For example, the Sierpinski triangle [113] is constructed from an equilateral triangle, eliminating its middle triangle and repeating this process on the resulting smaller triangles. This operation continues infinitely, removing the middle triangles from the newly generated smaller ones. The resulting shape is not one-dimensional, however; it is not two-dimensional either as its area is zero, so its intrinsic dimension lies between 1 and 2. This issue is resolved by considering fractional dimensionalities. For the Sierpinski triangle its *fractal* dimension is approximately 1.58 [308].

This concept may be applied to feature selection by considering the change in fractal dimension when certain features are removed from the original dataset under consideration. For example, the fractal dimension of the Lorenz dataset is approximately 2.05 [308]. As most of the data lie in a two-dimensional plane, the third dimension is effectively redundant. Removing this feature from the dataset will result in a slight, but acceptable, reduction in the fractal dimension. This forms the basis of the work carried out in [362]. The feature selection algorithm based on this can be seen in Figure 4.12. Given a dataset, the correlation fractal dimension is calculated, and attributes are removed via backward elimination until the removal of any further attribute reduces the fractal dimension too much.

One problem with this approach is that the calculation of the correlation fractal dimension requires datasets containing a large number of objects. If a dataset has too few instances, the procedure for calculating the fractal dimension will not be able to estimate the dimension of the dataset effectively. Another weakness of FDR is how to estimate the extent of the allowed reduction in fractal dimensionality ϵ. It is not clear how to determine this value beforehand. The fractal dimension of the full set of features is often used as an additional stopping criterion if the number of remaining features falls below this value (the intrinsic dimensionality of the dataset).

4.2.1.7 Feature Grouping Typically in feature selection the generation procedure incrementally adds or removes individual features. Recently there have been a couple of investigations into the potential utility of *grouping* features at each stage. This strategy can decrease the time taken in finding optimal subsets by selecting several features at once.

```
FDR(ℂ,ε)
Input: ℂ, the set of all conditional features;
ε, the allowed reduction in fractal dimension
Output: R, the feature subset
R ← ℂ; bestDim = 0; d_C = calculateFractalDim(ℂ)
while d_C − bestDim ≤ ε
   T ← R,
   foreach x ∈ R
      S ← R − {x}
      d_S = calculateFractalDim(S)
      if d_S > bestDim
         bestDim = d_S; w ← x
   R ← R − {w}
return R
```

Figure 4.12 FDR algorithm

An automatic feature grouping technique is proposed in [398] that uses k-means clustering [125] beforehand in order to generate groups. From these groups one or two features are pre-selected for a typical forward search FS method. No feature grouping takes place during the search, however. As yet, no results are available for this approach.

In group-wise feature selection (GFS) [255] the feature groups are again calculated beforehand. These groups are then used throughout the subset search instead of individual features. An overview of the algorithm can be seen in Figure 4.13. Here the effect of adding groups of features to the currently considered subset is evaluated at each stage. The group that produces the largest increase in performance is selected, and all features present within the group are added to the current subset. The process continues until the performance is of an appropriate quality.

In the evaluation a measurement cost is also considered, although this can be omitted if no measurement information is available or required. In addition to the GFS algorithm presented in Figure 4.13, an extension to it, GNFS, that performs a nested forward search within groups has also been proposed. Instead of selecting the best group, GNFS searches for the best subset within a group and adds this to the currently selected feature subset. Both algorithms perform comparably with their standard individual feature selection method, but with the benefit of reduced computation time.

All feature grouping approaches so far have relied on groups being defined beforehand. Group membership is not changed in the selection process leading to a dependence on a suitably accurate grouping mechanism for good results. In addition the *extent* of group membership is not considered at all; features either belong or do not belong to single groups. Fuzzy grouping can be used to handle this problem so that features can belong to more than one group with varying

```
GFS(G,ϵ)
Input: G, the set of feature groups; ϵ, required level
of subset performance
Output: R, the feature subset
     R ← {}; A ← {}; best = 0
     while evaluate(R) < ϵ
        foreach group Gi
            T ← all features from Gi
            ϵt = evaluate(R ∪ T)
            if ϵ > best
                   A ← T
                   best = ϵt
        R ← R ∪ A
     return R
```

Figure 4.13 Group-wise FS algorithm

degrees of membership. This additional membership information can then be used in the selection process. These ideas motivated the development of the new rough and fuzzy-rough set-based grouping FS technique, detailed in Section 10.1.

4.2.1.8 Other Approaches Besides the approaches outlined above, a filter method based on ideas from probabilistic reasoning and information theory is proposed in [180]. The central motivation behind this development is the observation that the goal of an induction algorithm is to estimate the probability distributions over the class values. In the same way, feature subset selection should attempt to remain as close as possible to these original distributions. The algorithm performs a backward elimination search, at each stage removing the feature that causes the least change between the distributions. The search stops when the desired number of features remain (specified by the user). An important problem with this method is that it requires the features in a dataset to be binary-valued. This constraint is added to avoid the bias toward many-valued features present in entropy-based measures.

Also worth mentioning is the Chi2 algorithm [211]. This is in effect a heuristic feature selector that discretizes continuous features and in the process removes irrelevant ones based on the χ^2 statistic [264]. Feature selection actually takes place only as a side effect of the discretization process; if a feature ends up with all its values mapped to a single discrete value, then it can be removed from the dataset without introduction of inconsistency.

4.2.2 Wrapper Methods

The Las Vegas wrapper (LVW) algorithm [213] is a wrapper method based on the earlier LVF algorithm [212] (described in Section 4.2.1.3); it can be outlined

```
LVW(ℂ, K, ε)
Input: ℂ, the set of conditional features; K, update threshold;
ε, error threshold
Output: R, the feature subset
R ← ℂ; k = 0
while ε not updated for K times
  T ← randomFeatureSubset()
  ε_t = learn(T)
  if (ε_t < ε) or (ε_t == ε and |T| < |R|)
     output T
     k = 0; ε = ε_t; R ← T
  k = k + 1
ε = learn(R); return R
```

Figure 4.14 LVW algorithm

as shown in Figure 4.14. This algorithm again uses a Las Vegas style of random subset creation that guarantees that given enough time, the optimal solution will be found. As with LVF, LVW produces intermediate solutions while working toward better ones that result in a lower classification error. The algorithm requires two threshold values to be supplied: ϵ, the classification error threshold, and the value K, used to determine when to exit the algorithm due to there being no recent updates to the best subset encountered so far.

Initially, the full set of conditional features are considered to be the best subset. The algorithm continues to generate random subsets and evaluates them using an inductive learning algorithm until no better subsets are encountered for a given number of attempts (the K criterion). Finally, training and testing is carried out on the resulting best feature subset. In the reported experimentation, C4.5 [281] is used as the learning algorithm because of its relatively fast induction time (a critical factor in designing wrapper-based systems).

In [312] a neural network-based wrapper feature selector is proposed that employs backward elimination in the search for optimal subsets. This algorithm is summarized in Figure 4.15. Initially, a three-layer feedforward network is trained using all features in the dataset. This is then evaluated using an error function that involves both the classification error and a measure of the network complexity. The attribute that gives the largest decrease in network accuracy is removed. This process repeats until no more attributes can be eliminated without exceeding the maximum error threshold.

A wrapper method was proposed in [171] where feature subsets are explored using heuristic search and evaluated using n-fold cross validation. The training data are split into n partitions and the induction algorithm run n times; the standard C4.5 package is used for decision tree induction. For each partition the algorithm uses $n - 1$ partitions for training and the remaining partition for testing. The average of the classification accuracy over all n runs is used as the estimated

```
NeuralNet(ℂ, ε_max)
Input: ℂ, the set of all conditional features; ε_max, network
classification error threshold
Output: R, the feature subset
R ← ℂ;  ε_w = 0
while ε_R ≤ ε_max
   T ← R,
   foreach x ∈ R
      S ← R − {x}
      ε_S = trainNet(S)
      if ε_S > ε_w
          ε_w = ε_S;  w ← x
   R ← R − {w}
trainNet(R)
```

Figure 4.15 Neural network feature selection

accuracy. The results show that although there is not much difference between the compared filter and wrapper approaches in terms of classification accuracy, the induced decision trees were smaller for the wrapper method in general.

4.2.3 Genetic Approaches

Genetic algorithms (GAs) [133] are generally effective for rapid search of large, nonlinear, and poorly understood spaces. Unlike classical feature selection strategies where one solution is optimized, a population of solutions can be modified at the same time [190,200,201,330]. This can result in several optimal (or close-to-optimal) feature subsets as output. Section 15.5 discusses the potential usefulness of this aspect of GAs in feature selection.

A feature subset is typically represented by a binary string with length equal to the number of features present in the dataset. A zero or one in the jth position in the chromosome denotes the absence or presence of the jth feature in this particular subset. The general process for feature selection using GAs can be seen in Figure 4.16.

An initial population of chromosomes is created; the size of the population and how they are created are important issues. From this pool of feature subsets, the typical genetic operators (crossover and mutation) are applied. Again, the choice of which types of crossover and mutation used must be carefully considered, as well as their probabilities of application. This process generates a new feature subset pool that may be evaluated in two different ways. If a filter approach is adopted, the fitness of individuals is calculated using a suitable criterion function $J(X)$. This function evaluates the "goodness" of feature subset X; a larger value of J indicates a better feature subset. Such a criterion function could be Shannon's entropy measure [281] or the dependency function from rough set theory [261].

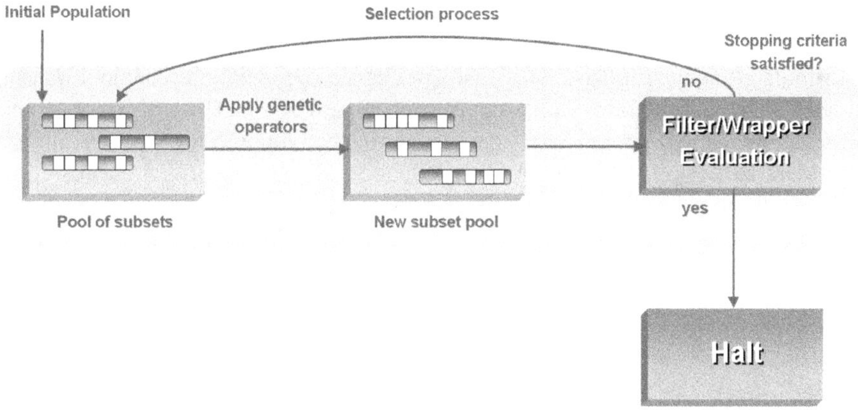

Figure 4.16 Feature selection with genetic algorithms

For the wrapper approach chromosomes are evaluated by inducing a classifier based on the feature subset, and obtaining the classification accuracy (or an estimate of it) on the data [347]. To guide the search toward minimal feature subsets, the subset size is also incorporated into the fitness function of both filter and wrapper methods. Indeed other factors may be included that are of interest, such as the cost of measurement for each feature. GAs may also learn rules directly, and in the process perform feature selection [60,170,392].

A suitable stopping criterion must be chosen. This is typically achieved by limiting the number of generations that take place or by setting some threshold that must be exceeded by the fitness function. If the stopping criterion is not satisfied, then individuals are selected from the current subset pool and the process described above repeats. Among the selection strategies that have been applied are roulette wheel selection [254] and rank-based selection [396,405]. In roulette wheel selection the probability of a chromosome being selected is proportional to its fitness. Rank selection sorts all the individuals by fitness and the probability that an individual will be selected is proportional to its rank in this sorted list.

As with all feature selection approaches, GAs can get caught in local minima, missing a dataset's true minimal feature subset. Also the fitness evaluation can be very costly as there are many generations of many feature subsets that must be evaluated. This is particularly a problem for wrapper approaches where classifiers are induced and evaluated for each chromosome.

4.2.4 Simulated Annealing Based Feature Selection

Annealing is the process by which a substance is heated (usually melted) and cooled slowly in order to toughen and reduce brittleness. For example, this process is used for a metal to reach a configuration of minimum energy (a perfect, regular crystal). If the metal is annealed too quickly, this perfect organization is unable to be achieved throughout the substance. Parts of the material will be regular, but these will be separated by boundaries where fractures are most likely to occur.

Simulated annealing (SA) [177] is a stochastic optimization technique that is based on the computational imitation of this process of annealing. It is concerned with the change of energy (cost) of a system. In each algorithmic step, an "atom" (a feature subset in FS) is given a small random displacement and the resulting change of energy, ΔE, is calculated. If $\Delta E \leq 0$, this new state is allowed and the process continues. However, if $\Delta E > 0$, the probability that this new state is accepted is

$$P(\Delta E) = e^{-(\Delta E/T)} \tag{4.4}$$

As the temperature T is lowered, the probability of accepting a state with a positive change in energy reduces. In other words, the willingness to accept a bad move decreases. The conversion of a combinatorial optimization problem into the SA framework involves the following:

- *Concise configuration description*. The representation of the problem to be solved should be defined in a way that allows solutions to be constructed easily and evaluated quickly.
- *Random move generator*. A suitable random transformation of the current state must be defined. Typically the changes allowed are small to limit the extent of search to the vicinity of the currently considered best solution. If this is not limited, the search degenerates to a random unguided exploration of the search space.
- *Cost function definition*. The cost function (i.e., the calculation of the state's energy) should effectively combine the various criteria that are to be optimized for the problem. This function should be defined in such a way that smaller function values indicate better solutions.
- *Suitable annealing schedule*. As with the real-world annealing process, problems are encountered if the initial temperature is too low, or if annealing takes place too quickly. Hence an annealing schedule must be defined that avoids these pitfalls. The schedule is usually determined experimentally.

To convert the feature selection task into this framework, a suitable representation must be used. Here the states will be feature subsets. The random moves can be produced by randomly mutating the current state with a low probability. The random transformations may also remove features from a given feature subset, allowing the search to progress both forward and backward. The cost function must take into account both the evaluated subset "goodness" (by a filter evaluation function or a wrapper classifier accuracy) and the subset size. The annealing schedule can be determined by experiment, although a good estimate may be $T(0) = |C|$ and $T(t+1) = \alpha * T(t)$, with $\alpha \geq 0.85$. Here t is the number of iterations and α determines the rate of cooling.

```
SAFS(T0, Tmin, α, Lk)
Input: T0, the initial temperature; Tmin, the minimum allowed
temperature; α, the extent of temperature decrease;
Lk, extent of local search
Output: R, the feature subset
    R← genInitSol()
    while T(t) > Tmin
       foreach i = 1...Lk
          S ← genSol(R)
          ΔE = cost(S) - cost(R)
          if ΔE ≤ 0
             M ← S
          else if P(ΔE) ≥ randNumber()
             M ← S
       R ← M
       T(t+1) = α*T(t)
    return R
```

Figure 4.17 SAFS algorithm

The SA-based feature selection algorithm can be seen in Figure 4.17. SAFS differs slightly from the general SA algorithm in that there is a measure of local search employed at each iteration, governed by the parameter L_k. An initial solution is created from which the next states are derived by random mutations and evaluated. The best state is remembered and used for processing in the next cycle. The chosen state may not actually be the best state encountered in this loop, due to the probability $P(\Delta E)$ that a state is chosen randomly (which will decrease over time). The temperature is decreased according to the annealing schedule and the algorithm continues until the lowest allowed temperature has been exceeded.

Problems with this approach include how to define the annealing schedule correctly. If α is too high, the temperature will decrease slowly, allowing more frequent jumps to higher energy states, slowing convergence. However, if α is too low, the temperature decreases too quickly, and the system will converge to local minima (equivalent to brittleness in the case of metal annealing). Also the cost function definition is critical—there must be a balancing of the importance assigned to the different evaluation criteria involved. Biasing one over another will have the effect of directing search toward solutions that optimize that criterion only.

4.3 SUMMARY

This chapter has reviewed the important problem of dimensionality reduction for datasets, with a focus on semantics-preserving reduction or feature selection. This has become a vital step in many areas such as machine learning, pattern recognition, and signal processing due to their inability to handle high-dimensional descriptions of input features. Additionally feature selection can provide a better understanding of the underlying concepts within data. It is estimated that every 20 months or so the amount of information in the world doubles, similarly applications for dealing with this vast amount of information must develop. Part of the solution to this must be drastic and efficient dimensionality reduction techniques. The extent of reduction needs to be severe to allow more detailed analysis of the data to take place by future processes. Although less of an issue for dimensionality reduction which usually takes place off-line, efficiency is still an important consideration. As one of the main goals of reduction is to enable further, more complex data processing that would otherwise be intractable, the reduction methods must not themselves be subject to the curse of dimensionality if possible.

Dimensionality reduction may be split into the two disjoint areas: transformation-based and selection-based reduction. Transformation-based approaches reduce dimensional data (which may exhibit linear or nonlinear relationships) by irreversibly transforming the data points, and as a result they destroy the original dataset semantics. Feature selection seeks to retain this information by selecting attributes as opposed to transforming them. This aspect is particularly useful when feature selection precedes other processes that require the original feature meanings to be intact, for example, rule induction where rules may need to be human-readable. Two general methods for this, filter and wrapper approaches, were shown and examples given.

This chapter proposed an area for research based on an alternative generation procedure, namely simulated annealing-based FS. Hill-climbing techniques often fail to find minimal feature subsets as they can be misled by initially promising features. Later on in the search, these features turn out to require the presence of many other less informative attributes, leading to oversized final subsets. By way of stochastic techniques, this problem may be countered to allow a degree of randomness in the search.

CHAPTER 5

ROUGH SET BASED APPROACHES TO FEATURE SELECTION

Many problems in machine learning involve high-dimensional descriptions of input features. It is therefore not surprising that much research has been carried out on dimensionality reduction [67,176,194,214,231]. However, existing work tends to destroy the underlying semantics of the features after reduction (e.g., transformation-based approaches [77]) or require additional information about the given data set for thresholding (e.g., entropy-based approaches [232]). A technique that can reduce dimensionality using information contained within the dataset and that preserves the meaning of the features (i.e., semantics-preserving) is clearly desirable. Rough set theory (RST) can be used as such a tool to discover data dependencies and to reduce the number of attributes contained in a dataset using the data alone, requiring no additional information [261,274,323].

Over the past 10 years RST has become a topic of great interest to researchers and has been applied to many domains. Given a dataset with discretized attribute values, it is possible to find a subset (termed a *reduct*) of the original attributes using RST that are the most informative; all other attributes can be removed from the dataset with minimal information loss (i.e., with no information loss in terms of the class attribute prediction accuracy for training patterns). From the dimensionality reduction perspective, informative features are those that are most predictive of the class attribute.

This chapter focuses on those recent techniques for feature selection that employ a rough set-based methodology for this purpose, highlighting current trends in this promising area. Rough set fundamentals are introduced with a simple example to illustrate its operation. Several extensions to this theory are

Computational Intelligence and Feature Selection: Rough and Fuzzy Approaches, by Richard Jensen and Qiang Shen

also presented that enable alternative approaches to feature selection. Many of these approaches are evaluated experimentally and compared. Some of the techniques described here can be found in rough set systems available online [291,292].

5.1 ROUGH SET ATTRIBUTE REDUCTION

Rough set attribute reduction (RSAR) [54,167] provides a filter-based tool by which knowledge may be extracted from a domain in a concise way; retaining the information content while reducing the amount of knowledge involved. The main advantage that rough set analysis has is that it requires no additional parameters to operate other than the supplied data [90]. In RSAR a subset with minimum cardinality is searched for.

Using the example dataset (Table 2.1 in Section 2.3), the dependencies for all possible subsets of $\mathbb{C}$ can be calculated:

$$\begin{aligned} \gamma_{\{a,b,c,d\}}(\{e\}) &= 8/8 & \gamma_{\{b,c\}}(\{e\}) &= 3/8 \\ \gamma_{\{a,b,c\}}(\{e\}) &= 4/8 & \gamma_{\{b,d\}}(\{e\}) &= 8/8 \\ \gamma_{\{a,b,d\}}(\{e\}) &= 8/8 & \gamma_{\{c,d\}}(\{e\}) &= 8/8 \\ \gamma_{\{a,c,d\}}(\{e\}) &= 8/8 & \gamma_{\{a\}}(\{e\}) &= 0/8 \\ \gamma_{\{b,c,d\}}(\{e\}) &= 8/8 & \gamma_{\{b\}}(\{e\}) &= 1/8 \\ \gamma_{\{a,b\}}(\{e\}) &= 4/8 & \gamma_{\{c\}}(\{e\}) &= 0/8 \\ \gamma_{\{a,c\}}(\{e\}) &= 4/8 & \gamma_{\{d\}}(\{e\}) &= 2/8 \\ \gamma_{\{a,d\}}(\{e\}) &= 3/8 \end{aligned}$$

Note that the given dataset is consistent, since $\gamma_{\{a,b,c,d\}}(\{e\}) = 1$. The minimal reduct set for this example is

$$R_{min} = \{\{b, d\}, \{c, d\}\}$$

If $\{b, d\}$ is chosen, then the dataset can be reduced as in Table 5.1 Clearly, each object can be uniquely classified according to the attribute values remaining.

The problem of finding a reduct of an information system has been the subject of much research [5,355]. The most basic solution to locating such a subset is to simply generate *all* possible subsets and retrieve those with a maximum rough set dependency degree. Obviously this is an expensive solution to the problem and is only practical for very simple datasets. Most of the time only one reduct is required as typically only one subset of features is used to reduce a dataset, so all the calculations involved in discovering the rest are pointless.

To improve the performance of the method above, an element of pruning can be introduced. By noting the cardinality of any pre-discovered reducts, the current

TABLE 5.1 Reduced dataset

$x \in \mathbb{U}$	b	d	$\Rightarrow$	e
0	R	T		R
1	S	S		T
2	R	S		S
3	S	T		T
4	R	R		S
5	T	S		S
6	S	S		T
7	S	0		S

```
QuickReduct(ℂ,𝔻)
Input: ℂ, the set of all conditional features; 𝔻, the set of
decision features
Output: R, the feature subset
    R ← {}
    while γ_R(𝔻) ≠ γ_ℂ(𝔻)
      T ← R
      foreach x ∈ (ℂ − R)
          if γ_{R∪{x}}(𝔻) > γ_T(𝔻)
              T ← R ∪ {x}
      R ← T
    return R
```

Figure 5.1 QuickReduct algorithm

possible subset can be ignored if it contains more elements. However, a better approach is needed, one that will avoid wasted computational effort.

The QuickReduct algorithm given in Figure 5.1 (adapted from [54]) attempts to calculate a reduct without exhaustively generating all possible subsets. It starts off with an empty set and adds in turn, one at a time, those attributes that result in the greatest increase in the rough set dependency metric, until this produces its maximum possible value for the dataset. Other such techniques may be found in [273].

According to the QuickReduct algorithm the dependency of each attribute is calculated, and the best candidate chosen. In Figure 5.2 this stage is illustrated using the example dataset. As attribute d generates the highest dependency degree, then that attribute is chosen and the sets $\{a, d\}$, $\{b, d\}$, and $\{c, d\}$ are evaluated. This process continues until the dependency of the reduct equals the consistency of the dataset (1 if the dataset is consistent). The generated reduct shows the way of reducing the dimensionality of the original dataset by eliminating those conditional attributes that do not appear in the set.

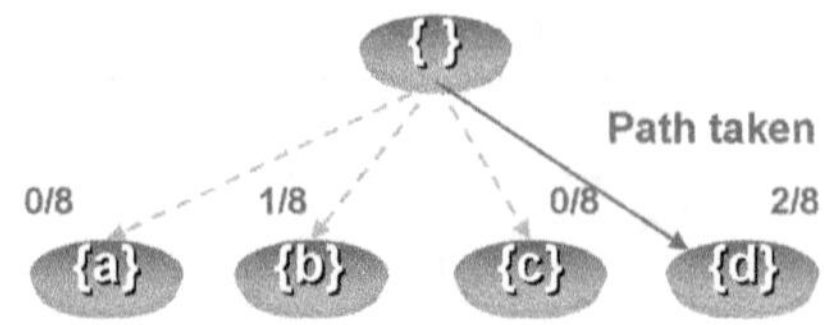

Figure 5.2 Branches of the search space

Determining the consistency of the entire dataset is reasonable for most datasets. However, it may be infeasible for very large data, so alternative stopping criteria may have to be used. One such criterion could be to terminate the search when there is no further increase in the dependency measure. This will produce exactly the same path to a reduct, due to the monotonicity of the measure [54], without the computational overhead of calculating the dataset consistency.

Other developments include ReverseReduct where the strategy is backward elimination of attributes as opposed to the current forward selection process. Initially all attributes appear in the reduct candidate; the least informative ones are incrementally removed until no further attribute can be eliminated without introducing inconsistencies. This procedure is not often used for large datasets, as the algorithm must evaluate large feature subsets (starting with the set containing *all* features), which is too costly, although the computational complexity is, in theory, the same as that of forward-looking QuickReduct. As both forward and backward methods perform well, it is thought that a combination of these within one algorithm would be effective. For instance, search could continue in a forward direction initially, and then resort to backward steps intermittently to remove less important features before continuing onward.

QuickReduct, however, is not guaranteed to find a *minimal* subset as has been shown in [56]. Using the dependency function to discriminate between candidates may lead the search down a nonminimal path. It is impossible to predict which combinations of attributes will lead to an optimal reduct based on changes in dependency with the addition or deletion of single attributes. It does result in a close-to-minimal subset, though, which is still useful in greatly reducing dataset dimensionality.

In [56] a potential solution to this problem has been proposed whereby the QuickReduct algorithm is altered, making it into an n-lookahead approach. However, even this cannot guarantee a reduct unless n is equal to the original number of attributes, but it reverts back to generate-and-test. The n-lookahead algorithm still suffers from the same problem as the original QuickReduct: it is impossible to tell at any stage whether the current path will be the shortest to a reduct.

It is interesting that the rough set degree of dependency measure is very similar to the consistency criterion used by the FOCUS algorithm and others [4,306]. In FOCUS a breadth-first search is employed such that any subset is rejected if this produces at least one inconsistency. If this is converted into a guided search using the consistency measure as a heuristic, it should behave exactly as QuickReduct. Consistency is defined as the number of discernible objects out of the entire object set—exactly that of the dependency measure.

5.1.1 Additional Search Strategies

The QuickReduct algorithm used in RSAR is a greedy hill-climbing approach. It uses the dependency function to choose the (locally) best feature subset and terminates when the maximum dependency has been reached for the dataset. Due to the monotonicity of the algorithm, there are no local minima. Of course, other strategies could be employed for this, several of which are briefly outlined here. See [298] for a fuller description of these search methods.

5.1.1.1 Breadth-First Search Breadth-first search is an uninformed search method that examines and expands all nodes of a graph in the search for a solution. The search is exhaustive: the entire graph is examined until the goal state is found. The method is also complete and optimal if the path cost is nondecreasing with depth. For RSAR the dependency function could be used as the search stopping criterion so that when a subset is encountered with maximum dependency, the search terminates. The resulting subset would be guaranteed to be minimal with this strategy. However, the time and space complexity are both $O(branchFactor^{depth})$, which is too costly for larger problems.

5.1.1.2 Depth-First Search Depth-first search is an uninformed search that progresses by expanding the first child node of the search tree that appears and thus going deeper and deeper until a goal state is found, or until it hits a node that has no children. At this point the search backtracks and continues from the next node. This strategy is not complete, as search may progress infinitely down a branch of the search tree. It is also not optimal as the first (not necessarily ideal) solution found is returned. This has the same time complexity as breadth-first search ($O(branchFactor^{depth})$) but better space complexity, $O(depth)$.

5.1.1.3 Iterative Deepening Iterative deepening visits each node in the search tree in the same order as the depth-first search but by gradually increasing the maximum depth limit of the search iteratively. At each iteration the maximum depth is increased and the search is re-run. This is repeated until the depth limit reaches the depth of the shallowest goal state. Like breadth-first search it is complete and optimal. The space complexity of iterative deepening is $O(branchFactor * depth)$. Time complexity of this method is $O(branchFactor^{depth})$.

5.1.1.4 Best-First Search Best-first search expands nodes according to some cost function. The search is greedy in that the best nodes (according to the evaluation function) are expanded first. This method is of benefit only when the evaluation function is suitably accurate. In this case the amount of search required may be drastically reduced. However, if the function is very expensive, the benefits of reducing the amount of search may be outweighed by the costs of assigning scores to nodes.

5.1.1.5 A* Search Best-first search is useful, but it does not take into account the cost of the path so far when choosing the next node. A* search is a strategy that seeks to find a solution that minimizes this total cost of solution path. It combines advantages of breadth-first search, where the shortest path is found first, with advantages of best-first search, where the node that is heuristically closest to the solution is explored next. This method is both complete and optimal.

5.1.2 Proof of QuickReduct Monotonicity

Theorem Suppose that R is a subset of the set of conditional attributes, x is an arbitrary conditional attribute that belongs to the dataset, and $\mathbb{D}$ is the set of decision attributes that dataset objects are classified into. Then $\gamma_{R\cup\{x\}}(\mathbb{D}) \geq \gamma_R(\mathbb{D})$.

Proof From expression (2.10), where the $\otimes$ operator is defined as a variation of the Cartesian product, it is known that

$$|\bigotimes\{q \in (R \cup \{x\}) : \mathbb{U}/\mathrm{IND}(\{q\})\}| \geq |\bigotimes\{q \in R : \mathbb{U}/\mathrm{IND}(\{q\})\}| \tag{5.1}$$

This relation, in conjunction with the definition of the positive region, leads to

$$|\bigcup\{A : A \in \mathbb{U}/\mathrm{IND}(R \cup \{x\}), A \subseteq (R \cup \{x\})\}| \\ \geq |\bigcup\{A : A \in \mathbb{U}/\mathrm{IND}(R), A \subseteq R\}| \tag{5.2}$$

This relation holds because the cardinality of a union of sets monotonically nondecreases as the cardinality of the sets increases. From this relation, and from the definition of the positive region, it follows that

$$|\bigcup_{A\in\mathbb{U}/\mathbb{D}} \underline{(R \cup \{x\})A}| \geq |\bigcup_{A\in\mathbb{U}/\mathbb{D}} \underline{R}A| \tag{5.3}$$

Given any nonempty universe $\mathbb{U}$, $|\mathbb{U}| \geq 1$. Thus

$$\frac{|\mathrm{POS}(R \cup \{x\})|}{|\mathbb{U}|} \geq \frac{|\mathrm{POS}(R)|}{|\mathbb{U}|} \tag{5.4}$$

That is, $\gamma_{R\cup\{x\}}(\mathbb{D}) \geq \gamma_R(\mathbb{D})$.

5.2 RSAR OPTIMIZATIONS

If the RSAR algorithm outlined previously is implemented using explicit set-theoretic representations, it will be extremely inefficient in terms of space and time requirements. More computationally feasible methods can be constructed.

5.2.1 Implementation Goals

A naive implementation of the gamma function would work directly with the various set-theoretic constructs involved. It would therefore seek to obtain an explicit representation of the positive region. Following the definitions, this would involve partitioning the dataset on both the decision attributes and candidate conditions. Then each equivalence class in the partition induced by the candidate conditions would be checked against every equivalence class in the partition induced by the decision attributes see whether the latter is a subset of the former. This operation is not only inherently time-consuming but may impose unreasonable spatial demands as the entire dataset must be loaded into memory in order to construct the partitions.

Implementing RSAR based on a library of generalized rough set operations can thus incur high computing costs, leading to unreasonable space and time requirements. It should be possible to simplify the reduction infrastructure by taking into account the specific properties of the problem at hand. This would not alleviate the inherent complexity of the problem. It should, however, minimize the coefficient of the second-order component of the complexity, making QuickReduct practically applicable to much more demanding domains. A number of requirements can be outlined:

- The implementation should be able to cope with datasets of extremely high dimensionality, reaching tens or hundreds of thousands of attributes. This makes RSAR suitable for applications such as text categorization problems.
- It should be able to cope effectively with datasets containing large numbers of samples.
- It should offer flexibility and efficiency, so that the implementation can be incorporated into a wider range of applications.

5.2.2 Implementational Optimizations

The following subsections introduce the key implementational enhancements to RSAR. The first two optimizations described, addressing data storage and discernibility vector dictionaries, are applied to all the experimental results in the next section.

5.2.2.1 Data Storage Given that RSAR should handle datasets with very high dimensionality and large numbers of samples, datasets are not explicitly read into main memory. They are accessed instead directly from peripheral memory.

All modern operating systems in use today have advanced caching features, which effectively transfer the data from peripheral memory onto main memory, if that is possible. This simplifies the implementation considerably as multiple passes are performed through the dataset (one per evaluation of γ).

Attributes may only have integer values. Strings and real numbers are not supported. Most integer operations are intrinsic on all modern computing platforms, and integers require less space than character strings or real numbers. RSAR only deals with discrete values; real values and discrete (string) tokens can be mapped to integer values using small preprocessing programs. This design decision also accelerates reading the dataset from peripheral memory, speeds up and simplifies parsing data into tokens (attribute values), simplifies the data structures needed to store the values (dynamic, 32-bit integer-indexed dictionaries are used instead of hashes), and thus increases overall processing speed. The implementation also includes a family of high-speed, optimized conventional set theory operators. These operate on bit strings using Boolean operations, intrinsic on all microprocessors of the past 20 years. This allows for efficient, compact storage of sets.

5.2.2.2 Discernibility Vector Dictionaries Discernibility vector dictionaries (DVD) form the basis of the optimized evaluator. They originate from an observation of the properties of the degree of dependency. The numerator in equation (2.16), $|POS_P(Q)|$, is equal to the number of dataset samples that are discernible with respect to the specified attribute subset. This reduces the problem of evaluating the denominator to counting the number of discernible samples. To this end the concept of indiscernibility is leveraged. Given a subset of attributes A and two data items $d, d' \in \mathbb{U}$, mapped to classes c, c', data vectors $\mathbf{v}, \mathbf{v}'$ are generated such that if $A = a_1, a_2, \ldots, a_n$, $\mathbf{v} = \langle d_{a_1}, d_{a_2}, \ldots, d_{a_n} \rangle$ and $\mathbf{v}' = \langle d'_{a_1}, d'_{a_2}, \ldots, d'_{a_n} \rangle$. Then d and d' are discernible iff $\mathbf{v} \neq \mathbf{v}' \vee c = c'$.

During the evaluation of γ for a particular attribute subset, vectors and their classes are implicitly obtained by reading the dataset. They are recorded in the DVD. DVD entries are pairs of the form $\mathbf{v} \mapsto \langle c, f, i \rangle$, where $\mathbf{v}$ is a data vector, c is the class it is labeled with, f is the frequency of $\mathbf{v}$, the number of times the pattern $\mathbf{v}$ was encountered in the dataset so far, and i is a Boolean flag denoting whether $\mathbf{v} \mapsto c$ and $\mathbf{v} \mapsto c'$, $c \neq c'$, have been found to be indiscernible. Each encountered data vector is added to this dictionary, which maybe implemented using a variety of different data structures. This may appear subject to combinatorial explosion as the domain of the DVDs index is a Cartesian product. However, datasets being finite by definition, a dataset of l samples will have at most l distinct data vectors. DVDs are temporary storage available only during the evaluation of γ for a particular attribute subset. This algorithm is outlined in Figure 5.3. After all dataset objects have been processed, dependency can be calculated by

$$\gamma = \frac{\sum f \,\forall \langle \mathbf{v}, f, \text{true} \rangle \in \text{DVD}}{\sum f \,\forall \langle \mathbf{v}, f, i \rangle \in \text{DVD}} \tag{5.5}$$

```
CHECK-DISCERNIBILITY(DVD, v, c)
Input: DVD, the current Discernibility Vector Dictionary;
v, a data vector to test; c, the equivalence class of v
Output: DVD, the updated dictionary
if ∃ DVD[v]
    ⟨c, f, i⟩ = DVD[v]
    if c = c′
        DVD[v] ← ⟨c, f+1, i⟩
        return DVD
    else
        DVD[v] ← ⟨c, f+1, false⟩
        return DVD
else
    DVD[v] ← ⟨c, 1, true ⟩
    return DVD
```

Figure 5.3 DVD update algorithm

5.2.2.3 Further Optimizations Additional speed may be gained by making further observations about the operation of RSAR. Increasingly sophisticated optimizations were implemented as different versions of the QuickReduct algorithm, numbered as follows:

QuickReduct. The original algorithm, augmented by using DVD-based calculation.

QuickReduct II, also known as ReverseReduct. A backward-eliminating version of QuickReduct.

QuickReduct III. Unlike the original QuickReduct, this version of the algorithm introduces a minor optimization: the reduction stops immediately as soon as the first reduct is found.

QuickReduct IV. Augmented version of QuickReduct III. In adding further features to the attribute subset, dataset samples are not examined if they are known to be discernible given the current attribute subset. Only heretofore indiscernible samples are used in the calculation of γ. This reduces the effective size of the dataset as QuickReduct IV approaches a solution. For example, assume a dataset of 150 samples is being processed, and the first evaluation of dependency yields 120 discernible samples. In adding further attributes to the attribute subset, γ will only be evaluated for the remaining 30 samples. This depends on the monotonicity of γ: if samples d and d' are discernible given the attribute set A, then they will also be discernible given the attribute set A', where $A \subseteq A'$

QuickReduct V. In selecting an attribute to be added to the candidate reduct, the attribute whose addition leads to the highest γ is used. This version of the algorithm immediately stops calculating γ if the result is guaranteed to be below the highest result so far. This is tested incrementally by summing the frequencies of indiscernible pairs in the DVD. This optimization saves considerable time, especially where datasets have a significant degree of redundancy and/or noise.

5.2.2.4 *Experimental Results* The aim of the following experiments is to gauge the implementation's performance when operating on worst-case and average-case scenarios. The worst-case scenario is one in which all remaining attributes have to be examined because they provide almost identical gains.

Given a dataset with n conditional attributes, γ is evaluated $n + (n-1) + \cdots + 2 + 1 = \frac{n}{2}(1+n)$ times, a $O(n^2)$ time complexity. Evaluating γ involves processing each dataset item using CHECK-DISCERNIBILITY—clearly, $O(l)$ with respect to the number of dataset items. The average case is described as a dataset in which high-signal and noise/redundant attributes are homogeneously mixed in a dataset. The best, nontrivial case is one where high signal attributes are the first to be chosen by the algorithm. For brevity and because of its simplicity, the last case is not addressed here. Experimental results were obtained by running different versions of the improved RSAR on artificially constructed datasets with varying dimensionality and number of samples to gauge efficiency and complexity. To smooth out timing errors, numerous runs were performed and averaged. Approximate curves are shown interpolating the experimental data.

The worst-case runtimes are shown in Figure 5.4, where RSAR performed FS on pure binary noise (each attribute is one bit of noise). The first graph shows the results of varying the number of attributes (with 20 dataset items); the second graph plots runtime over the number of dataset items (with 20 attributes). $O(n^2)$ and $O(l)$ complexities are obvious. As expected, QUICKREDUCT V is the best choice for this type of problem.

Two different average-case experiments were conducted. In both cases signal-rich attributes (drawn from the Iris dataset [38]) and attributes consisting of pure binary noise are combined so that the signal-rich attributes are separated by noise attributes. The two experiments differ in the layout of the datasets:

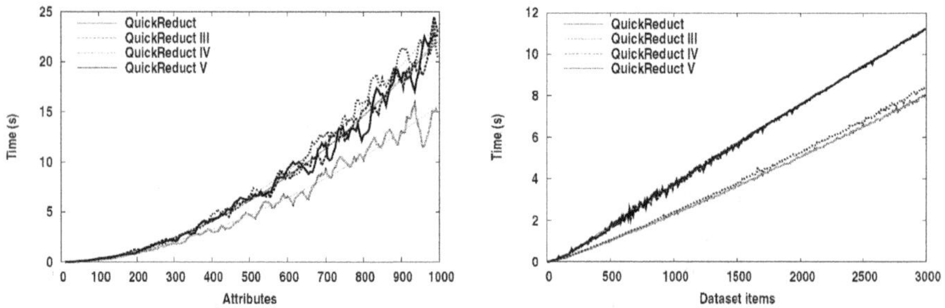

Figure 5.4 Evaluation of algorithms: Worst case

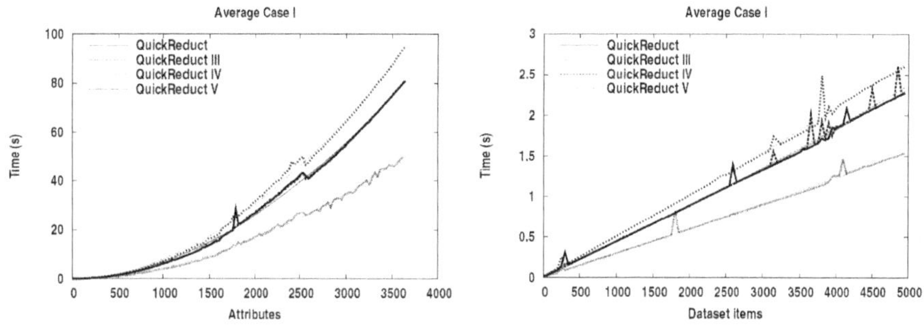

Figure 5.5 Evaluation of algorithms: Average case (pessimistic)

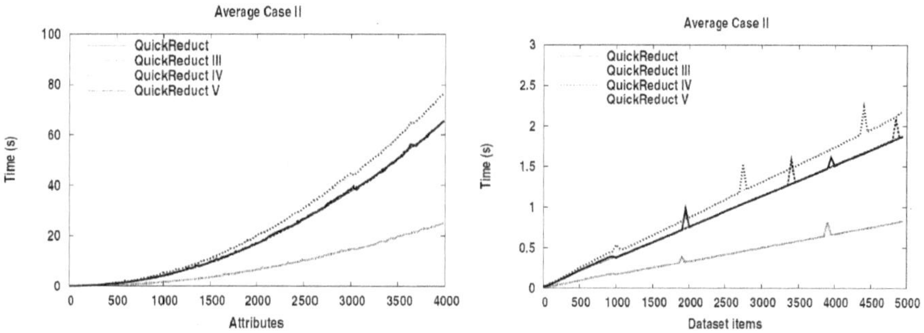

Figure 5.6 Evaluation of algorithms: Average case (optimistic)

average case I starts each dataset item with a noise attribute and ends the datum with the last information-rich attribute (pessimist case). Average case II begins with an information-rich attribute and ends the datum with noisy attributes (optimist case). There are four information-rich attributes in each generated dataset. Figures 5.5 and 5.6 illustrate the results. As expected, the optimistic case yields results considerably faster than the pessimistic case. Both cases demonstrate the efficiency of the algorithms, with QUICKREDUCT V locating the four signal attributes in a dataset of 4000 attributes within 20 seconds. It should be noted, that most application domains have less redundancy than these "average" cases.

5.3 DISCERNIBILITY MATRIX BASED APPROACHES

5.3.1 Johnson Reducer

The Johnson reducer is a simple greedy heuristic algorithm that is often applied to discernibility functions to find a single reduct [253]. Reducts found by this

```
JOHNSON(ℂ,f_D)
Input: ℂ, the set of all conditional features; f_D, the
discernibility function.
Output: R, the feature subset
R←{}, bestc = 0
while f_D not empty
  foreach a∈ℂ appearing in f_D
     c = heuristic(a)
     if c > bestc
         bestc = c; bestAttr ← a
  R←R∪{a}
  f_D = removeClauses(f_D,a)
return R
```

Figure 5.7 Johnson reducer

process have no guarantee of minimality but are generally of a size close to the minimal.

The algorithm in Figure 5.7 begins by setting the current reduct candidate, R, to the empty set. Then each conditional attribute appearing in the discernibility function is evaluated according to the heuristic measure. For the standard Johnson algorithm this is typically a count of the number of appearances an attribute makes within clauses; attributes that appear more frequently are considered to be more significant. The attribute with the highest heuristic value is added to the reduct candidate, and all clauses in the discernibility function containing this attribute are removed. As soon as all clauses have been removed, the algorithm terminates and returns the reduct R. R is assured to be a reduct as all clauses contained within the discernibility function have been addressed.

Variations of the algorithm involve alternative heuristic functions in an attempt to guide search down better paths [249,374]. However, no perfect heuristic exists, and hence there is still no guarantee of subset optimality.

5.3.2 Compressibility Algorithm

In [350] the authors present a method for the generation of all reducts in an information system by manipulating the clauses in discernibility functions. In addition to the standard simplification laws, the concept of strong compressibility is introduced and applied in conjunction with an expansion algorithm.

The strong compressibility simplification applies where clause attributes are either simultaneously present or absent in all clauses. In this situation the attributes may be replaced by a single representative attribute. As an example consider the formula

$$f_D = \{a \vee b \vee c \vee f\} \wedge \{b \vee d\} \wedge \{a \vee d \vee e \vee f\} \wedge \{d \vee c\}$$

The attributes a and f can be replaced by a single attribute as they are both present in the first and third clauses, and absent in the second and fourth. Replacing these with g results in

$$f_D = \{g \vee b \vee c\} \wedge \{b \vee d\} \wedge \{g \vee d \vee e\} \wedge \{d \vee c\}$$

If a reduct resulting from this discernibility function contains the new attribute g, then this attribute may be replaced by either a or f. Here $\{g, d\}$ is a reduct and so $\{a, d\}$ and $\{f, d\}$ are reducts of the original set of clauses. Hence fewer attributes are considered in the reduct-determining process with no loss of information. The complexity of this step is $O(a * c + a^2)$, where a is the number of attributes and c is the number of clauses.

Returning to the example above and following the compressibility algorithm (Figure 5.8) from step 4, the most commonly occurring attribute can be seen to be d. Hence the expansion law is applied with respect to this attribute to obtain

$$\begin{aligned} f_D &= f_1 \vee f_2 \\ &= (\{d\} \wedge \{g \vee b \vee c\}) \vee (\{g \vee b \vee c\} \wedge \{b\} \wedge \{g \vee e\} \wedge \{c\}) \\ &= (\{d\} \wedge \{g \vee b \vee c\}) \vee (\{b\} \wedge \{g \vee e\} \wedge \{c\}) \end{aligned}$$

As all components are in simple form, step 6 is carried out, where all strongly equivalent classes are replaced by their equivalent attributes:

$$\begin{aligned} f_D &= f_1 \vee f_2 \\ &= (\{d\} \wedge \{a \vee f \vee b \vee c\}) \vee (\{b\} \wedge \{a \vee f \vee e\} \wedge \{c\}) \end{aligned}$$

```
COMPRESSIBILITY(f_D)
Input: f_D, the discernibility function
Output: all minimal reducts
(1)   while f_D not in simple form
(2)     applyAbsorptionLaws(f_D) //remove supersets
(3)     replaceStronglyCompressibleAttributes(f_D)
(4)     a ← mostFrequentAttribute(f_D)
(5)     applyExpansionLaw(a,f_D)
(6)   substituteStronglyCompressibleClasses(f_D)
(7)   Reds ← calculateReducts(f_D)
(8)   return minimalElements(Reds)
```

Figure 5.8 Compressibility algorithm

The corresponding reducts for the function components are as follows (step 7):

$$\text{Red}(f_1) = (\{a, d\}, \{d, f\}, \{b, d\}, \{c, d\})$$
$$\text{Red}(f_2) = (\{b, a, c\}, \{b, f, c\}, \{b, e, c\})$$

The reducts for the system are generated by taking the union of the components and determining the minimal elements:

$$\text{Red}(f_D) = (\{a, d\}, \{d, f\}, \{b, d\}, \{c, d\}, \{b, a, c\}, \{b, f, c\}, \{b, e, c\})$$

5.4 REDUCTION WITH VARIABLE PRECISION ROUGH SETS

Variable precision rough sets (VPRS) [419] attempts to improve upon rough set theory by relaxing the subset operator. VPRS was proposed to analyze and identify data patterns that represent statistical trends rather than functional. The main idea of VPRS is to allow objects to be classified with an error smaller than a certain predefined level.

This approach is arguably easiest to be understood within the framework of classification. Let $X, Y \subseteq \mathbb{U}$. The relative classification error is defined by

$$c(X, Y) = 1 - \frac{|X \cap Y|}{|X|}$$

Observe that $c(X, Y) = 0$ if and only if $X \subseteq Y$. A degree of inclusion can be achieved by allowing a certain level of error, β, in the classification:

$$X \subseteq_\beta Y \text{ iff } c(X, Y) \leq \beta, \qquad 0 \leq \beta < 0.5$$

If $\subseteq_\beta$ is used instead of $\subseteq$, the β-upper and β-lower approximations of a set X can be defined as

$$\underline{R}_\beta X = \bigcup\{[x]_R \in \mathbb{U}/R | [x]_R \subseteq_\beta X\}$$
$$\overline{R}_\beta X = \bigcup\{[x]_R \in \mathbb{U}/R | c([x]_R, X) < 1 - \beta\}$$

Note that $\underline{R}_\beta X = \underline{R}X$ for $\beta = 0$. The positive, negative, and boundary regions in the original rough set theory can now be extended to

$$POS_{R,\beta}(X) = \underline{R}_\beta X \tag{5.6}$$
$$NEG_{R,\beta}(X) = \mathbb{U} - \overline{R}_\beta X \tag{5.7}$$
$$BND_{R,\beta}(X) = \overline{R}_\beta X - \underline{R}_\beta X \tag{5.8}$$

Returning to the example dataset in Table 2.1, equation (5.6) can be used to calculate the β-positive region for $R = \{b, c\}$, $X = \{e\}$ and $\beta = 0.4$. Setting β

to this value means that a set is considered to be a subset of another if they share about half the number of elements. The partitions of the universe of objects for R and X are

$$\mathbb{U}/R = \{\{2\}, \{0, 4\}, \{3\}, \{1, 6, 7\}, \{5\}\}$$
$$\mathbb{U}/X = \{\{0\}, \{1, 3, 6\}, \{2, 4, 5, 7\}\}$$

For each set $A \in \mathbb{U}/R$ and $B \in \mathbb{U}/X$, the value of $c(A,B)$ must be less than β if the equivalence class A is to be included in the β-positive region. Considering $A = \{2\}$ gives

$$c(\{2\}, \{0\}) = 1 > \beta$$
$$c(\{2\},\{1, 3, 6\}) = 1 > \beta$$
$$c(\{2\},\{2, 4, 5, 7\}) = 0 < \beta$$

So object 2 is added to the β-positive region as it is a β-subset of $\{2, 4, 5, 7\}$ (and is in fact a traditional subset of the equivalence class). For $A = \{1, 6, 7\}$, a more interesting case is encountered:

$$c(\{1, 6, 7\}, \{0\}) = 1 > \beta$$
$$c(\{1, 6, 7\}, \{1, 3, 6\}) = 0.3333 < \beta$$
$$c(\{1, 6, 7\}, \{2, 4, 5, 7\}) = 0.6667 > \beta$$

Here the objects 1, 6, and 7 are included in the β-positive region as the set $\{1, 6, 7\}$ is a β-subset of $\{1, 3, 6\}$. Calculating the subsets in this way leads to the following β-positive region:

$$POS_{R,\beta}(X) = \{1, 2, 3, 5, 6, 7\}$$

Compare the region above with the positive region generated previously: $\{2, 3, 5\}$. Objects 1, 6, and 7 are now included due to the relaxation of the subset operator. Consider a decision table $(\mathbb{U}, \mathbb{C} \cup \mathbb{D})$, where $\mathbb{C}$ is the set of conditional attributes and $\mathbb{D}$ the set of decision attributes. The β-positive region of an equivalence relation Q on $\mathbb{U}$ may be determined by

$$POS_{R,\beta}(Q) = \bigcup_{X \in \mathbb{U}/Q} \underline{R}_{\beta}X$$

where R is also an equivalence relation on $\mathbb{U}$. This function can then be used to calculate dependencies and thus determine β-reducts. The dependency function becomes

$$\gamma_{R,\beta}(Q) = \frac{|POS_{R,\beta}(Q)|}{|\mathbb{U}|}$$

It can be seen that the QuickReduct algorithm outlined previously can be adapted to incorporate the reduction method built upon VPRS theory. By supplying a suitable β value to the algorithm, the β-lower approximation, β-positive region, and β-dependency can replace the traditional calculations. This more approximate final reduct may be a better generalization when encountering unseen data. Additionally, setting β to 0 forces such a method to behave exactly like standard rough set theory.

Extended classification of reducts in the VPRS approach may be found in [27,28,189]. However, the variable precision approach requires the additional parameter β that has to be specified from the start. By repeated experimentation, this parameter can be suitably approximated. Nevertheless, problems arise when searching for true reducts as VPRS incorporates an element of imprecision in determining the number of classifiable objects.

5.5 DYNAMIC REDUCTS

Reducts generated from an information system are sensitive to changes in the system. This sensitivity can be seen when removing a randomly chosen set of objects from the original object set. The reducts frequently occurring in random subtables can be considered to be stable; it is these reducts that are encompassed by *dynamic reducts* [23]. Let $\mathcal{A} = (\mathbb{U}, \mathbb{C} \cup d)$ be a decision table; then any system $\mathcal{B} = (\mathbb{U}', \mathbb{C} \cup d)$ $(\mathbb{U}' \subseteq \mathbb{U})$ is called a subtable of $\mathcal{A}$. If $\mathcal{F}$ is a family of subtables of $\mathcal{A}$, then

$$DR(\mathcal{A}, \mathcal{F}) = Red(\mathcal{A}, d) \cap \{\bigcap_{\mathcal{B} \in \mathcal{F}} Red(\mathcal{B}, d)\}$$

defines the set of $\mathcal{F}$-dynamic reducts of $\mathcal{A}$. From this definition it follows that a relative reduct of $\mathcal{A}$ is dynamic if it is also a reduct of all subtables in $\mathcal{F}$. In most cases this is too restrictive, so a more general notion of dynamic reducts is required.

By introducing a threshold, $0 \leq \epsilon \leq 1$, the concept of $(\mathcal{F}, \epsilon)$-dynamic reducts can here be defined:

$$DR_\epsilon(\mathcal{A}, \mathcal{F}) = \{C \in Red(\mathcal{A}, d) : s_F(C) \geq \epsilon\}$$

where

$$s_F(C) = \frac{|\{\mathcal{B} \in \mathcal{F} : C \in Red(\mathcal{B}, d)\}|}{|\mathcal{F}|}$$

is the $\mathcal{F}$-stability coefficient of C. This lessens the previous restriction that a dynamic reduct must appear in *every* generated subtable. Now a reduct is considered to be dynamic if it appears in a certain proportion of subtables, determined by the value ϵ. For example, by setting ϵ to 0.5 a reduct is considered to be

```
DynamicRed(𝒜,ϵ,its)
Input: 𝒜, the original decision table; ϵ, the dynamic reduct
threshold; its, the number of iterations
Output: dynamic reducts
    R ← {}
    A ← calculateAllReducts(𝒜)
    foreach j = 1...its
        𝒜_j ← deleteRandomRows(𝒜)
        R ← R ∪ calculateAllReducts(𝒜_j)
    foreach C ∈ A
        if s_F(C, R) ≥ ϵ then output C
```

Figure 5.9 Dynamic reduct algorithm

dynamic if it appears in at least half of the subtables. Note that if $\mathcal{F} = \{\mathcal{A}\}$, then $DR(\mathcal{A}, \mathcal{F}) = Red(\mathcal{A}, d)$. Dynamic reducts may then be calculated according to the algorithm given in Figure 5.9. First, all reducts are calculated for the given information system, $\mathcal{A}$. Then, the new subsystems $\mathcal{A}_j$ are generated by randomly deleting one or more rows from $\mathcal{A}$. All reducts are found for each subsystem, and the dynamic reducts are computed using $s_F(C, R)$, which denotes the significance factor of reduct C within all reducts found, R.

Returning to the example decision table (call this $\mathcal{A}$), the first step is to calculate all its reducts. This produces the set of all reducts $A = \{\{b, d\}, \{c, d\}, \{a, b, d\}, \{a, c, d\}$, and $\{b, c, d\}\}$. The reduct $\{a, b, c, d\}$ is not included as this will always be a reduct of any generated subtable (it is the full set of conditional attributes). The next step randomly deletes a number of rows from the original table $\mathcal{A}$. From this, all reducts are again calculated. For one subtable the result might be $R = \{\{b, d\}, \{b, c, d\}, \{a, b, d\}\}$. In this case the subset $\{c, d\}$ is not a reduct (though it was for the original dataset). If the number of iterations is set to just one, and if ϵ is set to a value less than 0.5 (implying that a reduct should appear in half of the total number of discovered reducts), then the reduct $\{c, d\}$ is deemed not to be a dynamic reduct.

Intuitively, the hope in finding stable reducts is they will be more representative of the real world, meaning it is more likely that they will be reducts for unseen data. A comparison of dynamic and nondynamic approaches can be found in [22], where various methods were tested on extracting laws from decision tables. In the experiments the dynamic method and the conventional RS method both performed well. In fact it appears that the RS method has on average a lower error rate of classification than the dynamic RS method.

A disadvantage of this dynamic approach is that several subjective choices have to be made before the dynamic reducts can be found (e.g., the choice of the value of ϵ); these values are not contained in the data. Also the huge complexity of finding all reducts within subtables forces the use of heuristic techniques such as genetic algorithms to perform the search. For large datasets this step may well be too costly.

5.6 RELATIVE DEPENDENCY METHOD

In [123] a feature selection method based on an alternative dependency measure is presented. The technique was originally proposed to avoid the calculation of discernibility functions or positive regions, which can be computationally expensive without optimizations

The authors replace the traditional rough set degree of dependency with an alternative measure, the relative dependency, defined as follows for an attribute subset R:

$$\kappa_R(D) = \frac{|U/IND(R)|}{|U/IND(R \cup D)|} \tag{5.9}$$

The authors then show that R is a reduct iff $\kappa_R(D) = \kappa_C(D)$ and $\forall X \subset R$, $\kappa_X(D) \neq \kappa_C(D)$.

Two algorithms are constructed for feature selection based on this measure. The first (Figure 5.10) performs backward elimination of features where attributes are removed from the set of considered attributes if the relative dependency equals 1 upon their removal. Attributes are considered one at a time, starting with the first, evaluating their relative dependency. The second algorithm initially ranks the individual attributes beforehand using an entropy measure before the backward elimination is performed.

Returning to the example dataset, the backward elimination algorithm initializes R to the set of conditional attributes, $\{a, b, c, d\}$. Next the elimination of attribute a is considered:

$$\kappa_{\{b,c,d\}}(D) = \frac{|\mathbb{U}/IND(\{b,c,d\})|}{|\mathbb{U}/IND(\{b,c,d,e\})|} = \frac{|\{\{0\},\{1,6\},\{2\},\{3\},\{4\},\{5\},\{7\}\}|}{|\{\{0\},\{1,6\},\{2\},\{3\},\{4\},\{5\},\{7\}\}|} = 1$$

```
RELATIVEREDUCT(ℂ,𝔻)
Input: ℂ, the set of all conditional features; 𝔻, the set of
decision features
Output: R, the feature subset
(1)   R←ℂ
(2)   foreach a∈ℂ
(3)      if κR−{a}(𝔻) = = 1
(4)         R←R−{a}
(5)   return R
```

Figure 5.10 Backward elimination based on relative dependency

As the relative dependency is equal to 1, attribute a can be removed from the current reduct candidate $R \leftarrow \{b, c, d\}$. The algorithm then considers the elimination of attribute b from R:

$$\kappa_{\{c,d\}}(D) = \frac{|\mathbb{U}/IND(\{c, d\})|}{|\mathbb{U}/IND(\{c, d, e\})|} = \frac{|\{\{0\}, \{1, 6\}, \{2, 5\}, \{3\}, \{4\}, \{7\}\}|}{|\{\{0\}, \{1, 6\}, \{2, 5\}, \{3\}, \{4\}, \{7\}\}|} = 1$$

Again, the relative dependency of $\{c, d\}$ evaluates to 1, so attribute b is removed from R, ($R = \{c, d\}$). The next step evaluates the removal of c from the reduct candidate:

$$\kappa_{\{d\}}(D) = \frac{|\mathbb{U}/IND(\{d\})|}{|\mathbb{U}/IND(\{d, e\})|} = \frac{|\{\{0, 3\}, \{1, 2, 5, 6\}, \{4, 7\}\}|}{|\{\{0\}, \{3\}, \{1, 6\}, \{2, 5\}, \{4, 7\}\}|} = \frac{3}{5}$$

As this does not equal 1, attribute d is not removed from R. The algorithm then evaluates the elimination of attribute d from R ($R = \{c, d\}$):

$$\kappa_{\{c\}}(D) = \frac{|\mathbb{U}/IND(\{c\})|}{|\mathbb{U}/IND(\{c, e\})|} = \frac{|\{\{0, 4\}, \{1, 6, 7\}, \{2, 3, 5\}\}|}{|\{\{0\}, \{4\}, \{1, 6\}, \{7\}, \{2, 5\}, \{3\}\}|} = \frac{3}{6}$$

Again, the relative dependency does not evaluate to 1; hence attribute d is retained in the reduct candidate. As there are no further attributes to consider, the algorithm terminates and outputs the reduct $\{c, d\}$.

5.7 TOLERANCE-BASED METHOD

Another way of attempting to handle imprecision is to introduce a measure of similarity of attribute values and define the lower and upper approximations based on these similarity measures.

5.7.1 Similarity Measures

In this approach suitable similarity relations must be defined for each attribute, although the same definition can be used for all attributes if applicable. A standard measure for this purpose, given in [351], is

$$SIM_a(x, y) = 1 - \frac{|a(x) - a(y)|}{|a_{\max} - a_{\min}|} \tag{5.10}$$

where a is the attribute under consideration, and $a_{\max}$ and $a_{\min}$ denote the maximum and minimum values respectively for this attribute.

When considering more than one attribute, the defined similarities must be combined to provide a measure of the overall similarity of objects. For a subset of attributes, P, this can be achieved in many ways; two commonly adopted approaches are

$$(x, y) \in SIM_{P,\tau} \quad \text{iff} \prod_{a \in P} SIM_a(x, y) \geq \tau \tag{5.11}$$

$$(x, y) \in SIM_{P,\tau} \quad \text{iff} \frac{\sum_{a \in P} SIM_a(x, y)}{|P|} \geq \tau \tag{5.12}$$

where τ is a global similarity threshold. This framework allows for the specific case of traditional rough sets by defining a suitable similarity measure (e.g., equality of attribute values and equation (5.11)) and threshold ($\tau = 1$). Further similarity relations are investigated in [249], but are omitted here.

From the framework above the so-called tolerance classes generated by a given similarity relation for an object x are defined as

$$SIM_{P,\tau}(x) = \{y \in \mathbb{U} | (x, y) \in SIM_{P,\tau}\} \tag{5.13}$$

5.7.2 Approximations and Dependency

Lower and upper approximations are defined in a similar way to traditional rough set theory:

$$\underline{P_\tau}X = \{x | SIM_{P,\tau}(x) \subseteq X\} \tag{5.14}$$

$$\overline{P_\tau}X = \{x | SIM_{P,\tau}(x) \cap X \neq \emptyset\} \tag{5.15}$$

The tuple $\langle \underline{P_\tau}X, \overline{P_\tau}X \rangle$ is called a tolerance rough set [335]. Positive region and dependency functions then become

$$POS_{P,\tau}(Q) = \bigcup_{X \in \mathbb{U}/Q} \underline{P_\tau}X \tag{5.16}$$

$$\gamma_{P,\tau}(Q) = \frac{|POS_{P,\tau}(Q)|}{|\mathbb{U}|} \tag{5.17}$$

From these definitions, methods for reduct search can be constructed that use the tolerance-based degree of dependency, $\gamma_{P,\tau}(Q)$, to gauge the significance of attribute subsets (in a similar way as QuickReduct). The resulting algorithm can be found in Figure 5.11.

```
TOLERANCEREDUCT(ℂ,𝔻,τ)
Input: ℂ, the set of all conditional features; 𝔻, the set of
decision features; τ, the similarity threshold
Output: R, the feature subset
    R ← {}; γ^τ_best = 0;
    while γ^τ_best ≠ γ^τ_prev
        T ← R
        γ^τ_prev = γ^τ_best
        foreach x ∈ (ℂ − R)
            if γ_{R∪{x},τ}(𝔻) > γ_{T,τ}(𝔻)
                T ← R ∪ {x}
                γ^τ_best = γ_{T,τ}(𝔻)
        R ← T
    return R
```

Figure 5.11 Tolerance QUICKREDUCT

```
SELECT(ℂ, 𝔻, O, ε)
Input: ℂ, the set of all conditional features; 𝔻, the set of
decision features; O, the set of objects (instances); ε, reduct
threshold
Output: R, the feature subset
    R ← calculateCore()
    while γ_R(𝔻) < ε
        O ← O − POS_R(𝔻) //optimization
        foreach a ∈ ℂ − R
            v_a = |POS_{R∪{a}}(𝔻)|
            m_a = |largestEquivClass(POS_{R∪{a}}(𝔻))|
        Choose a with largest v_a*m_a
        R ← R ∪ {a}
    return R
```

Figure 5.12 Heuristic filter-based algorithm

5.8 COMBINED HEURISTIC METHOD

In [417] a filter-based approach is presented that adopts a heuristic employing a combination of rough set constructs. The algorithm proposed, as reformalized in Figure 5.12, begins with the core of the dataset (those attributes that cannot be removed without introducing inconsistencies) and incrementally adds attributes based on a heuristic measure. Additionally a threshold value is required as a stopping criterion to determine when a reduct candidate is "near enough" to being a reduct. On each iteration the objects that are consistent with the current reduct candidate are removed (an optimization that can be used with RSAR). As the process starts with the core of the dataset, "the core" has to be calculated beforehand. Using the discernibility matrix for this purpose can be quite impractical for datasets of large dimensionality. However, there are other methods that can calculate the core in an efficient manner [261]. For example, this can be done by calculating the degree of dependency of the full feature set and the corresponding dependencies of the feature set minus each attribute. The features that result in a dependency decrease are core attributes. There are also alternative methods available that allow the calculation of necessary information about the discernibility matrix without the need to perform operations directly on it [250].

5.9 ALTERNATIVE APPROACHES

Other approaches to generating reducts from information systems have been developed and can be found in [37,338,384]. Among the first rough set-based approaches is the PRESET algorithm [236], which is another feature selector that uses rough set theory to rank heuristically the features, assuming a noise-free binary domain. Since PRESET does not try to explore all combinations of the features, it is certain that it will fail on problems whose attributes are highly correlated. There have also been investigations into the use of different reduct quality measures (see [273] for details).

In [415], a new rough set-based feature selection heuristic, parameterized average support heuristic (PASH), is proposed. Unlike the existing methods, PASH is based on a special parameterized lower approximation that is defined to include all predictive instances. Predictive instances are instances that may produce predictive rules that hold true with a high probability but are not necessarily always true. The traditional model could exclude predictive instances that may produce such rules. The main advantage of PASH is that it considers the overall quality of the potential rules, thus producing a set of rules with balanced support distribution over all decision classes. However, it requires a parameter to be defined by the user that adjusts the level of approximation. One of the benefits of rough set theory is that it does not require such additional information.

5.10 COMPARISON OF CRISP APPROACHES

In order to evaluate several of the mainstream approaches to rough set-based feature selection, an investigation into how these methods perform in terms of resulting subset optimality has been carried out here. Several real and artificial datasets are used for this purpose. In particular, it is interesting to compare those methods that employ an incremental-based search strategy with those that adopt a more complex stochastic/probabilistic mechanism.

5.10.1 Dependency Degree Based Approaches

Five techniques for finding crisp rough set reducts are tested here on 13 datasets. These techniques are RSAR (using QuickReduct), EBR (using the same search mechanism as QuickReduct), GenRSAR (genetic algorithm based), AntRSAR (ant-based), and SimRSAR (simulated annealing-based).

5.10.1.1 Experimental Setup Before the experiments are described, a few points must be made about the latter three approaches, GenRSAR, AntRSAR, and SimRSAR.

GenRSAR employs a genetic search strategy in order to determine rough set reducts. The initial population consists of 100 randomly generated feature subsets, the probabilities of mutation and crossover are set to $0.4/|\mathbb{C}|$ and 0.6, respectively, and the number of generations is set to 100. The fitness function considers both the size of subset and its evaluated suitability, and it is defined as follows:

$$fitness(R) = \gamma_R(\mathbb{D}) * \frac{|\mathbb{C}| - |R|}{|\mathbb{C}|} \tag{5.18}$$

AntRSAR follows the mechanism described in [160] and Section 10.2. This "work" demonstrates that the general ant based FS framework presented in this book can be applied to crisp rough set based selection. Here the precomputed heuristic desirability of edge traversal is the entropy measure, with the subset evaluation performed using the rough set dependency heuristic (to guarantee that true rough set reducts are found). The number of ants used is set to the number of features, with each ant starting on a different feature. Ants construct possible solutions until they reach a rough set reduct. To avoid fruitless searches, the size of the current best reduct is used to reject those subsets whose cardinality exceeds this value. Pheromone levels are set at 0.5, with a small random variation added. Levels are increased by only those ants who have found true reducts. The global search is terminated after 250 iterations; α is set to 1 and β is set to 0.1.

SimRSAR employs a simulated annealing-based feature selection mechanism [163]. The states are feature subsets, with random state mutations set to changing three features (either adding or removing them). The cost function attempts to maximize the rough set dependency (γ) while minimizing the subset cardinality. For these experiments the cost of subset R is defined as

$$cost(R) = \left[\frac{\gamma_{\mathbb{C}}(\mathbb{D}) - \gamma_R(\mathbb{D})}{\gamma_{\mathbb{C}}(\mathbb{D})}\right]^a + \left[\frac{|R|}{|\mathbb{C}|}\right]^b \tag{5.19}$$

where a and b are defined in order to weight the contributions of dependency and subset size to the overall cost measure. In the experiments "presented" here, $a = 1$ and $b = 3$. The initial temperature of the system is estimated as $2 * |\mathbb{C}|$ and the cooling schedule is $T(t + 1) = 0.93 * T(t)$.

The experiments were carried out on three datasets from [285], namely *m-of-n*, *exactly*, and *exactly2*. The remaining datasets are from the machine learning repository [38]. Those datasets containing real-valued attributes have been discretized to allow all methods to be compared fairly.

5.10.1.2 Experimental Results Table 5.2 presents the results of the 5 methods on the 13 datasets. It shows the size of reduct found for each method, as well as the size of the optimal (minimal) reduct. RSAR and EBR produced the same subset every time, unlike AntRSAR and SimRSAR, which often found different subsets and sometimes different subset cardinalities. On the whole, it appears to be the case that AntRSAR and SimRSAR outperform the other three methods. This is at the expense of the time taken to discover these reducts, as can be seen in Figure 5.13 (results for RSAR and EBR do not appear as they are consistently faster than the other methods). In all experiments the rough ordering of techniques with respect to time is RSAR $<$ EBR $\leq$ SimRSAR $\leq$ AntRSAR $\leq$ GenRSAR. AntRSAR and SimRSAR perform similarly throughout—for some datasets, AntRSAR is better (e.g., Vote) and for others SimRSAR is best (e.g., LED). The performance of these two methods may well be improved by fine-tuning the parameters to each individual dataset.

From these results it can be seen that even for small- and medium-sized datasets, incremental hill-climbing techniques often fail to find minimal subsets. For example, RSAR is misled early in the search for the LED dataset, resulting in its choosing seven extraneous features. Although this fault is due to the non-optimality of the guiding heuristic, a perfect heuristic does not exist rendering these approaches unsuited to problems where a minimal subset is essential. However, for most real-world applications the extent of reduction achieved via such methods is acceptable. For systems where the minimal subset is required (perhaps due to the cost of feature measurement), stochastic feature selection should be used. The performance of GA-based methods can be improved by increasing their computation time, such as by increasing the number of generations and/or the population size.

5.10.2 Discernibility Matrix Based Approaches

Three techniques that use the discernibility matrix to locate reducts are evaluated here on the same datasets used previously. HC is a simple hill climber that selects the next attribute based on its frequency in the clauses appearing in the discernibility matrix, following a similar strategy to that of the reduction method based on Johnson's algorithm given in the rough set exploration system (RSES) [292]. NS follows a similar strategy to HC but also uses information about the size of the clauses in the guiding heuristic.

Clause-based search (CS), introduced here, performs search in a breadth-first manner. The process starts with an empty list, *Subsets*, which keeps a record of all current feature subsets. Clauses from the discernibility matrix are considered one at a time in order of their size, with those of the smallest cardinality

TABLE 5.2 Subset sizes found for five techniques

Index	Dataset	Features	Optimal	RSAR	EBR	AntRSAR	SimRSAR	GenRSAR
0	M-of-N	13	6	8	6	6	6	6(6) 7(12)
1	Exactly	13	6	9	8	6	6	6(10) 7(10)
2	Exactly2	13	10	13	11	10	10	10(9) 11(11)
3	Heart	13	6	7	7	6(18) 7(2)	6(29) 7(1)	6(18) 7(2)
4	Vote	16	8	9	9	8	8(15) 9(15)	8(2) 9(18)
5	Credit	20	8	9	10	8(12) 9(4) 10(4)	8(18) 9(1) 11(1)	10(6) 11(14)
6	Mushroom	22	4	5	4	4	4	5(1) 6(5) 7(14)
7	LED	24	5	12	5	5(13) 6(4) 7(3)	5	6(1) 7(3) 8(16)
8	Letters	25	8	9	9	8	8	8(8) 9(12)
9	Derm	34	6	7	6	6(17) 7(3)	6(12) 7(8)	10(6) 11(14)
10	Derm2	34	8	10	10	8(3) 9(17)	8(3) 9(7)	10(2) 11(8)
11	WQ	38	12	14	14	12(2) 13(7) 14(11)	13(16) 14(4)	16
12	Lung	56	4	4	4	4	4(7) 5(12) 6(1)	6(8) 7(12)

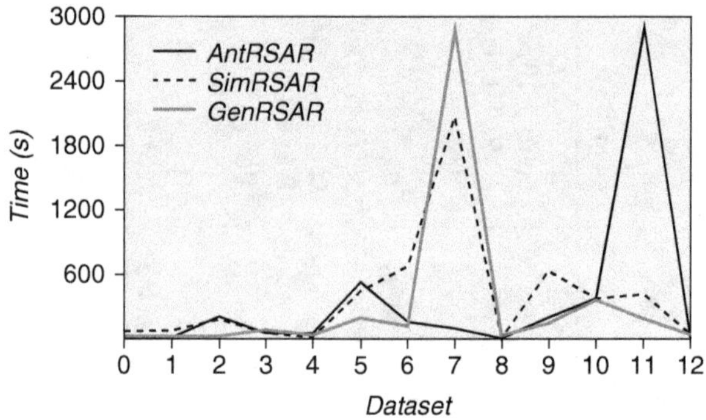

Figure 5.13 Average runtimes for AntRSAR, SimRSAR and GenRSAR

chosen first. When a clause is selected, the features appearing within the clause are added to every set in *Subsets*. For example, if *Subsets* contains $\{a, b\}$ and $\{c, d\}$, and the next considered clause is $\{d \vee e\}$, then each appearing attribute is added. The *Subsets* list will now contain $\{a, b, d\}$, $\{a, b, e\}$, $\{c, d\}$, and $\{c, d, e\}$. This guarantees that each set in *Subsets* satisfies all the clauses that have been encountered so far. If one of these subsets satisfies all clauses, the algorithm terminates as a reduct has been found. If not, then the process continues by selecting the next clause and adding these features. This process will result in a minimal subset, but it has an exponential time and space complexity.

The results of the application of these three methods to the 13 datasets can be found in Table 5.3. HC and NS perform similarly throughout, differing only in their results for the Letters and WQ datasets. CS will always find the smallest valid feature subset, although it is too costly to apply to larger datasets in its present form. On the whole, all three methods perform as well as or better

TABLE 5.3 Subset sizes found for discernibility matrix-based techniques

Dataset	Features	HC	NS	CS
M-of-N	13	6	6	6
Exactly	13	6	6	6
Exactly2	13	10	10	10
Heart	13	6	6	6
Vote	16	8	8	8
Credit	20	10	10	8
Mushroom	22	4	4	4
LED	24	5	5	5
Letters	25	9	10	8
Derm	34	6	6	6
Derm2	34	9	9	8
WQ	38	14	13	12
Lung	56	4	4	4

than the dependency-based methods. However, HC, NS, and CS all require the calculation of the discernibility matrix beforehand.

5.11 SUMMARY

Feature selection seeks to reduce data while retaining semantics by selecting attributes as opposed to transforming them. This aspect is particularly useful when feature selection precedes other processes that require the original feature meanings to be intact, for example, rule induction where rules may need to be human-comprehensible. This chapter focused on some of the recent developments in rough set theory for the purpose of feature selection.

Several approaches to discovering rough set reducts were experimentally evaluated and compared. The results highlighted the shortcomings of conventional hill-climbing approaches to feature selection. These techniques often fail to find minimal data reductions. Some guiding heuristics are better than others, but as no perfect heuristic exists, there can be no guarantee of optimality. From the experimentation it appears that the entropy-based measure is a more useful hill-climbing heuristic than the rough set based measure. However, the entropy measure is a more costly operation than that of dependency evaluation, and this may be an important factor when processing large datasets. Because of the failure of hill-climbing methods and the fact that exhaustive searches are not feasible for even medium-sized datasets, stochastic approaches provide a promising feature selection mechanism.

User-supplied information is essential to many existing algorithms for feature selection in the literature. This is a significant drawback. Some feature selectors require noise levels to be specified by the user beforehand; some simply rank features, leaving the users to choose their own subsets. There are those feature selectors that require the user to state how many features are to be chosen, or they must supply a threshold that determines when the algorithm should terminate. All of these require users to make decisions based on their own (possibly faulty) judgments.

It is most often the case that the values of attributes are both crisp and *real valued*, and this is where many feature selectors, particularly those based on traditional rough set theory, encounter a problem. It is not possible to say whether two attribute values are similar and to what extent they are the same; for example, two close values may only differ as a result of noise, but in RST they are considered to be as different as two values of a different order of magnitude. According to RST, the values -0.1 and -0.11 are as different as -0.1 and 300.

One answer to this problem has been to discretize the dataset beforehand, producing a new dataset with crisp values. This is often still inadequate, however, as the degrees of membership of values to discretized values are not considered at all. For example, two values may both be mapped to the same label "Negative," but one may be much more negative than the other. The values -0.1 and -2000 could both be mapped to this class, although they are significantly different.

This is a source of information loss, which is against the rough set ideology of retaining information content. Further discussion of this issue can be found in [29].

It is clear that there is a need for a method that will provide the means of data reduction for crisp and real-value attributed datasets and utilize the extent to which values are similar. Fuzzy sets [408] and the process of fuzzification provide a mechanism by which real-valued features can be effectively managed. By allowing values to belong to more than one label, with various degrees of membership, the vagueness present in data can be modeled. This information can then be exploited by further fuzzy methods to enable reasoning under uncertainty.

For feature selection this could be achieved through the use of *fuzzy-rough* sets [86]. Fuzzy-rough set theory is an extension of crisp rough set theory, allowing all memberships to take values in the range [0,1]. This permits a higher degree of flexibility compared to the strict requirements of crisp rough sets that only deal with full or zero set membership.

Also there are few feature selection methods that can handle data with continuous decision attributes. This type of data is often encountered in regression/function approximation problems. For example, in QSAR [46] such datasets are encountered where the measurements (the decision feature is molecular activity) are all continuous. The use of FS can discover the key variables that influence the decision quantity, without transforming the feature space.

CHAPTER 6

APPLICATIONS I: USE OF RSAR

This chapter examines several systems where an RSAR step has been employed to reduce dimensionality. The applications considered here are different in nature: medical image classification, text categorization, and algae population estimation. Because of the domain-independence of the RSAR technique, it can be applied to any problem with no additional information required about the data being processed. In addition other applications of rough set theory are presented briefly for the interested reader.

6.1 MEDICAL IMAGE CLASSIFICATION

Neural network-based classifiers have been successful in many application areas. However, complex application problems such as real-life medical image modeling and analysis have emphasized the issues of feature set dimensionality reduction and feature semantics preservation. To capture the essential characteristics of a given real image, many features may have to be extracted without explicit knowledge of what properties might best represent the image a priori. Of course, generating more features increases computational complexity, while not all such features are essential and some may even reduce the overall representational power of the feature set due to measurement noise. A method that can determine

Computational Intelligence and Feature Selection: Rough and Fuzzy Approaches, by Richard Jensen and Qiang Shen

the most significant features, based on sample measurements, is therefore highly desirable to support neural network-based classification.

The use of a rough set-based tool allows the induction of low-dimensionality feature sets from sample descriptions of feature patterns of a higher dimensionality. It causes no information loss, insofar as this is possible in the domain at hand [319,321]. As only selected features are employed to perform actual classification, the reduction of feature sets reduces the structural complexity of the neural network classifiers.

The RSAR approach has an advantageous side effect in that it removes redundancy from feature extraction on line, and helps minimize the noise associated with the extra feature measurements otherwise involved. More significant, however, is the fact that rough set feature selection preserves the semantics of the surviving features after removing redundant ones. This is critical in satisfying the requirement of user understandability of the generated classifiers, ensuring the direct interpretation of what characteristics have led to the identified classes. This feature is particularly important in medical applications, which forms a particular motivation of the present research.

6.1.1 Problem Case

Comparing normal and abnormal blood vessel structures, via the analysis of cell images, is central to pathology and medicine [317]. At the core of this analysis is the capture of the underlying features of such images. Many feature extraction methods are available to yield various characteristic descriptions of a given image. However, little knowledge is available as to what features are most helpful to provide the discrimination power between normal and abnormal cells and between their types, while it is computationally impractical to generate many features and then to perform classification based on these features for rapid diagnosis. Generating a good number of features and selecting from them the most informative ones off line, and then using those selected on line is the usual way to avoid this difficulty. The problem is that the features produced ought to have an embedded meaning, and such meaning should not be altered during the selection process. Therefore this domain presents a challenging case to test the potential of RSAR.

The samples of subcutaneous blood vessels used in [317] were taken from patients suffering critical limb ischaemia immediately after leg amputation. The level of amputation was always selected to be in a non-ischaemic area. The vessel segments obtained from this area represented internal proximal (normal) arteries, while the distal portion of the limb represented ischaemic (abnormal) ones. Images were collected using an inverted microscope, producing an image database of 318 cell images, each sized 512×512 pixels with gray levels ranging from 0 to 255. Examples of the three types of cell image taken from non-ischaemic and ischaemic resistance arteries are shown in Figure 6.1. Note that many of these images seem similar to the eye. It is therefore a difficult task for visual inspection and classification.

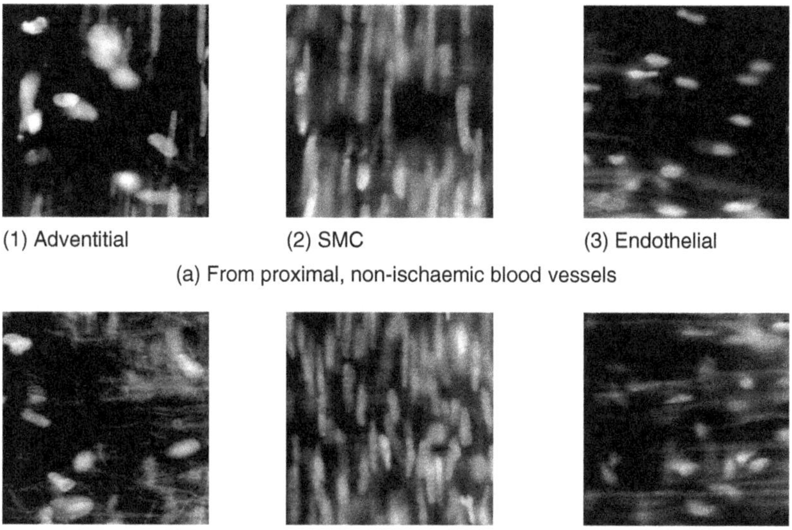

(1) Adventitial (2) SMC (3) Endothelial

(a) From proximal, non-ischaemic blood vessels

(1) Adventitial (2) SMC (3) Endothelial

(b) From distal, ischaemic blood vessels

Figure 6.1 Section cell images. The first, second and third columns, respectively, show adventitial, smooth muscle, and endothelial cells in proximal non-ischaemic and distal ischaemic subcutaneous blood vessels, taken from a human lower limb.

6.1.2 Neural Network Modeling

In this medical image classification work, each image classifier is implemented using a traditional multi-layer feedforward artificial neural network (MFNN). To capture and represent many possible and essential characteristics of a given image, fractal models [113,220] are used. Note that although these particular techniques are herein adopted to perform their respective task, the work described does not rely on them; it is generally applicable when other classification and feature extraction methods are employed.

An MFNN-based classifier accomplishes classification by mapping input feature patterns onto their underlying image classes. The design of each MFNN classifier used for the present work is specified as follows: The number of nodes in its input layer is set to that of the dimensionality of the given feature set (before or after feature reduction), and the number of nodes within its output layer is set to the number of underlying classes of interest. The internal structure of the network is designed to be flexible and may contain one or two hidden layers.

The training of the classifier is essential to its runtime performance, and is here carried out using the backpropagation algorithm [296]. Feature patterns that represent different images, coupled with their respective underlying image class indexes, are selected as the training data, with the input features being

TABLE 6.1 Features and Their Reference Number

Feature Number	Feature Meaning	Feature Number	Feature Meaning
1	0° direction	7	3rd finest resolution
2	45° direction	8	4th finest resolution
3	90° direction	9	5th finest resolution
4	135° direction	10	Mean
5	Finest resolution	11	STD
6	2nd finest resolution		

normalized into the range of 0 to 1. Each feature pattern consists of 9 fractal features (including 5 isotropic fractals measured on the top 5 finest resolutions and 4 directional fractals [317]) and the mean and standard deviation (STD), with their reference numbers listed in Table 6.1. Note that when applying the trained classifier, only those features selected during the learning phase are required to be extracted, so no discretization is needed but real-valued features are directly fed to the classifier.

6.1.3 Results

Eighty-five images selected from the image database are used for training and the remaining 233 images are employed for testing. For simplicity, only MFNNs with one hidden layer are considered.

Table 6.2 lists the results of using RSAR and the original full set of features. The error rate of using the five selected features is lower than that of using the full feature set. This improvement of performance is obtained by a structurally much simpler network of 10 hidden nodes, as opposed to the classifier that requires 24 hidden nodes to achieve the optimal learning. This is indicative of the power of RSAR in helping reduce not only redundant feature measures but also the noise associated with such measurement. Also the classifier using those five RSAR-selected features considerably outperforms those using five randomly selected features, with the average error of the latter reaching 19.1%.

Again, a comparison against a widely recognized benchmark method should help reflect the success of the system. The results of rough feature selection are systematically compared to those obtained via the use of principal component analysis (PCA) [77], as summarized in Table 6.3. Note that PCA is perhaps the most adopted dimensionality reduction technique. Although efficient, it irreversibly destroys the underlying semantics of the feature set. Therefore, in

TABLE 6.2 Results of using Rough Selected and Original Full Set of Features

Method	Dimensionality	Features	Structure	Error
Rough	5	1, 4, 9, 10, 11	$5 \times 10 + 10 \times 6$	7.55%
Original	11	1, 2, 3, 4, 5, 6, 7, 8, 9, 10, 11	$11 \times 24 + 24 \times 6$	9.44%

TABLE 6.3 Results of Using Rough and PCA-Selected Features

Method	Dimensionality	Features	Structure	Error
Rough	**5**	**1, 4, 9, 10, 11**	**$5 \times 10 + 10 \times 6$**	**7.7%**
PCA	1	1	$1 \times 12 + 12 \times 6$	57.1%
	2	1, 2	$2 \times 12 + 12 \times 6$	32.2%
	3	1, 2, 3	$3 \times 12 + 12 \times 6$	31.3%
	4	1, 2, 3, 4	$4 \times 24 + 24 \times 6$	28.8%
	5	**1, 2, 3, 4, 5**	**$5 \times 20 + 20 \times 6$**	**18.9%**
	6	1, 2, 3, 4, 5, 6	$6 \times 18 + 18 \times 6$	15.4%
	7	1, 2, 3, 4, 5, 6, 7	$7 \times 24 + 24 \times 6$	11.6%
	8	1, 2, 3, 4, 5, 6, 7, 8	$8 \times 24 + 24 \times 6$	13.7%
	9	1, 2, 3, 4, 5, 6, 7, 8, 9	$9 \times 12 + 12 \times 6$	9.9%
	10	1, 2, 3, 4, 5, 6, 7, 8, 9, 10	$10 \times 20 + 20 \times 6$	7.3%
	11	1, 2, 3, 4, 5, 6, 7, 8, 9, 10, 11	$11 \times 8 + 8 \times 6$	7.3%

this table, for the results of using PCA, feature number $i, i \in \{1, 2, \ldots, 11\}$, stands for the ith principal component, namely the transformed feature that is corresponding to the ith largest variance.

The advantages of using RSAR are clear. Of the same dimensionality (i.e., 5), the classifier using the features selected by the rough set approach has a substantially higher classification accuracy, and this is achieved via a considerably simpler neural network. When increasing the dimensionality of principal features, the error rate generally gets reduced, but the classifier generally underperforms until almost the full set of principal features is used. The overall structural complexity of all these classifiers is more complex than that of the classifier using the five RSAR-selected features. In addition the use of those classifiers that use PCA-selected features require many more feature measurements to achieve comparable classification results.

6.2 TEXT CATEGORIZATION

6.2.1 Problem Case

Many techniques involving electronic text categorization (TC) describe documents by using vectors of real numbers that exhibit extremely high dimensionality: typically one value per word or pair of words in a document or corpus of documents [54]. These vector ordinates are used as preconditions to rules or similarity metrics that decide what category the document belongs to. There are several tens of thousands of such ordinates in all but the simplest applications, making effective text categorization an extremely hard, if not intractable, problem for even the most powerful computers, unless the task is simplified.

Applying rough set attribute reduction to this problem domain provides assistance in locating those parts of text datasets that possess the necessary information, thereby reducing the amount of data to be handled by text classifiers. Additionally it allows humans to inspect the models generated, be they fuzzy or conventional, and thus understand the rules underlying the categorization task. This provides much needed transparency to a domain that has traditionally been characterized by opaqueness.

Data for this case study consists of folders (classes) of human-classified email messages used as training examples. In this context a feature is the perceived importance of words and phrases within messages. The concept of importance is defined by different metrics [365], although they all are rather simple in order to keep the overall computational complexity under control. Different types of model, including fuzzy models, were built using the training data and metrics.

6.2.2 Metrics

Four metrics are used to measure the importance of keywords that appear in text, namely the Boolean existential model, frequency metric, TF-IDF metric, and fuzzy relevance.

- The Boolean existential model metric assigns a weight of one to all keywords that exist in the document. Absent keywords have an implied weight of zero. These weights can ultimately be used by a crisp, Boolean inference engine.
- The frequency metric makes term weights equal to the terms' frequency in the document.
- The term frequency-inverse document frequency (TF-IDF) metric [303] assigns higher weights to keywords that are frequent in the current document, yet uncommon in most others.
- The fuzzy relevance metric (FRM) is a highly experimental attempt at judging the relevance of terms based on their place in the frequency histogram. This is, in essence, a distribution-based metric: it applies a bell curve to assign more relevance to keywords that are more frequent than the average, but not the most frequent keywords.

6.2.3 Datasets Used

Six email folders were used as experimentation datasets. The folders were chosen because they offered a wide spectrum of different features to be tested: different natural languages, homogeneous and heterogeneous writing styles and formats, different content, and similar linguistic elements in different contexts. Folders ranged in size from 159 to 380 messages, with an average of 225 messages. In terms of memory footprint, they ranged from 265 to 1251 kbytes, with an average of 634 kbytes. There are combinations of extremely long and extremely short messages in the folders.

As this is a real-world set of email folders, spelling mistakes are inevitable, a fact that makes it difficult for language-dependent systems to operate. All folders have been hand-classified by the authors with some reasonable rationale: personal folders contain messages from the same person and mailing list folders contain messages from the same list. Other email users may partition their email messages differently, for instance, based on content matter. It is expected that the system should be able to generalize rules to classify new messages according to this partitioning. It is, of course, not expected that the system will be able to distinguish between semantic differences in texts. For instance, it is unlikely to be successful to train the system to detect humorous messages, unless such messages use a different vocabulary from that of other types of training messages.

6.2.4 Dimensionality Reduction

Numerous sets of two to six folders were selected at random. The folders of each set were split into training and testing sets at a 1:4 ratio. The training sets were used to create sets of classification means using each of the four keyword acquisition metrics, as described previously. The results of dimensionality reduction are shown in Table 6.4. The first two numeric columns show the average dimensionality before and after reduction. The reduction itself is dramatic, decreasing the width of datasets by approximately 3.5 orders of magnitude (i.e., reduction by a factor of around 3162). So the text classification domain has a very high degree of redundancy. The redundancy can be taken advantage of in an effort to improve the efficiency of such systems by the use of the present approach.

As expected, the results are very similar for all four techniques, which give a measure of the information content of a term. As long as this measure is reasonably accurate, RSAR is able to remove redundant terms from the dataset. It is interesting to observe that postreduction dimensionality is not proportional to dimensionality before reduction. RSAR does not simply remove a certain proportion of attributes; rather, it attempts to remove *all* redundant or useless information from the data. The reduction factor is more or less constant. This implies that around one in every 3162 document terms can be used as a coordinate keyword in email messages.

TABLE 6.4 Average Dimensionality Reduction for Different Keyword Acquisition Metrics

Metric	Average Attributes Before	Average Attributes After	Average Reduction (Orders of Magnitude)
Existential	33,305	10.16	3.52
Frequency	35,328	9.63	3.56
TF-IDF	33,327	8.97	3.57
FRM	35,328	9.63	3.56

The average postreduction dimensionality displays an interesting pattern: it provides an indirect measure of the success of each of the four metrics at locating coordinate keywords. The better the metric, the smaller is the postreduction dimensionality (i.e., the more informative the surviving individual attributes). In this regard, TF-IDF yields the best results, followed closely by the frequency and FRM metrics and then the existential metric. This is generally expected. Since the existential metric is Boolean, it granulates the keyword relevance more than the other techniques. This requires more keywords to make up for the lesser information content of the weights. The similarity of the frequency and FRM metrics can also be easily explained: FRM is a function of the term frequency, representing frequency using fuzzy memberships.

For complex fuzzy applications the FRM could hence be avoided: normalized term frequency would be enough to provide the required information. This conforms to the assertion of [365] that even simply using the frequency as a relevance metric yields good text categorization results. It may therefore be argued that the frequency metric is preferable to TF-IDF wherever efficiency is the primary concern—it provides almost the same results as TF-IDF but without the computational cost of the latter.

6.2.5 Information Content of Rough Set Reducts

Having established that the RSAR method significantly reduces the dimensionality of keyword datasets, it should be established whether the reduct set of keywords provides more information for the classification of email messages than any other arbitrary set of keywords. For this illustration three message folders were chosen at random. They were again split into training and test sets at a 1:4 ratio. The reduct set was obtained for the training set and used to classify the email messages in the test dataset. For simplicity the Boolean existential metric was used as the keyword acquisition metric, and the BIM classifier was employed to classify the messages. RSAR reduced the dataset to just eight keywords.

Random sets of 1 to 20 keywords were chosen from the training set, and thus 20 keyword datasets containing the corresponding attributes were constructed. These datasets were then used to classify the same set of three folders as above using the same combination of metric and inference modules. One hundred runs of this procedure were performed to provide a representative sample. It would have been desirable to evaluate classification accuracies for all combinations of attributes, but this is completely infeasible, given the pre-reduction dimensionality of the keyword datasets. (The results are shown in Figure 6.3.) The rough set reduct (of eight keywords) obtained an accuracy of 75%. However, the random choices of attributes exhibited significantly reduced classification accuracy, at most managing 40% but with the lowest accuracy dropping to zero. Note that as demonstrated in the next experiment, these results are not typical for the present methodology. They were due to the arbitrary choice of keyword acquisition metric and inference method. A better choice for these components can yield accuracy as high as 98% to 99.5%.

It might be expected that the average classification rate would increase slightly as the dimensionality was increased. This did not happen, but 20 attributes are a fraction of the initial dimensionality of around 30,000, so any potential trends are not evident at this scale.

As this sampling is only of the set of all combinations of attributes, it may be argued that freak cases of high accuracy cannot be ruled out. However, for lack of a heuristic, the RSAR technique will have to do; an exhaustive, brute force search of all combinations is, to say the least, intractable.

6.2.6 Comparative Study of TC Methodologies

Results presented above have shown the success of the present work in reducing the dimensionality of datasets used in TC tasks, all the while retaining the information content of the original datasets. To measure the effect of the work on the actual classification task, a comparative study of the integration of the RSAR technique with different keyword acquisition methods and inference algorithms was carried out. All 12 combinations of the 4 keyword acquisition metrics and the 3 inference engines were investigated. Each experiment involves selecting random sets of 2 to 6 folders and training the system to classify previously unseen messages into these categories.

Traditionally IF and IR systems are benchmarked based on their *precision* and *recall*. Precision is defined as the ratio of the number of retrieved documents that are relevant to a user's query over the total number of documents retrieved as a response to the query. Recall is defined as the ratio of the number of relevant documents retrieved over the total number of relevant documents indexed by the system.

However, a *relevant/irrelevant* partition is a dual-class categorization task. The present work attempts to classify documents into an arbitrary number of classes. Hence the generation of precision/recall graphs would convey considerably less information about the success of the system. A more conventional metric, classification accuracy, is used instead. This is valid because the datasets used have similar sizes for the underlying document classes.

Each set of folders was split into training and test data, at the usual 1:4 ratio. Keyword datasets were acquired using one of the four keyword acquisition subsystems; the datasets were reduced and a classification test was performed using the test dataset. Classification accuracy and dimensionality reduction statistics were gathered to provide an insight into the working of the corresponding pairing of keyword acquisition and inference method, in combination with the RSAR.

The system was run 20 times for each set of folders within each of the 12 experiments in order to obtain a more representative statistical sample. The results are presented in Figure 6.2. Six graphs are shown overall; the left column shows average classification accuracy for each of the three types of classifier or inference engine. The four keyword acquisition metrics are plotted in different colors. The vertical axis represents classification accuracy; the horizontal axis represents the number of classes (the number of email folders used). The right column of

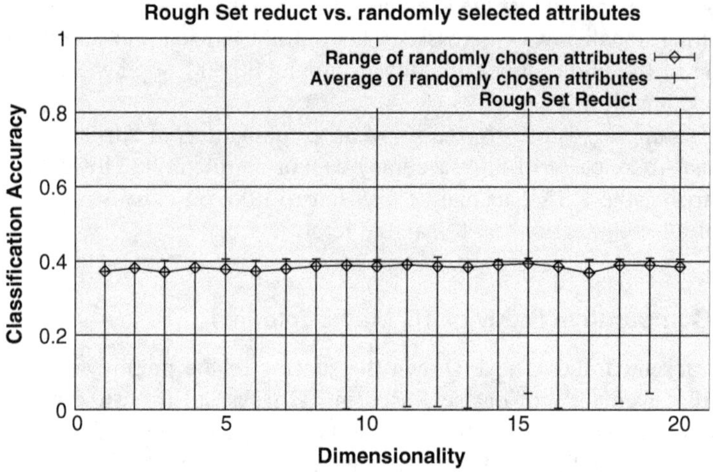

Figure 6.2 Verifying the information content of the Rough Set reduct. The thick line represents the accuracy of the Rough Set reduct. Minimum, maximum and average classification accuracies for randomly selected keyword datasets of one to twenty dimensions are also shown.

the figure gives average dimensionality reduction plotted on a logarithmic scale against the number of folders.

The most obvious observation is that dimensionality reduction is similar in all 12 combinations of techniques. It increases as the number of folders increases. This implies that despite the increase in total number of acquired keywords, there is little or no increase in the number of information-rich keywords. Hence the RSAR technique is expected to provide consistent results independently of the number of corpora of documents.

These results show that the TF-IDF metric performs best in almost all tests. The Boolean classifier gave good results with the FRM. The FRM was more or less consistent in its results with this classifier, with a rough average accuracy of 75%: not excellent, but acceptable, although certain individual runs went very close to perfect classification. As the number of folders increased, the Boolean and frequency metrics gave better and more consistent results. Classifying all six folders with a combination of the frequency metric and the Boolean classifier yielded excellent results, around 95% accuracy.

The VSM classifier proved more predictable: runs with the traditional combination of TF-IDF and the VSM offered good accuracy in the range 60% to 99%. Again, overall accuracy seems to increase as more corpora are added to the problem. The fuzzy classifier proved to give excellent classification using the TF-IDF metric: 90% to 100% accuracy with consistent behavior and very little variance. The FRM did acceptably well but was outperformed by the frequency metric, coming relatively close to the TF-IDF. As expected, the Boolean metric

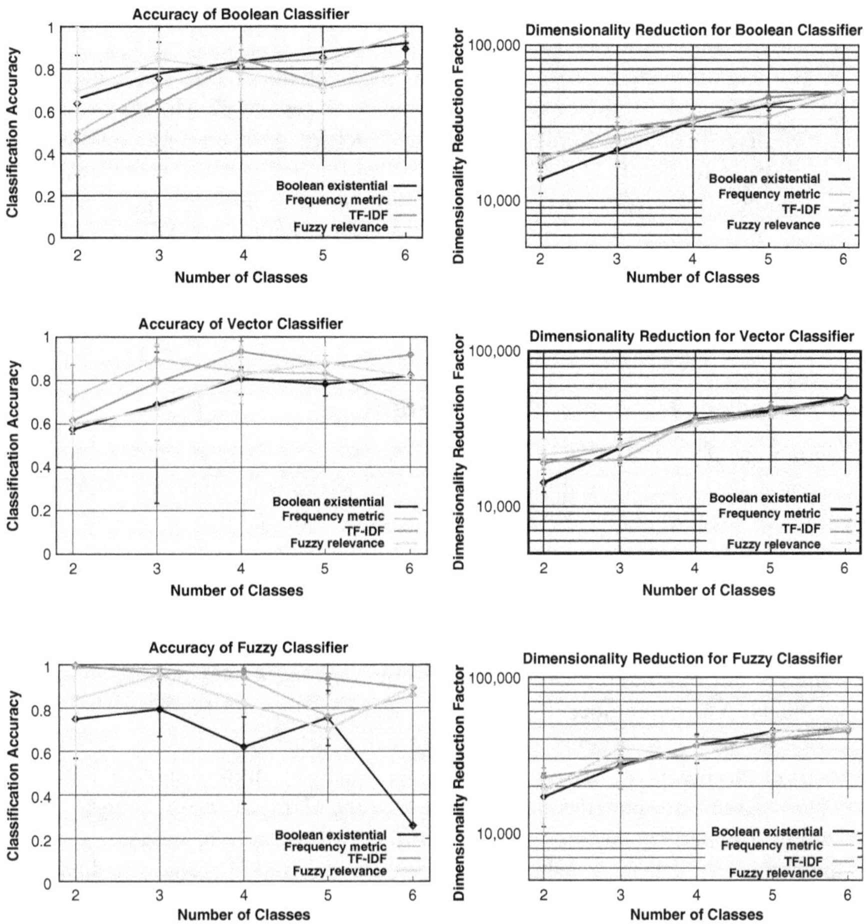

Figure 6.3 Comparison of the success of the four keyword acquisition metrics used with the three different inference engines. The left column shows classification accuracy; the right dimensionality reduction.

failed to provide consistent results, displaying widely varying accuracy. Overall, the fuzzy classifier has so far provided the best results for this system, especially when used with the TF-IDF metric.

An attempt was made to compare the rough set-assisted systems with more conventional TC systems that did not incorporate strong dimensionality reduction techniques, by bypassing RSAR and executing the inference engines on the unreduced datasets. Unfortunately, this experiment failed to the point where no results have been obtained. The unreduced datasets had, on average, around 33,000 keywords to be used by the inference engines as rule conditions or document vector ordinates. With approximately 500 messages in the training set, pre-reduction inference required unacceptably long periods of time and had to be terminated before any usable results could be gathered.

A 500-example dataset of 33,000 keywords contains 33,001 space delimited single- or double-digit numbers and measures in excess of 32 Mbytes of peripheral memory. Evaluating complex similarity metrics on each and every one of those 500 examples of 33,000 keywords to measure their similarity against a new document takes a rather long time. Fortunately, an average dimensionality decrease from 33,000 to 10 attributes manages to reduce the dataset to a few hundreds of bytes of uncompressed ASCII keyword weights. CPU processing times decrease similarly, which is especially significant for the more complicated systems, like those incorporating the VSM.

6.2.7 Efficiency Considerations of RSAR

Although the RSAR method alleviates many of the efficiency considerations of TC approaches, such as runtime and main and peripheral memory footprints, it would be prudent to know RSAR's own efficiency. RSAR's complexity should be lesser than that of TC techniques for it to be effective on a theoretical basis. For practical purposes, the algorithm's actual runtime should also be small enough for its use to offer speed gains.

To test this, two experiments were performed. In the first, a dataset of four conditional attributes and one decision attribute was used. The dataset contained 150 objects. A large number of RSAR runs were carried out. In each, RSAR was required to reduce a larger version of the same dataset, formed by repeatedly increasing the length of the dataset. This was done by duplicating the dataset's objects (without altering the information content of the data, of course). Each reduction was performed three times, with the average time calculated. A graph of the number of dataset objects against the time required to reduce the dataset is shown in Figure 6.4 (left graph). Barring a few irregularities (most likely caused by the algorithm running on a multi-tasking UNIX system), the graph exhibits

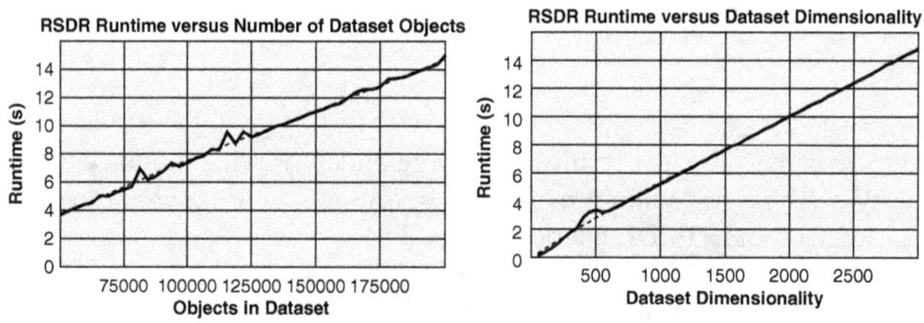

Figure 6.4 RSAR's runtime efficiency with respect to the number of dataset objects (dataset length) and dataset dimensionality (dataset width). The streamlined, experimental version of the algorithm manages to obtain very nearly linear operation. The dashed thin lines show projected linear runtime and clearly match the observed runtimes of the algorithm. All times are in seconds.

linearity. Reduction times are also very satisfactory, with around 7.5 seconds needed per 100,000 objects.

The second experiment involved the dimensionality of the dataset. The dataset contained 210 objects, with dimensionality increasing progressively by the addition of columns of noisy data. Numerous different datasets were constructed with 50,000 to 200,000 attributes. Three runs per dataset were performed, with average time shown in the right graph in Figure 6.4. Runtime was once more close to linear. Although an intuitive understanding of QuickReduct implies that, for a dimensionality of n, $(n^2 + n)/2$ evaluations of γ may be performed for the worst-case dataset, the actual observed runtime remains apparently linear at the scales shown here. Thus, although the *worst-case* runtime complexity is $O((n^2 + n)/2)$, the average case requires considerably less effort. In fact, if all the information-rich attributes are in the first few columns of the dataset, then it does not matter to QuickReduct how many other attributes exist in the dataset. Therefore the best-case complexity is close to $O(1)$. Average complexity was experimentally determined to be approximately $O(n)$.

6.2.8 Generalization

To accelerate and simplify the training process, the proposed system should be able to generalize using as few training examples as possible. This aids in cases where there is a lack of training examples. To test the system's ability to generalize, two folders were chosen at random. The folders were split into training and test datasets, the system was trained, and its classification accuracy tested. TF-IDF was used to measure the relevance of acquired keywords and the fuzzy reasoner was used to classify the unseen email messages. A large number of runs of the experiment were performed with varying training/testing ratios. There were

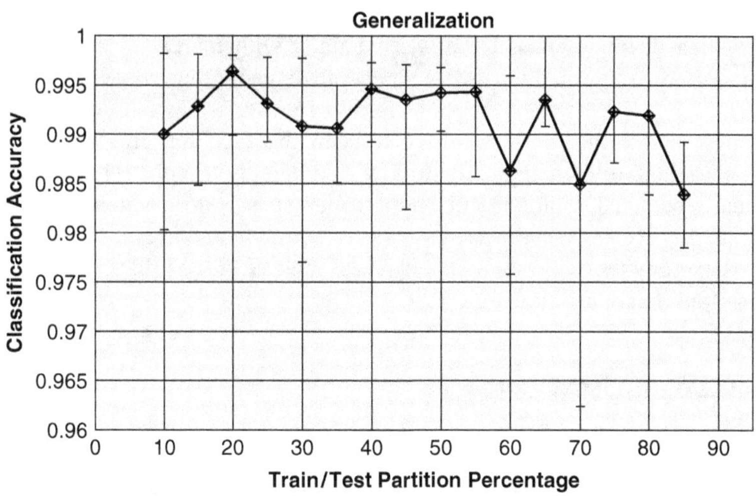

Figure 6.5 System's ability to generalize given different training/test partition ratios

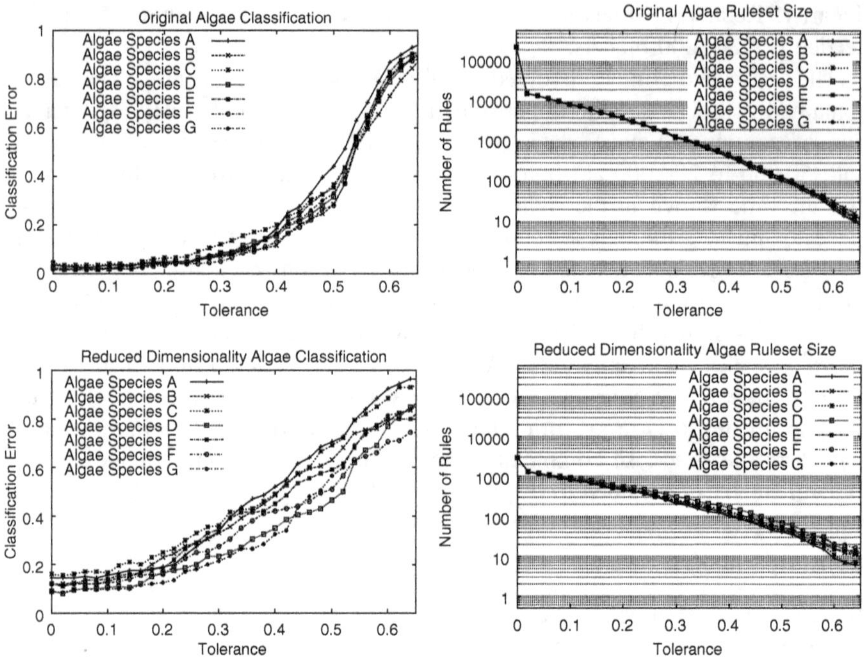

Figure 6.6 Algae estimation accuracy before (*top*) and after (*bottom*) dimensionality reduction. Note that ruleset sizes are given on a logarithmic scale.

five runs per ratio, with ratios ranging from 1:10 to 8.5:10. The results are shown in Figure 6.6.

These results are surprising in that accuracy is consistently high, with no clearly discernible trend, although there exists a reasonable amount of variation between different runs. Even when trained with merely 10% of messages in a folder, the system managed to generalize enough to be 98% to 99.5% accurate.

The reason for this high accuracy rests in the fact that email messages in human-classified folders typically follow a relatively simple set of rules. For example, a personal folder might contain messages from a single sender. The RSAR quickly locates what messages have in common and uses it for the classification. Assuming the training sample is representative of the folder, generalization of the system becomes almost perfect.

6.3 ALGAE ESTIMATION

6.3.1 Problem Case

The task of this application is to provide a system to estimate populations of various alga species (families of ubiquitous single-celled plants), say, for environment protection purposes. A more detailed analysis of this domain is carried out in Chapter 13 where the impact of fuzzy-rough feature selection is investigated. To derive the rules required for estimation, training samples were taken from different European rivers over the period of one year. These samples involve the following feature measurements [94]: the time of year the sample was taken, river size, water flow rate, and eight chemical concentrations known to influence alga growth patterns.

It is relatively easy to locate relations between one or two of these quantities and a species of alga. However, the process of identifying relations between different chemical elements and the population of different alga species requires expertise in chemistry and biology. It also involves well-trained personnel and microscopic examination that cannot be automated given the state of the art. Thus the process becomes expensive and slow, even for a subset of the quantities involved here. There are complex relations at work between the variables of this application domain: algae may influence other algae, as well as be influenced by the concentration of chemicals. As such, there is expected to be some redundancy in the data. This forms an important reason for the present explanation of the RSAR technique.

The dataset available for training includes 200 instances. The first three features of each instance (season, river size, and flow rate) are represented as fuzzy linguistic variables. Chemical concentrations and algae population estimates are represented as continuous quantities, which are later fuzzified to generate the fuzzy model. The dataset includes a few samples with missing values. Of the 200 instances, two exhibiting mostly unknown values were removed from the dataset because of their extremely low quality.

The same modeling method as used in the last case study is employed here. For convenience and easy interpretability, each of the seven alga species was processed separately to produce seven independent models.

6.3.2 Results

It is, first of all, interesting to investigate what effects dimensionality reduction may have on the runtime performance of this particular application. To show whether feature reduction has an impact on overall accuracy, fuzzy rulesets were induced from the entire, unreduced algae dataset [94], one per species. The results are shown on the top row of Figure 6.6. Then RSAR was employed to reduce the dimensionality of the dataset. This resulted in a 7 feature dataset selected from the original, 11 feature one. The results of testing the rulesets induced from this reduced dataset are illustrated on the bottom row of Figure 6.6.

The exact selected features were different for each alga species, although certain features were present in all seven reduct sets, namely the season and concentrations 1, 4, and 7. There is a certain drop in accuracy (approximately 10%) after feature selection, which may indicate that the attribute reduction process has removed some of the necessary information. Two empirical conclusions can be drawn from the conducted experiments: first, not all features contribute the same information; second, the results obtained from random sets of features are worse than those obtained from the reduct set. The latter conclusion demonstrates that RSAR does indeed locate a relatively high-quality reduct. At the same time, by retaining the original semantics of the data and by selecting important features, the readability of the generated models is increased significantly, allowing for rules to be inspected and understood by humans.

6.4 OTHER APPLICATIONS

This section provides a brief overview of some of the many applications of rough set theory. There are several properties of rough sets that make the theory an obvious choice for use in dealing with real problems; for example, it handles uncertainty present in real data through approximations and also does not require threshold information in order to operate (as is the case with many current techniques).

6.4.1 Prediction of Business Failure

Attempts to develop business failure prediction models began seriously sometime in the late 1960s and continue through today. Although there has been much research in this area, there is still no unified well-specified theory of how and why corporations fail. Financial organizations need these predictions for evaluating firms of interest.

Many methods have been used for the purpose of bankruptcy prediction, such as logit analysis, discriminant analysis, and probit analysis [357]. A comprehensive review of the various approaches to modeling and predicting this is presented in [80]. Although some of these methods led to satisfactory models, they suffered from limitations, often due to unrealistic statistical assumptions. As a result the rough set model, with its aim of keeping model assumptions to a minimum, appeared to be a highly useful approach for the analysis of financial information tables.

Rough set-based failure prediction was investigated in [344] and [79]. In these investigations the rough set approach was evaluated against several other methods, including C4.5 [281], discriminant analysis, and logit analysis. For the rough approach, decision rules were generated from the reducts produced by analysis of the financial information. All methods were then evaluated on data from the previous three years. The rough set model was found to be more accurate than discriminant analysis by an average of 6.1% per case, using a minimal set of reduced rules. It also outperformed C4.5 but performed similarly to logit analysis.

A comparative study of the rough sets model versus multivariable discriminant analysis (MDA) can be found in [356]. It was demonstrated that through the use of rough set theory, the prediction of corporate bankruptcy was 97.0% accurate—an improvement over MDA, which achieved an accuracy of 96.0%.

6.4.2 Financial Investment

Trading systems have been built using rough set approaches. In [115,114,420], the rough set model was applied to discover strong trading rules that reflect highly repetitive patterns in data. Historical data from the Toronto stock exchange in 1980 was used for the extraction of trading rules for five companies. Experts confirmed that the extracted rules described the stock behavior and market sensitivity of these companies. Depending on a roughness parameter, the rules generated were either "general" or "exact." The general rules were all recognized relationships in the investment industry, whereas the exact rules made less sense.

In the work reported in [23] the problem of how to deduce rules that map the financial indicators at the end of a month to the stock price changes a month later was addressed. This was based on 15 market indicators. From this study only a satisfactory performance was achieved with many issues still to be tackled, such as data filtration and how to handle missing data. In [17] research was carried out into rough set reduct analysis and rule construction for forecasting the total index of the Oslo stock exchange. This also achieved satisfactory results, with a highest accuracy of 45%.

Research has been carried out on building trading systems for the S&P index [294]. Here a hybrid system was developed that incorporated both neural networks and rough sets. Rules generated by rough sets were used to supervise neural networks to correct for possible errors in predictions. This system reduced drawdown by 25% to 50% and increased the average winner/loser ratio by 50% to 100%.

Rough sets have also been applied to financial decision analysis and explanation for an industrial development bank, ETEVA [341,342]. The bank was interested in investing its capital in firms, while reducing the risk involved in such an investment. To achieve this, a rough set-based firm assessment system was constructed that decided, based on a number of financial ratios, whether a company was acceptable, unacceptable, or uncertain. An information table was constructed with the help of the financial manager of ETEVA. From this table the rough set-generated rules revealed the financial policy applied in the selection of firms. The rules can also be used to evaluate new firms that seek financing from the bank.

6.4.3 Bioinformatics and Medicine

A common and diagnostically challenging problem facing emergency department personnel in hospitals is that of acute abdominal pain in children. There are many potential causes for this pain—most are usually nonserious. However, the pain may be an indicator that a patient has a serious illness, requiring immediate treatment and possibly surgery. Experienced doctors will use a variety of relevant historical information and physical observations to assess children. Such attributes occur frequently in recognizable patterns, allowing a quick and efficient diagnosis. Inexperienced physicians, on the other hand, may lack the knowledge and information to be able to recognize these patterns. The techniques developed in [96] provide a rough set-based clinical decision model to assist such inexperienced physicians. In this research rough sets are used to support diagnosis by distinguishing among three disposition categories: discharge, observation/further investigation, and consult. Preliminary results show that the system gives an accuracy comparable to physicians, although it is dependent on a suitably high data quality.

Rough set data analysis is also applied to the problem of extracting protein–protein interaction sentences in biomedical literature [112]. Due to the abundance of published information relevant to this area, manual information extraction is a formidable task. This approach develops decision rules of protein names, interaction words, and their mutual positions in sentences. To increase the set of potential interaction words, a morphological model is developed, generating spelling and inflection variants. The performance of the method is evaluated using a hand-tagged dataset containing 1894 sentences, producing a precision-recall breakeven performance of 79.8% with leave-one-out cross-validation.

Automated classification of calculated electroencephalogram (EEG) parameters has been shown to be a promising method for detection of intraoperative awareness. In [252] rough set-based methods were employed to generate classification rules, resulting in satisfactory accuracy rates of approximately 90%.

Gene expression experiments, where the genetic content of samples is obtained with high-throughput technologies, result in high-dimensional data. For useful information to be discovered from this data (usually comprising of thousands of genes), automated methods must be able to either cope with this dimensionality or reduce it intelligently. Typically the latter option is chosen as this has the additional benefit of making the extracted knowledge more readable. Many rough set-based methods have been applied to this task—both for feature reduction and classification rule discovery [147,230].

6.4.4 Fault Diagnosis

A rough set approach for the diagnosis of valve faults in a multicylinder diesel engine is investigated in [325]. The use of rough sets enabled the diagnosis of several fault categories in a generic manner. A decision table was constructed from attributes extracted from the vibration signals, with four operational states studied among the signal characteristics: normal, intake valve clearance too small, intake valve clearance too large, and exhaust valve clearance too large. Three sampling points were selected for the collection of vibration signals. The results demonstrated that the system is effective for such fault diagnosis, and the extracted rules correspond well with prior knowledge of the system.

In [222] a rough set-based method for continuous failure diagnosis in assembly systems is presented. Sensor measurements were used to construct a diagnosis table from which rough set rules were extracted.

6.4.5 Spacial and Meteorological Pattern Classification

Sunspot observation, analysis, and classification form an important part in furthering knowledge about the sun, the solar weather, and its effect on earth. Certain categories of sunspot groups are associated with solar flares. Observatories around the world track all visible sunspots in an effort to early detect flares. Sunspot recognition and classification are currently manual and labor-intensive processes that could be automated if successfully learned by a machine. The approach presented in [251] employs a hierarchical rough set-based learning method for sunspot classification. It attempts to learn the modified Zurich classification scheme through rough set-based decision tree induction. The resulting system is evaluated on sunspots extracted from satellite images, with promising results.

In [270] a new application of rough set theory for classifying meteorological radar data is introduced. Volumetric radar data is used to detect storm events responsible for severe weather. Classifying storm cells is a difficult problem as they exhibit a complex evolution throughout their life span. Also the high dimensionality and imprecision of the data can be prohibitive. Here a rough set approach is employed to classify a number of meteorological storm events.

6.4.6 Music and Acoustics

A dominance-based rough set approach, an extension of rough sets to preference-ordered information systems, was used in [159] to generate preference models for violin quality grading. A set of violins were submitted to a violin-maker's competition and evaluated by a jury according to several assessment criteria. The sound of the instruments was recorded digitally and then processed to obtain sound attributes. These features, along with jury assessments, were analyzed by the rough set method, generating preference models. It was shown that the jury's rankings were well approximated by the automated approach.

In [199] an approach to classifying swallowing sound signals is given, utilizing rough set theory. This approach has been developed to facilitate the detection of patients at risk of aspiration, or choking. The waveform dimension is used to describe sound signal complexity and major changes in signal variance. From swallow sound data tables, decision rules were derived, via rough sets. The algorithms yielded a high classification accuracy while producing a comparatively small ruleset.

A decision system employing rough sets and neural networks is presented in [186]. The aim of the study was to automatically classify musical instrument sounds on the basis of a limited number of parameters, and to test the quality of musical sound parameters that are included in the MPEG-7 standard. The use of wavelet-based parameters led to better audio retrieval efficiency.

The classification of musical works is considered in [128], based on the inspection of standard music notations. A decision table is constructed, with features representing various aspects of musical compositions (objects), such as rhythm disorder, beat characteristics, and harmony. From this table classification rules are induced (via rough set rule induction) and used to classify unseen compositions.

6.5 SUMMARY

This chapter has presented three main applications of rough set attribute reduction, within the very different domains of medical image classification, text categorization, and algae population estimation. Classification systems for these tasks benefit greatly from the application of RSAR both in terms of induction time speedup and resulting knowledge readability. A brief overview of other applications of rough set theory was presented, showing its utility in a wide range of tasks.

CHAPTER 7

ROUGH AND FUZZY HYBRIDIZATION

This chapter provides a broad overview of logical and black box approaches to fuzzy and rough hybridization. The logical approaches include theoretical, supervised learning, feature selection, and unsupervised learning. The black box approaches consist of neural and evolutionary computing. Since both theories originated in the expert system domain, there are a number of research proposals that combine rough and fuzzy concepts in supervised learning. However, continuing developments of rough and fuzzy extensions to clustering, neurocomputing, and genetic algorithms make hybrid approaches in these areas a potentially rewarding research opportunity as well.

7.1 INTRODUCTION

Fuzzy set theory has entered the fifth decade of research and development efforts [408]. Rough set theory celebrates its twenty-fifth anniversary this year [260]. The two theories complement each other and as such constitute important components of soft computing. Researchers have explored a variety of different ways in which these two theories interact with each other. The origins of both theories were essentially logical. Hence much of the hybridization between fuzzy and rough set theory is logically based. Moreover rough set theory was proposed for supervised learning. Therefore there is a significant body of work that combines supervised classification using fuzzy and rough set theory. This chapter begins with a review

Computational Intelligence and Feature Selection: Rough and Fuzzy Approaches, by Richard Jensen and Qiang Shen

of fuzzy and rough hybridization in supervised learning, information retrieval, and feature selection.

Both fuzzy and rough set theories are attracting attention among researchers for a more flexible representation of clusters. There are various extensions of fuzzy and rough set theory for clustering. These include the fuzzy and rough extensions of popular clustering algorithms including K-means, Kohonen self-organizing maps, evolutionary unsupervised learning, and support vector clustering. In addition to a brief look at these extensions, this chapter looks at attempts to combine fuzzy and rough set theory in unsupervised clustering. These research efforts include truly unsupervised hybridizations as well as semisupervised data-mining techniques that use the rule generation aspect of rough set theory.

The two theories have also made forays into black box techniques such as neurocomputing and evolutionary computing. The fuzzy and rough extensions of supervised and unsupervised neural networks have been in existence for more than a decade. Consequently there has also been many fuzzy and rough hybridized neurocomputing work. Moreover some innovative combinations of logical foundations of rough and fuzzy set theory have been integrated into neural networks. The fuzzy and rough evolutionary computing is relatively nascent. There are some examples of hybridization of fuzzy and rough sets in evolutionary computing. In some cases the genetic algorithms are used to aid other hybridization techniques such as rough-fuzzy neural networks.

7.2 THEORETICAL HYBRIDIZATION

There have been two main lines of thought in the hybridization of fuzzy and rough sets, the constructive approach and the axiomatic approach. A general framework for the study of fuzzy-rough sets from both of these viewpoints is presented in [403]. For the constructive approach, generalized lower and upper approximations are defined based on fuzzy relations. Initially these were fuzzy similarity/equivalence relations [86], but they have since been extended to arbitrary fuzzy relations [403]. The axiomatic approach is primarily for the study of the mathematical properties of fuzzy-rough sets [386]. Here various classes of fuzzy-rough approximation operators are characterized by different sets of axioms that guarantee the existence of types of fuzzy relations producing the same operators.

Dubois and Prade defined the fuzzy P-lower and P-upper approximations as follows [86]:

$$\mu_{\underline{P}X}(F_i) = \inf_x \max\{1 - \mu_{F_i}(x), \mu_X(x)\} \qquad \forall i \tag{7.1}$$

$$\mu_{\overline{P}X}(F_i) = \sup_x \min\{\mu_{F_i}(x), \mu_X(x)\} \qquad \forall i \tag{7.2}$$

where F_i is a fuzzy equivalence class and X is the (fuzzy) concept to be approximated. The tuple $\langle \underline{P}X, \overline{P}X \rangle$ is called a fuzzy-rough set. Also defined in the literature are rough-fuzzy sets [85,349], which can be seen to be a particular case of fuzzy-rough sets. A rough-fuzzy set is a generalization of a rough set, derived from the approximation of a fuzzy set in a crisp approximation space. In [400] it is argued that to be consistent, the approximation of a crisp set in a fuzzy approximation space should be called a fuzzy-rough set, and the approximation of a fuzzy set in a crisp approximation space should be called a rough-fuzzy set, making the two models complementary. In this framework the approximation of a fuzzy set in a fuzzy approximation space is considered to be a more general model, unifying the two theories. However, most researchers consider the traditional definition of fuzzy-rough sets in [86] as standard. The specific use of min and max operators in the definitions above is expanded in [282], where a broad family of fuzzy-rough sets is constructed, each member represented by a particular implicator and t-norm. The properties of three well-known classes of implicators (S-, R-, and QL-implicators) are investigated. Further investigations in this area can be found in [72,359,388,403].

In [39,237] an axiomatic approach is taken, but restricted to fuzzy T-similarity relations (and hence fuzzy T-rough sets). Wu et al. [385] investigated the properties of generalized fuzzy-rough sets, defining a pair of dual generalized fuzzy approximation operators based on arbitrary fuzzy relations. The approach presented in [228] introduces definitions for generalized fuzzy lower and upper approximation operators determined by a residual implication. Assumptions are found that allow a given fuzzy set-theoretic operator to represent a lower or upper approximation from a fuzzy relation. Different types of fuzzy relations produce different classes of fuzzy-rough set algebras.

The work in [283] generalizes the fuzzy-rough set concept through the use of residuated lattices. An arbitrary residuated lattice is used as a basic algebraic structure, and several classes of L-fuzzy-rough sets and their properties are investigated. In [53] a complete completely distributive (CCD) lattice is selected as the foundation for defining lower and upper approximations in an attempt to provide a unified framework for rough set generalizations. It is demonstrated that the existing fuzzy-rough sets are special cases of the approximations on a CCD lattice for T-similarity relations.

The relationships between fuzzy-rough set models and fuzzy ([0, 1]-) topologies on a finite universe have been investigated. The first such research was reported in [39], where it was proved that the lower and upper approximation operators were fuzzy interior and closure operators, respectively, for fuzzy T-similarity relations. The work carried out in [403] investigated this for arbitrary fuzzy relations. In [278,387] it was shown that a pair of dual fuzzy-rough approximation operators can induce a topological space if and only if the fuzzy relation is reflexive and transitive. The sufficient and necessary condition that a fuzzy interior (closure) operator derived from a fuzzy topological space can associate with a fuzzy reflexive and transitive relation such that the induced fuzzy lower (upper) approximation operator is the fuzzy interior (closure) operator is also examined.

In addition to the previous approaches to hybridization, other generalizations are possible. One of the first attempts at hybridizing the two theories is reported in [389], where rough sets are expressed by a fuzzy membership function to represent the negative, boundary, and positive regions. All objects in the positive region have a membership of one, and those belonging to the boundary region have a membership of 0.5. Those that are contained in the negative region (and therefore do not belong to the rough set) have zero membership. In so doing, a rough set can be expressed as a fuzzy set, with suitable modifications to the rough union and intersection operators. In [244] a definition of fuzzy-rough sets is given based on an algebraic approach to rough sets proposed by Iwinski, where a rough set is defined as a pair of subsets from a sub-Boolean algebra without reference to the universe. The lower and upper bounds of an Iwinski rough set are then fuzzified. As stated in [400], the precise meaning of the upper and lower bounds may not be clear.

7.3 SUPERVISED LEARNING AND INFORMATION RETRIEVAL

In [304] a fuzzy-rough nearest neighbor classification approach is presented that attempts to handle both fuzzy uncertainty (caused by overlapping classes) and rough uncertainty (through insufficient features). A fuzzy-rough ownership function is constructed that attempts to incorporate these two factors, and it also allows a possibilistic class membership assignment. All training patterns influence the ownership function, and hence no decision is required as to the number of neighbors to consider, although there are other parameters that must be defined for its successful operation. It should be noted that the algorithm does not use fuzzy lower or upper approximations to determine class membership.

This work is extended in [377]. Here the classification task is divided into four stages. First, the fuzzy-rough ownership is calculated (via a leave-one-out strategy) for each training pattern in the dataset for all classes. This value indicates the degree to which other patterns support each data item. Next, the training set is filtered to eliminate inconsistencies—where the highest fuzzy-rough ownership value for a data item suggests a class other than the known class. Then, representative points are selected from the processed training set, and fuzzy-rough ownership values are updated based on mountain clustering. Finally, test patterns are classified using only the representative points determined previously by the fuzzy-rough algorithm in [304]. Bian and Mazlack [35] incorporate rough uncertainty into the fuzzy K-NN classifier through the use of fuzzy upper and lower approximations as defined in [86]. The K nearest neighbors are used to determine the membership degree of the test pattern to the fuzzy lower and upper approximations for every class.

The induction of gradual decision rules, based on fuzzy-rough hybridization, is given in [117,118]. For this approach new definitions of fuzzy lower and upper approximations are constructed that avoid the use of fuzzy logical connectives altogether, and hence bypass this arbitrary choice. Decision rules are induced from lower and upper approximations defined for positive and negative relationships between credibility of premises and conclusions. In addition a new scheme of inference is proposed based on a generalized fuzzy-rough modus ponens.

Fuzzy decision tree induction is an area of much interest, due mainly to its ability to model vagueness, and fuzzy decision tree construction seems to be a logical area of application for fuzzy-rough sets. However, so far very little work has been carried out. Initial research is presented in [32] where a method for fuzzy decision tree construction is given that employs the fuzzy-rough ownership function from [304]. This is used to define both an index of fuzzy-roughness and a measure of fuzzy-rough entropy as a node-splitting criterion. Traditionally fuzzy entropy (or its extension) has been used for this purpose. In [165] a fuzzy decision tree algorithm is proposed, based on fuzzy ID3, that incorporates the fuzzy-rough dependency function defined in [163] as a splitting criterion. It is shown that the fuzzy-rough method performs comparably to fuzzy ID3 for fuzzy datasets, and better than it for crisp data. A fuzzy-rough rule induction method is proposed in [136] for generating certain and possible rulesets from hierarchical data.

Fuzzy-rough ownership is used to measure the average ruggedness of data for time series in [305]. The rate of average fuzzy-roughness change is examined when the resolution of the time series is changed. By this method, a fractal dimension that quantifies the regularity of the time series is determined. The proposed dimension is relatively insensitive to translation, scaling, noise, and nonstationarity.

A characteristic of natural language is its inherent vagueness and uncertainty, and hence is an ideal area of application for fuzzy-rough set hybrids. In [73] an approach to Web query expansion is proposed using fuzzy-rough set theory as a framework for query refinement. A thesaurus for this task, defining an approximation space, can be approximated from the upper and the lower side. The use of fuzzy-rough sets is motivated by the extra flexibility afforded by graded thesauri and weighted queries—a thesaurus may be represented as a fuzzy relation and the query as a fuzzy set. Query expansion and refinement occurs through the use of fuzzy lower approximations of fuzzy upper approximations of the query itself. Automatic text summarization is a process that condenses a source document into a much shorter text with the aim of retaining its core content. The authors in [143] propose a fuzzy-rough set aided summarization method to extract key sentences from a text as a document summary. Sentences are given a relevance ranking based on fuzzy-rough approximations based on a fuzzy relevance clustering.

7.4 FEATURE SELECTION

One of the primary and most successful applications of crisp rough sets is to the area of feature selection. It is natural, then, to apply its hybridization to this area [191]. Such work has been carried out in [163,168,322], where a reduction method was proposed based on fuzzy extensions to the positive region and dependency function based on fuzzy lower approximations. A greedy hill-climber is used to perform subset search, using the fuzzy dependency function both for subset evaluation and as a stopping criterion. The method was used successfully within a range of problem domains, including Web content classification and complex system monitoring [168].

Optimizations are given in [33,168] to improve the performance of the method. In [34] a compact computational domain is proposed to reduce the computational effort required to calculate fuzzy lower approximations for large datasets, based on some of the properties of fuzzy connectives. Fuzzy entropy is used in [218] to guide the search toward smaller reducts. In [33] an alternative search algorithm is presented that alleviates some of the problems encountered with a greedy hill-climber approach. This problem is also tackled in [164] via the use of a novel ant colony optimization based framework for feature selection. A genetic algorithm is used in [418] for search based on the fuzzy dependency function within a face recognition system with promising results.

The work in [361,378] improves upon these developments by formally defining relative reductions for fuzzy decision systems. A discernibility matrix is constructed for the computation of all such reductions. As the resulting discernibility matrix is crisp, some information could have been lost in this process. Additionally there are complexity issues when computing discernibility matrices for large datasets. However, in the crisp rough set literature there have been methods proposed that avoid this, and similar constructions may be applicable here.

Feature selection algorithms, based on the generalization of fuzzy approximation spaces to fuzzy probability approximation spaces, are introduced in [141]. This is achieved through the introduction of a probability distribution on the universe. Information measures for fuzzy indiscernibility relations are presented in [140] for the computation of feature importance. Reducts are computed through the use of a greedy selection algorithm.

7.5 UNSUPERVISED LEARNING AND CLUSTERING

Rough set theory has made substantial progress as a classification tool in data mining [261]. The basic concept of representing a set as lower and upper bounds can be used in a broader context. Clustering in relation to rough set theory is attracting increasing interest among researchers [129,269,368]. One of the first proposals of unsupervised learning in the context of rough set theory came in 1996 [204], with the concept of lower and upper bounds of numeric ranges used to create a rough Kohonen neural network. A similar concept was also used for the clustering of traffic patterns based on rough genetic algorithms [206]. These early proposals did not exploit the set-theoretic concept of lower and upper bounds. The rough set representation of clusters can be found in [205], which described how a rough set-theoretic classification scheme can be represented using a rough set genome. In subsequent publications [210,209], modifications of K-means and Kohonen self-organizing maps (SOM) were proposed to create intervals of clusters based on rough set theory. Recently [383] showed how K nearest neighbor classification can be augmented with rule induction. Asharaf et al. [9] created rough boundaries around irregular shaped clusters based on support vector clustering.

Unsupervised clustering based on fuzzy set theory can be traced back more than 25 years to Bezdek's proposal of the fuzzy c-means algorithm [31]. Hoppner et al. [138] provide a comprehensive overview of fuzzy clustering. The fuzzy clustering techniques discussed in their book include the fuzzy c-means, the Gustafson–Kessel, and the Gath-and-Geva algorithm for classification problems. There have been a few studies that compare, complement, and combine rough and fuzzy set based clustering. Lingras [207] compared the results of fuzzy clustering that included creation of rough set representation of clusters using fuzzy memberships from the fuzzy c-means algorithm. Asharaf and Murty [8] describe a hybridized fuzzy-rough approach to clustering. Chimphlee et al. [55] described how feature selection based on independent component analysis can be used for hybridized rough-fuzzy clustering of Web user sessions.

There are a number of approaches that combine the traditional supervised classification from rough set theory with unsupervised fuzzy clustering. For example, [416] use a rough set-based fuzzy clustering algorithm in which the objects of fuzzy clustering are initial clusters obtained in terms of equivalence relations. The preponderance of many small classes is countered through secondary clustering on the basis of defining fuzzy similarity between two initial clusters. Wang et al. [379] integrated fuzzy clustering and variable precision rough set theory. This approach was used to effectively discover association rules in process planning.

Traditional hybridization of fuzzy and rough sets can also be seen in [271], where a texture separation algorithm is proposed to solve the problem of unsupervised boundary localization in textured images using rough fuzzy sets and hierarchical clustering. Pal [256] in a book chapter describes how rough-fuzzy initialization can be used for clustering with the example of multi-spectral image segmentation. In another chapter in the same book, he describes how rough set-based rule extraction can be combined with self-organizing maps.

7.6 NEUROCOMPUTING

While most of the rough-fuzzy hybridization efforts are focused on classification, such hybridization is finding increasing popularity in neurocomputing. Use of rough set theory in neurocomputing is relatively recent, when in 1996 Lingras [203,204] showed how upper and lower bounds of ranges of numbers may be able to reduce training time and improve prediction performance. Lingras [205] also showed how rough neurons can be combined with neo-fuzzy neurons.

A slightly different knowledge-based approach can be found in a book by Pal and Mitra [257], where they report the usefulness of rough and fuzzy hybridization in knowledge-based networks. Han et al. [122] used a combination of rough-fuzzy neural computing for classifying faults in high voltage power systems. They also showed how the combination of rough and fuzzy sets compares favorably with fuzzy neural computing. Pal et al. [259] used an innovative combination of multiple technologies—including rough sets, fuzzy sets, neurocomputing, and genetic algorithms—that provided accelerated training and a compact network suitable for generating a minimum number of rules with high certainty values. Zhang [413] combined the logical view of rough set theory with fuzzy neurocomputing. Rough set theory was used to reduce the rule set in the initial fuzzy system by eliminating inconsistent and redundant rules. The resulting rule set is used to create and train a simple fuzzy neural network. A similar hybridization can also be found in [142], where particle swarm optimization is used to refine network parameters.

The traditional rough set strength of rule set management was combined with neuro-fuzzy systems by [6] for time series prediction based on actual stock market data. The proposed model creates rules with reasonable interpretability and leads to higher yields. The use of upper and lower bounds of numeric ranges is also shown to be effective in classification of multi-spectral images in [226], where a rough-fuzzy Hopfield net (RFHN) is proposed. The approach used upper and lower bounds of gray levels captured from a training vector in multi-spectral images instead of all the information in the image. The resulting RFHN reduces training time because of the use of $2/N$ pixels for an N-dimensional multi-spectral image. The resulting classification is also shown to be an improvement over the conventional approach. A similar improvement in performance with the use of rough-fuzzy neural networks over fuzzy neural networks was reported in [98] for prediction of short-term electricity load forecasting. They also used genetic algorithms for the selection of the best set of inputs for the prediction. The application of rough-fuzzy neurocomputing continues to diversify as indicated by the rough-fuzzy neural network controller used by [52] for wastewater management.

7.7 EVOLUTIONARY AND GENETIC ALGORITHMS

There is a 15-year history of hybridization of fuzzy and genetic algorithms (GAs). The first 10-year history of such hybridization is reported in [61]. Hybridization of GAs with rough set theory was first based on lower and upper bounds of numeric ranges in the form of rough genetic algorithms in 1998 [206]. Another hybridization between GAs and rough set theory used rough sets representation in the genetic encoding [210] for creating rough set representations of clusters. Mitra and Mitra [233] proposed a hybrid decision support system for detecting the different stages of cervical cancer. Hybridization included the evolution of knowledge-based subnetwork modules with GAs using rough set theory and the ID3 algorithm.

Genetic algorithms have been used with hybridized rough and fuzzy neural research by a number of authors. For example, [98] used GAs for selecting the best set of inputs for rough-fuzzy neurocomputing. Similarly [259] used GAs in rough and fuzzy set based rule generation. These genetic algorithms were not specifically extended based on either rough or fuzzy set theory. One example of an extended GA based on rough set theory combined with fuzzy set theory can be found in [316]. The authors applied rough GAs based on lower and upper bounds of numeric ranges to the classification problem in an undetermined environment based on a fuzzy distance function by calculating attribute weights.

7.8 SUMMARY

This chapter reviewed fuzzy and rough hybridization efforts. The review can be broadly categorized into logical and black box approaches. The logical approaches can be further subdivided into theoretical, supervised learning, feature selection, and unsupervised learning. The black box approaches consist of neural and evolutionary computing. There is significant theoretical work on hybridization of fuzzy and rough set theories, as well as its usage in classification and similar supervised learning techniques. There is also an increasing interest in extension of these two theories in unsupervised clustering and in semisupervised learning.

Neurocomputing extensions of fuzzy and rough set theory have been in existence for more than a decade. The hybridization of fuzzy and rough set theories can be seen in the neurocomputing world at various levels. These two theories not only provide different modeling of input and output neurons but are also used for introducing logical reasoning in neural networks. While rough and fuzzy extensions of genetic algorithms are being proposed, the use of hybridized rough and fuzzy sets in genetic computing is relatively limited. However, a number of researchers have used evolutionary computing in the more traditional logical hybridization of these two theories. Since fuzzy and rough set theory complement each other, one expects further work in their hybridization in all the areas discussed in this chapter.

CHAPTER 8

FUZZY-ROUGH FEATURE SELECTION

The RSAR process described previously can only operate effectively with datasets containing discrete values. It is not possible in the original theory to say whether two attribute values are similar and to what extent they are the same; for example, two close values may only differ as a result of noise. However, in RST attribute values affected by noise are considered to be as different as two values of a different order of magnitude. Additionally, there is no way of handling noisy data. As most datasets contain real-valued features, it is necessary to perform a discretization step beforehand. This is typically implemented by standard fuzzification techniques [321], enabling linguistic labels to be associated with attribute values. Fuzzification also aids uncertainty modeling by allowing the possibility of the membership of a value to more than one fuzzy label. However, membership degrees of feature values to fuzzy sets are not exploited in the process of dimensionality reduction.

As a result extensions to the original theory have been proposed, for example, based on similarity or tolerance relations [335,343,352]. One promising approach is the use of *fuzzy-rough* sets. Fuzzy-rough sets encapsulate the related but distinct concepts of vagueness (for fuzzy sets [408]) and indiscernibility (for rough sets), both of which occur as a result of uncertainty in knowledge [86]. Vagueness arises due to a lack of sharp distinctions or boundaries in the data. This is typical of human communication and reasoning. Rough sets can be said to model ambiguity resulting from a lack of information through set approximations.

Computational Intelligence and Feature Selection: Rough and Fuzzy Approaches, by Richard Jensen and Qiang Shen

8.1 FEATURE SELECTION WITH FUZZY-ROUGH SETS

Fuzzy-rough sets encapsulate the related but distinct concepts of vagueness (for fuzzy sets [408]) and indiscernibility (for rough sets), both of which occur as a result of uncertainty in knowledge [86]. A fuzzy-rough set is defined by two fuzzy sets, fuzzy lower and upper approximations, obtained by extending the corresponding crisp rough set notions. In the crisp case, elements that belong to the lower approximation (i.e., have a membership of 1) are said to belong to the approximated set with absolute certainty. In the fuzzy-rough case, elements may have a membership in the range [0, 1], allowing greater flexibility in handling uncertainty.

Fuzzy-rough feature selection (FRFS) [168] provides a means by which discrete or real-valued noisy data (or a mixture of both) can be effectively reduced without the need for user-supplied information. Additionally this technique can be applied to data with continuous or nominal decision attributes, and as such can be applied to regression as well as classification datasets. The only additional information required is in the form of fuzzy partitions for each feature that can be automatically derived from the data. In the work presented here, excluding the simple examples, *all* fuzzy sets are derived solely from the data. This avoids the need for domain experts to provide information on the data involved and ties in with the advantage of rough sets in that it requires no information other than the data. However, if such experts are readily available, it is beneficial to capture their knowledge in the form of fuzzy data partitions to improve the transparency of the selection process and any other future processes (e.g., rule induction).

8.2 FUZZY-ROUGH REDUCTION PROCESS

FRFS builds on the notion of fuzzy lower approximation to enable reduction of datasets containing real-valued features. As will be shown, the process becomes identical to the crisp approach when dealing with nominal well-defined features.

The crisp positive region in traditional rough set theory is defined as the union of the lower approximations. By the extension principle [409], the membership of an object $x \in \mathbb{U}$, belonging to the fuzzy positive region can be defined by

$$\mu_{POS_{P}(Q)}(x) = \sup_{X \in \mathbb{U}/Q} \mu_{\underline{P}X}(x) \tag{8.1}$$

Object x will not belong to the positive region only if the equivalence class it belongs to is not a constituent of the positive region. This is equivalent to the crisp version where objects belong to the positive region only if their underlying equivalence class does so. Similarly the negative and boundary regions can be defined. For this particular feature selection method, the upper approximation is not used, although this may be useful for other methods.

Using the definition of the fuzzy positive region, the new dependency function can be defined as follows:

$$\gamma'_P(Q) = \frac{|\mu_{POS_{P(Q)}}(x)|}{|\mathbb{U}|} = \frac{\sum_{x \in \mathbb{U}} \mu_{POS_{P(Q)}}(x)}{|\mathbb{U}|} \tag{8.2}$$

As with crisp rough sets the dependency of Q on P is the proportion of objects that are discernible out of the entire dataset. In the present approach this corresponds to determining the fuzzy cardinality of $\mu_{POS_{P(Q)}}(x)$ divided by the total number of objects in the universe.

The definition of dependency degree covers the crisp case as its specific instance. This can be easily shown by recalling the definition of the crisp dependency degree given in (2.16). If a function $\mu_{POS_{P(Q)}}(x)$ is defined that returns 1 if the object x belongs to the positive region, 0 otherwise, then the above definition may be rewritten as

$$\gamma_P(Q) = \frac{\sum_{x \in \mathbb{U}} \mu_{POS_{P(Q)}}(x)}{|\mathbb{U}|} \tag{8.3}$$

which is identical to (8.2).

If the fuzzy-rough reduction process is to be useful, it must be able to deal with multiple features, finding the dependency among various subsets of the original feature set. For example, it may be necessary to be able to determine the degree of dependency of the decision feature(s) with respect to $P = \{a, b\}$. In the crisp case, $\mathbb{U}/P$ contains sets of objects grouped together that are indiscernible according to both features a and b. In the fuzzy case, objects may belong to many equivalence classes, so the Cartesian product of $\mathbb{U}/IND(\{a\})$ and $\mathbb{U}/IND(\{b\})$ must be considered in determining $\mathbb{U}/P$. In general,

$$\mathbb{U}/IND(P) = \otimes\{\mathbb{U}/IND(\{a\}) | a \in P\}, \tag{8.4}$$

where

$$A \otimes B = \{X \cap Y \mid X \in A, Y \in B, X \cap Y \neq \emptyset\} \tag{8.5}$$

Each set in $\mathbb{U}/P$ denotes an equivalence class. For example, if $P = \{a, b\}$, $\mathbb{U}/IND(\{a\}) = \{N_a, Z_a\}$ and $\mathbb{U}/IND(\{b\}) = \{N_b, Z_b\}$, then

$$\mathbb{U}/P = \{N_a \cap N_b, N_a \cap Z_b, Z_a \cap N_b, Z_a \cap Z_b\}$$

The extent to which an object belongs to such an equivalence class is therefore calculated by using the conjunction of constituent fuzzy equivalence classes, say F_i, $i = 1, 2, \ldots, n$:

$$\mu_{F_1 \cap \ldots \cap F_n}(x) = \min(\mu_{F_1}(x), \mu_{F_2}(x), \ldots, \mu_{F_n}(x)) \tag{8.6}$$

8.3 FUZZY-ROUGH QUICKREDUCT

A problem may arise when the fuzzy-rough approach is compared to the crisp approach. In conventional RSAR, a reduct is defined as a subset R of the features that have the same information content as the full feature set A. In terms of the dependency function this means that the values $\gamma(R)$ and $\gamma(A)$ are identical and equal to 1 if the dataset is consistent. However, in the fuzzy-rough approach this is not necessarily the case as the uncertainty encountered when objects belong to many fuzzy equivalence classes results in a reduced total dependency.

A possible way of combating this additional uncertainty would be to determine the degree of dependency of a set of decision features $\mathbb{D}$ upon the full feature set and use this as the denominator rather than $|\mathbb{U}|$ (for normalization), allowing γ' to reach 1. With these issues in mind, a new QuickReduct algorithm has been developed as given in Figure 8.1. It employs the new dependency function γ' to choose which features to add to the current reduct candidate in the same way as the original QuickReduct process (see Figure 5.1). The algorithm terminates when the addition of any remaining feature does not increase the dependency (such a criterion could be used with the original QuickReduct algorithm).

As the new degree of dependency measure is nonmonotonic, it is possible that the QuickReduct-style search terminates having reached only a local optimum. The global optimum may lie elsewhere in the search space. This motivates the adoption of an alternative search mechanism, presented in Section 10.2. However, the algorithm as presented in Figure 8.1 is still highly useful in locating good subsets quickly.

It is also possible to reverse the search process in a manner identical to that of ReverseReduct, that is, start with the full set of features and incrementally

```
FRQuickReduct(ℂ,𝔻)
Input: ℂ, the set of all conditional features; 𝔻, the set of
decision features
Output: R, the feature subset
    R ← {}; γ'best = 0; γ'prev = 0
    while γ'best ≠ γ'prev
       T ← R
       γ'prev = γ'best
       foreach x ∈ (ℂ − R)
          if γ'R∪{x}(𝔻) > γ'T(𝔻)
             T ← R ∪ {x}
             γ'best = γ'T(𝔻)
       R ← T
    return R
```

Figure 8.1 Fuzzy-rough QuickReduct algorithm

remove the least informative features. This process continues until no more features can be removed without reducing the total number of discernible objects in the dataset. Again, this tends not to be applied to larger datasets as the cost of evaluating these larger feature subsets is too great.

8.4 COMPLEXITY ANALYSIS

An intuitive understanding of QUICKREDUCT implies that for a dimensionality of n, $(n^2 + n)/2$ evaluations of the dependency function may be performed for the worst-case dataset. However, as FRFS is used for dimensionality reduction prior to any involvement of the system that will employ those features belonging to the resultant reduct, this operation has no negative impact upon the runtime efficiency of the system.

In fact, as feature selection can only take place when $n \geq 2$, the base case is $n = 2$. Suppose that the set of conditional features in this case is $\{a_1, a_2\}$; the QUICKREDUCT algorithm makes two initial dependency evaluations (for a_1 and a_2) and a final evaluation for $\{a_1, a_2\}$ (in the worst case). Hence the order of complexity of the algorithm is 3 (or $(n^2 + n)/2$) for $n = 2$. Suppose that for $n = k$ the order of complexity of the algorithm is

$$\frac{(k^2 + k)}{2} \tag{8.7}$$

For $k + 1$ features, $\{a_1, \ldots, a_k, a_{k+1}\}$, QUICKREDUCT makes $k + 1$ initial evaluations of the dependency function to determine the best feature (call this a_i). Once a_i is chosen, for the remaining features there are $(k^2 + k)/2$ more evaluations in the worst case according to (8.7). Hence the total number of evaluations for $n = k + 1$ is

$$\frac{k^2 + k}{2} + (k + 1) = \frac{k^2 + 3k + 2}{2} = \frac{(k + 1)^2 + (k + 1)}{2}$$

The complexity of the algorithm is therefore $O((n^2 + n)/2)$ in the worst case. In the best case the first feature considered results in the maximum degree of dependency for the data, and hence only one heuristic evaluation is performed. In crisp RSAR the average complexity is seen to be close to linear with the inclusion of several optimizations [56].

8.5 WORKED EXAMPLES

To illustrate the operation of FRFS, two small example datasets are considered. The first contains real-valued conditional attributes with nominal decisions. In crisp RSAR the dataset would be discretized using the nonfuzzy sets. However, in the fuzzy-rough approach membership degrees are used in calculating the fuzzy lower approximations and fuzzy positive regions. The second dataset is defined by fuzzy membership values only, with corresponding fuzzy decisions.

In addition to demonstrating the operation of FRFS, a further purpose of presenting these two datasets is to show the applicability of the development to different types of dataset. FRFS is equally applicable to datasets where the conditional values are crisp and decisions are fuzzy, and also to the situation where both conditional and decision attributes are crisp. In the latter case the operation of FRFS is exactly that of the original crisp RSAR.

8.5.1 Crisp Decisions

Table 8.1 contains three real-valued conditional attributes and a crisp-valued decision attribute. To begin with, the fuzzy-rough QuickReduct algorithm initializes the potential reduct (i.e., the current best set of attributes) to the empty set.

Using the fuzzy sets defined in Figure 8.2 (for all conditional attributes), and setting $A = \{a\}$, $B = \{b\}$, $C = \{c\}$ and $Q = \{q\}$, the following equivalence classes are obtained:

$$\mathbb{U}/A = \{N_a, Z_a\}$$

$$\mathbb{U}/B = \{N_b, Z_b\}$$

TABLE 8.1 Example dataset: Crisp decisions

Object	a	b	c	q
1	−0.4	−0.3	−0.5	no
2	−0.4	0.2	−0.1	yes
3	−0.3	−0.4	−0.3	no
4	0.3	−0.3	0	yes
5	0.2	−0.3	0	yes
6	0.2	0	0	no

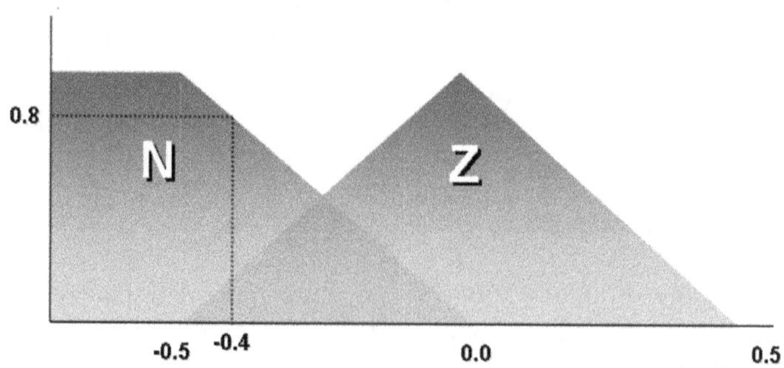

Figure 8.2 Fuzzifications for conditional features

TABLE 8.2 Membership values of objects to corresponding fuzzy sets

Object	a		b		c		q	
	N_a	Z_a	N_b	Z_b	N_c	Z_c	{1, 3, 6}	{2, 4, 5}
1	0.8	0.2	0.6	0.4	1.0	0.0	1.0	0.0
2	0.8	0.2	0.0	0.6	0.2	0.8	0.0	1.0
3	0.6	0.4	0.8	0.2	0.6	0.4	1.0	0.0
4	0.0	0.4	0.6	0.4	0.0	1.0	0.0	1.0
5	0.0	0.6	0.6	0.4	0.0	1.0	0.0	1.0
6	0.0	0.6	0.0	1.0	0.0	1.0	1.0	0.0

$$\mathbb{U}/C = \{N_c, Z_c\}$$

$$\mathbb{U}/Q = \{\{1, 3, 6\}, \{2, 4, 5\}\}$$

The first step is to calculate the lower approximations of the sets A, B, and C, using equation (2.25) in Section 2.4.2. To clarify the calculations involved, Table 8.2 contains the membership degrees of objects to fuzzy equivalence classes. For simplicity, only A will be considered here, that is, by using A to approximate Q. For the first decision equivalence class $X = \{1, 3, 6\}$, $\mu_{\underline{A}\{1,3,6\}}(x)$ needs to be calculated:

$$\mu_{\underline{A}\{1,3,6\}}(x) = \sup_{F \in \mathbb{U}/A} \min(\mu_F(x), \inf_{y \in \mathbb{U}} \max\{1 - \mu_F(y), \mu_{\{1,3,6\}}(y)\})$$

Consider the first fuzzy equivalence class of A, N_a:

$$\min(\mu_{N_a}(x), \inf_{y \in \mathbb{U}} \max\{1 - \mu_{N_a}(y), \mu_{\{1,3,6\}}(y)\})$$

For object 2 this can be calculated as follows: From Table 8.2 it can be seen that the membership of object 2 to the fuzzy equivalence class N_a, $\mu_{N_a}(2)$, is 0.8. The remainder of the calculation involves finding the smallest of the following values:

$$\max(1 - \mu_{N_a}(1), \mu_{\{1,3,6\}}(1)) = \max(0.2, 1.0) = 1.0$$

$$\max(1 - \mu_{N_a}(2), \mu_{\{1,3,6\}}(2)) = \max(0.2, 0.0) = 0.2$$

$$\max(1 - \mu_{N_a}(3), \mu_{\{1,3,6\}}(3)) = \max(0.4, 1.0) = 1.0$$

$$\max(1 - \mu_{N_a}(4), \mu_{\{1,3,6\}}(4)) = \max(1.0, 0.0) = 1.0$$

$$\max(1 - \mu_{N_a}(5), \mu_{\{1,3,6\}}(5)) = \max(1.0, 0.0) = 1.0$$

$$\max(1 - \mu_{N_a}(6), \mu_{\{1,3,6\}}(6)) = \max(1.0, 1.0) = 1.0$$

From the calculations above, the smallest value is 0.2; hence

$$\min(\mu_{N_a}(x), \inf_{y \in \mathbb{U}} \max\{1 - \mu_{N_a}(y), \mu_{\{1,3,6\}}(y)\}) = \min(0.8, \inf\{1, 0.2, 1, 1, 1, 1\})$$
$$= 0.2$$

Similarly for Z_a,

$$\min(\mu_{Z_a}(x), \inf_{y \in \mathbb{U}} \max\{1 - \mu_{Z_a}(y), \mu_{\{1,3,6\}}(y)\})$$
$$= \min(0.2, \inf\{1, 0.8, 1, 0.6, 0.4, 1\}) = 0.2$$

Thus

$$\mu_{\underline{A}\{1,3,6\}}(2) = 0.2$$

Calculating the A-lower approximation of $X = \{1, 3, 6\}$ for every object gives

$$\mu_{\underline{A}\{1,3,6\}}(1) = 0.2 \quad \mu_{\underline{A}\{1,3,6\}}(2) = 0.2$$
$$\mu_{\underline{A}\{1,3,6\}}(3) = 0.4 \quad \mu_{\underline{A}\{1,3,6\}}(4) = 0.4$$
$$\mu_{\underline{A}\{1,3,6\}}(5) = 0.4 \quad \mu_{\underline{A}\{1,3,6\}}(6) = 0.4$$

The corresponding values for $X = \{2, 4, 5\}$ can also be determined:

$$\mu_{\underline{A}\{2,4,5\}}(1) = 0.2 \quad \mu_{\underline{A}\{2,4,5\}}(2) = 0.2$$
$$\mu_{\underline{A}\{2,4,5\}}(3) = 0.4 \quad \mu_{\underline{A}\{2,4,5\}}(4) = 0.4$$
$$\mu_{\underline{A}\{2,4,5\}}(5) = 0.4 \quad \mu_{\underline{A}\{2,4,5\}}(6) = 0.4$$

It is a coincidence that $\mu_{\underline{A}\{2,4,5\}}(x) = \mu_{\underline{A}\{1,3,6\}}(x)$ for this example. From these values the fuzzy positive region for each object can be calculated via using

$$\mu_{POS_A(Q)}(x) = \sup_{X \in \mathbb{U}/Q} \mu_{\underline{A}X}(x)$$

This results in

$$\mu_{POS_A(Q)}(1) = 0.2 \quad \mu_{POS_A(Q)}(2) = 0.2$$
$$\mu_{POS_A(Q)}(3) = 0.4 \quad \mu_{POS_A(Q)}(4) = 0.4$$
$$\mu_{POS_A(Q)}(5) = 0.4 \quad \mu_{POS_A(Q)}(6) = 0.4$$

The next step is to determine the degree of dependency of Q on A:

$$\gamma'_A(Q) = \frac{\sum_{x \in \mathbb{U}} \mu_{POS_A(Q)}(x)}{|\mathbb{U}|} = 2/6$$

Calculating for B and C gives

$$\gamma'_B(Q) = \frac{2.4}{6}, \quad \gamma'_C(Q) = \frac{1.6}{6}$$

From this result it can be seen that attribute b will cause the greatest increase in dependency degree. This attribute is chosen and added to the potential reduct. The process iterates and the two dependency degrees calculated are

$$\gamma'_{\{a,b\}}(Q) = \frac{3.4}{6}, \quad \gamma'_{\{b,c\}}(Q) = \frac{3.2}{6}$$

Adding attribute a to the reduct candidate causes the larger increase of dependency, so the new candidate becomes $\{a, b\}$. Last, attribute c is added to the potential reduct:

$$\gamma'_{\{a,b,c\}}(Q) = \frac{3.4}{6}$$

As this action causes no increase in dependency, the algorithm stops and outputs the reduct $\{a, b\}$. The steps taken by the fuzzy-rough QuickReduct algorithm in reaching this decision can be seen in Figure 8.3. The dataset can now be reduced to only those attributes appearing in the reduct. When crisp RSAR is performed on this dataset (after using the same fuzzy sets to discretize the real-valued attributes), the reduct generated is $\{a, b, c\}$, namely the full conditional attribute set [162]. Unlike crisp RSAR, the true minimal reduct was found using the information on degrees of membership. It is clear from this example alone that the information lost by using crisp RSAR can be important when trying to discover the smallest reduct from a dataset.

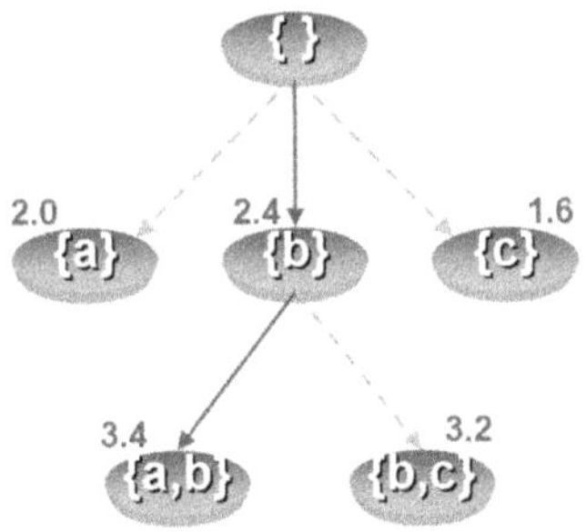

Figure 8.3 Path taken by the fuzzy-rough QuickReduct algorithm

TABLE 8.3 Example dataset containing fuzzy values

Object	*a*			*b*			*c*		*Plan*		
	*A*1	*A*2	*A*3	*B*1	*B*2	*B*3	*C1*	*C*2	X	Y	Z
1	0.3	0.7	0.0	0.2	0.7	0.1	0.3	0.7	0.1	0.9	0.0
2	1.0	0.0	0.0	1.0	0.0	0.0	0.7	0.3	0.8	0.2	0.0
3	0.0	0.3	0.7	0.0	0.7	0.3	0.6	0.4	0.0	0.2	0.8
4	0.8	0.2	0.0	0.0	0.7	0.3	0.2	0.8	0.6	0.3	0.1
5	0.5	0.5	0.0	1.0	0.0	0.0	0.0	1.0	0.6	0.8	0.0
6	0.0	0.2	0.8	0.0	1.0	0.0	0.0	1.0	0.0	0.7	0.3
7	1.0	0.0	0.0	0.7	0.3	0.0	0.2	0.8	0.7	0.4	0.0
8	0.1	0.8	0.1	0.0	0.9	0.1	0.7	0.3	0.0	0.0	1.0
9	0.3	0.7	0.0	0.9	0.1	0.0	1.0	0.0	0.0	0.0	1.0

8.5.2 Fuzzy Decisions

With the fuzzy-rough QuickReduct algorithm, Table 8.3 can be reduced in size. First of all the lower approximations need to be determined. Consider the first feature in the dataset: setting $P = \{a\}$ produces the fuzzy partitioning $\mathbb{U}/P = \{A1, A2, A3\}$. Additionally setting $Q = \{Plan\}$ produces the fuzzy partitioning $\mathbb{U}/Q = \{X, Y, Z\}$. To determine the fuzzy P-lower approximation of Plan X ($\mu_{\underline{P}X}(x)$), each $F \in \mathbb{U}/P$ must be considered. For $F = A1$,

$$\min(\mu_{A1}(x), \inf_{y\in\mathbb{U}} \max\{1 - \mu_{A1}(y), \mu_X(y)\}) = \min(\mu_{A1}(x), 0.6)$$

Similarly, for $F = A2$, $\min(\mu_{A2}(x), 0.3)$ and $F = A3$, $\min(\mu_{A3}(x), 0.0)$. To calculate the extent to which an object x in the dataset belongs to the fuzzy P-lower approximation of X, the union of these values is calculated. For example, object 1 belongs to $\underline{P}X$ with a membership of

$$sup\{\min(\mu_{A1}(1), 0.6), \min(\mu_{A2}(1), 0.3), \min(\mu_{A3}(1), 0.0)\} = 0.3.$$

Likewise, for Y and Z,

$$\mu_{\underline{P}Y}(1) = 0.2, \mu_{\underline{P}Z}(1) = 0.3$$

The extent to which object 1 belongs to the fuzzy positive region can be determined by considering the union of fuzzy P-lower approximations:

$$\mu_{POS_P(Q)}(1) = \sup_{S\in\mathbb{U}/Q} \mu_{\underline{P}S}(1) = 0.3$$

Similarly, for the remaining objects,

$$\mu_{POS_P(Q)}(2) = 0.6 \quad \mu_{POS_P(Q)}(3) = 0.3$$
$$\mu_{POS_P(Q)}(4) = 0.6 \quad \mu_{POS_P(Q)}(5) = 0.5$$

$$\mu_{POS_{P(Q)}}(6) = 0.3 \quad \mu_{POS_{P(Q)}}(7) = 0.6$$

$$\mu_{POS_{P(Q)}}(8) = 0.3 \quad \mu_{POS_{P(Q)}}(9) = 0.3$$

With these values the new degree of dependency of Q on $P = \{a\}$ can be calculated:

$$\gamma'_P(Q) = \frac{\sum_{x\in\mathbb{U}} \mu_{POS_{P(Q)}}(x)}{|\{1, 2, 3, 4, 5, 6, 7, 8, 9\}|} = \frac{3.8}{9}$$

The fuzzy-rough QuickReduct algorithm uses this decision process to evaluate subsets of features in an incremental fashion. The algorithm starts with an empty set and considers the addition of each individual feature:

$$\begin{aligned}\gamma'_{\{a\}}(Q) &= 3.8/9\\ \gamma'_{\{b\}}(Q) &= 2.1/9\\ \gamma'_{\{c\}}(Q) &= 2.7/9\end{aligned}$$

As feature a causes the greatest increase in dependency degree, it is added to the reduct candidate and the search progresses:

$$\begin{aligned}\gamma'_{\{a,b\}}(Q) &= 4.0/9\\ \gamma'_{\{a,c\}}(Q) &= 5.7/9\end{aligned}$$

Here c is added to the reduct candidate as the dependency is increased. There is only one feature addition to be checked at the next stage, namely

$$\gamma'_{\{a,b,c\}}(Q) = 5.7/9$$

This causes no dependency increase, resulting in the algorithm terminating and outputting the reduct $\{a, c\}$. Hence the original dataset can be reduced to these features with minimal information loss (according to the algorithm).

8.6 OPTIMIZATIONS

There are several optimizations that can be implemented to speed up the FRFS process. The original definition of the fuzzy positive region, given in equation (8.1), can be more explicitly defined as

$$\mu_{POS_{P(Q)}}(x) = \sup_{X\in\mathbb{U}/Q} \left\{ \sup_{F\in\mathbb{U}/P} \min(\mu_F(x), \inf_{y\in\mathbb{U}} \max\{1 - \mu_F(y), \mu_X(y)\}) \right\} \tag{8.8}$$

TABLE 8.4 Experimental comparison of the two formulations for the calculation of the positive region

Dataset	Number of Features	Equation (8.8) (seconds)	Equation (8.9) (seconds)	Optimized (seconds)
Glass	10	29.5	26.7	7.18
Wine	14	5.41	3.05	2.20
Olitos	26	47.6	21.9	13.0
JobSat	27	19.2	5.75	2.72
Ionosphere	35	204.5	107.8	76.9
Selwood	54	57.5	15.9	5.64
Isolet	618	368.4	131.9	47.2
Phenetyl	629	740.7	145.0	70.3
Caco	714	2709.3	213.5	114.1

where P is a subset of the conditional attributes, Q the decision attribute(s). In order to speed up computation time, equation (8.8) can be rewritten as

$$\mu_{POS_{P(Q)}}(x) = \sup_{F \in \mathbb{U}/P} \left\{ \min(\mu_F(x), \sup_{X \in \mathbb{U}/Q} \{\inf_{y \in \mathbb{U}} \max(1 - \mu_F(y), \mu_X(y))\}) \right\} \tag{8.9}$$

This reformulation helps to speed up the calculation of the fuzzy positive region by considering each fuzzy equivalence class F in $\mathbb{U}/P$ first. If the object x is found not to belong to F, the remainder of the calculations for this class need not be evaluated, due to the use of the min operator. This can save substantial time, as demonstrated in Table 8.4, where the two definitions of the positive region are used to determine reducts from several small to large datasets. The times here are the times taken for each version of FRFS to find a reduct. Each version of FRFS will follow exactly the same route and will locate identical reducts, hence the results are comparable. All the datasets are from the Machine Learning Repository [38] and contain real-valued conditional features with nominal classifications.

Additionally in Table 8.4 average runtimes are given for the optimized implementation of the fuzzy-rough feature selector. This includes the use of the algorithm presented in Figure 8.4, which is designed to result in the faster computation of the fuzzy-rough metric for small feature subsets. Excess computation is avoided at lines (4) and (6), which exploit the nature of t-norms and s-norms.

8.7 EVALUATING THE FUZZY-ROUGH METRIC

In order to evaluate the utility of the new fuzzy-rough measure of feature significance, a series of artificial datasets were generated and used for comparison with 5 other leading feature ranking measures. The datasets were created by

```
CALCULATEGAMMA'(ℂ,𝔻,P)
Input: ℂ, the set of all conditional features; 𝔻, the set of
decision features; P, feature subset for evaluation
Output: γ'_P(𝔻)
m[], γ' ← 0
foreach F ∈ 𝕌/P
    deg = sup_{X∈𝕌/𝔻} {inf_{y∈𝕌} max{1 − μ_F(y), μ_X(y)}}
    if deg ≠ 0
        foreach o ∈ 𝕌
            if m[o] ≠ 1 and deg > m[o]
                m[o] = max(min(μ_F(o), deg), m[o])
foreach o ∈ 𝕌
    γ' += m[o]
return γ'
```

Figure 8.4 Optimized γ' calculation for small subsets

generating around 30 random feature values for 400 objects. Two or three features (referred to as x, y, or z) are chosen to contribute to the final Boolean classification by means of an inequality. For example, in Table 8.6, if the inequality $(x + y)^2 > 0.25$ holds for an object, then it is classified as 1, with a classification of 0 otherwise. The task for the feature rankers was to discover those features that are involved in the inequalities, ideally rating the other irrelevant features poorly in contrast.

The tables presented in the metric comparison section show the ranking given to the features that are involved in the inequality that determines the classification. The final row indicates whether all the other features are given a ranking of zero. The full results can be seen in Appendix A. For the data presented in Table 8.5, the first feature, x, is used to determine the classification. The values of features y and z are derived from x: $y = \sqrt{x}$, $z = x^2$.

8.7.1 Compared Metrics

The metrics compared are the fuzzy-rough measure (FR), Relief-F (Re), Information Gain (IG), Gain Ratio (GR), OneR (1R), and the statistical measure χ^2.

TABLE 8.5 Feature evaluation for $x > 0.5$, $y = \sqrt{x}$, $z = x^2$

Feature	FR	Re	IG	GR	1R	χ^2
x	0.5257	0.31758	0.997	1.0	99.5	200
y	0.5296	0.24586	0.997	1.0	99.5	200
z	0.5809	0.32121	0.997	1.0	99.5	200
Others	$= 0$	$\neq 0$	$= 0$	$= 0$	$\neq 0$	$= 0$

Metrics other than the fuzzy-rough measure were obtained from [382]. A brief description of each is presented next.

8.7.1.1 *Information Gain* The Information Gain (IG) [146] is the expected reduction in entropy resulting from partitioning the dataset objects according to a particular feature. The entropy of a labeled collection of objects S is defined as

$$Entropy(S) = \sum_{i=1}^{c} -p_i \log_2 p_i \tag{8.10}$$

where p_i is the proportion of S belonging to class i. Based on this, the IG metric is

$$IG(S, A) = Entropy(S) - \sum_{v \in values(A)} \frac{|S_v|}{|S|} Entropy(S_v) \tag{8.11}$$

where $values(A)$ is the set of values for feature A, S the set of training examples, S_v the set of training objects where A has the value v. This metric is the one used in ID3 [279] for selecting the best feature to partition the data.

8.7.1.2 *Gain Ratio* One limitation of the IG measure is that it favors features with many values. The gain ratio (GR) seeks to avoid this bias by incorporating another term, split information, that is sensitive to how broadly and uniformly the attribute splits the considered data:

$$Split(S, A) = -\sum_{i=1}^{c} \frac{|S_i|}{|S|} \log_2 \frac{|S_i|}{|S|} \tag{8.12}$$

where each S_i is a subset of objects generated by partitioning S with the c-valued attribute A. GR is then defined as follows:

$$GR(S, A) = \frac{IG(S, A)}{Split(S, A)} \tag{8.13}$$

8.7.1.3 *χ^2 Measure* In the χ^2 method [211], features are individually evaluated according to their χ^2 statistic with respect to the classes. For a numeric attribute, the method first requires its range to be discretized into several intervals. The χ^2 value of an attribute is defined as

$$\chi^2 = \sum_{i=1}^{m} \sum_{j=1}^{k} \frac{(A_{ij} - E_{ij})^2}{E_{ij}} \tag{8.14}$$

where m is the number of intervals, k the number of classes, A_{ij} the number of samples in the ith interval, jth class, R_i the number of objects in the ith interval,

C_j the number of objects in the jth class, N the total number of objects, and E_{ij} the expected frequency of A_{ij} ($E_{ij} = R_i * C_j/N$). The larger the χ^2 value, the more important is the feature.

8.7.1.4 Relief-F This is the Relief-F measure, based on the original RELIEF measure described in Section 4.2.1.1. Relief evaluates the worth of an attribute by repeatedly sampling an instance and considering the value of the given attribute for the nearest instance of the same and different class. Relief-F extends this idea to dealing with multi-class problems as well as handling noisy and incomplete data.

8.7.1.5 OneR The OneR (1R) classifier [135] learns a one-level decision tree; that is, it generates a set of rules that test one particular attribute. One branch is assigned for every value of a feature; each branch is assigned the most frequent class. The error rate is then defined as the proportion of instances that do not belong to the majority class of their corresponding branch. Features with the higher classification rates are considered to be more significant than those resulting in lower accuracies.

8.7.2 Metric Comparison

The tables presented here are summaries of those given in appendix A. From the results in Table 8.5 it can be observed that all metrics successfully rank the influential features highest. IG, GR, 1R, and χ^2 rank these features equally, whereas Re and FR rank feature z higher. Only FR, IG, GR, and χ^2 rate all the other features as zero.

As can be seen from the results in Table 8.5, feature rankers can discover the influential features but on their own are incapable of determining multiple feature interactions. The table could be reduced to one feature only (either x, y, or z) without any loss of information as only these contribute to the classification. However, the rankers all rate these features highly and would only provide enough information to reduce the data to at least these three attributes. Here the rankers have found the predictive (or relevant) features but have been unable to determine which of these are redundant.

Table 8.6 shows the results for the inequality $(x + y)^2 > 0.25$. Both features x and y are required for deciding the classification. All feature rankers evaluated detect this. FR, IG, GR, 1R, and χ^2 also rank the tenth feature highly—probably due to a chance correlation with the decision. The results in Table 8.7 are for

TABLE 8.6 Feature evaluation for $(x + y)^2 > 0.25$

Feature	FR	Re	IG	GR	1R	χ^2
x	0.2330	0.1862	0.2328	0.1579	86.75	128.466
y	0.2597	0.1537	0.1687	0.1690	87.75	71.971
Others	$\neq 0$	$\neq 0$	$\neq 0$	$\neq 0$	$\neq 0$	$\neq 0$

TABLE 8.7 Feature evaluation for $(x+y)^2 > 0.5$

Feature	FR	Re	IG	GR	1R	χ^2
x	0.2090	0.140067	0.241	0.156	79.0	119.562
y	0.2456	0.151114	0.248	0.165	78.25	122.336
Others	$= 0$	$\neq 0$	$= 0$	$= 0$	$\neq 0$	$= 0$

TABLE 8.8 Feature evaluation for $(x+y)^3 < 0.125$

Feature	FR	Re	IG	GR	1R	χ^2
x	0.2445	0.1486	0.134	0.134	87.75	57.455
y	0.2441	0.1659	0.159	0.164	87.25	73.390
Others	$= 0$	$\neq 0$	$= 0$	$= 0$	$\neq 0$	$= 0$

TABLE 8.9 Feature evaluation for $x * y * z > 0.125$

Feature	FR	Re	IG	GR	1R	χ^2
x	0.1057	0.0750547	0.169	0.123	64.25	73.653
y	0.0591	0.1079423	0.202	0.226	66.75	88.040
z	0.1062	0.0955878	0.202	0.160	67.50	84.283
Others	$= 0$	$\neq 0$	$= 0$	$= 0$	$\neq 0$	$= 0$

a similar inequality, with all the feature rankers correctly rating the important features. FR, IG, GR, and χ^2 evaluate the remaining features as having zero significance.

In Table 8.8 all metrics apart from 1R locate the relevant features. For this dataset, 1R chooses 22 features as being the most significant, while ranking features x and y last. This may be due to the discretization process that must precede the application of 1R. If the discretization is poor, then the resulting feature evaluations will be affected.

Tables 8.9 shows the results for data classified by $x * y * z > 0.125$. All feature rankers correctly detect these variables. However, in Table 8.10 the results can be seen for the same inequality but with the impact of variable z increased. All metrics determine that z has the most influence on the decision, and almost all choose x and y next. Again, the 1R measure fails and chooses features 15, 19, and 24 instead.

Only the FR and Re metrics are applicable to datasets where the decision feature is continuous. Results from the application of these two measures to such data can be found in Appendix B. Both methods find the features that are involved in generating the decision values.

This short investigation into the utility of the new fuzzy-rough measure has shown that it is comparable with the leading measures of feature importance. Indeed its behavior is similar to that of information gain and gain ratio metrics. This is interesting as both of these measures are entropy-based. As mentioned in

TABLE 8.10 Feature evaluation for $x * y * z^2 > 0.125$

Feature	FR	Re	IG	GR	1R	χ^2
x	0.1511	0.0980	0.1451	0.0947	76.5	65.425
y	0.1101	0.0557	0.0909	0.1080	78.0	35.357
z	0.2445	0.1474	0.2266	0.2271	79.75	93.812
Others	$= 0$	$\neq 0$	$= 0$	$= 0$	$\neq 0$	$= 0$

Section 4.2.1.5, a feature subset with a maximum (crisp) rough set dependency has a corresponding entropy of 0. Unlike these metrics, the fuzzy-rough measure can also be applied to datasets containing real-valued decision features.

8.7.3 Application to Financial Data

The comparisons carried out in Section 8.7 concerning the fuzzy-rough metric used artificial data where the relevancy of features was known beforehand. To demonstrate the utility of the new measure in determining relevant features from real-world data, a financial dataset (the trust crisis dataset) was chosen for analysis. An expert was asked to define a ranking of feature importance for the data in order to compare this with the automated ranking produced by the metrics.

8.7.3.1 The Trust Crisis Dataset An investigation was carried out in [193] in an attempt to analyse the split capital investment trust crisis of 2001 and 2002. This analysis sought to examine what factors were behind the failure of a number of these trusts. Many retail investors, thinking that they were investing in low-risk financial products, were left facing substantial losses in their investments. This ultimately led to a hearing by the Treasury Select Committee. Although trust managers apportioned blame to falling equity markets, leading analysts contended that this was an accident waiting to happen.

In addition to other analyzes, the study investigated the effects of 18 different financial-related aspects (features) of the trusts to determine how these may have contributed to trust failure. It is therefore interesting to compare an expert-defined ordering of feature influence on failure with the automated ordering of feature ranking. The features involved in the analysis are given in Table 8.11, with their corresponding ordering of importance. According to the expert, the most influential features are Split & High Yielding Trusts, though Equities, Fixed Interest & Convertible Stock, and Cash are correlated with this. Bank Debt is the next most influential feature. Although an ordering is given to Management Fees, Charge to Capital, and Admin Fees, it is difficult to differentiate among them.

8.7.3.2 Results The results of the application of the feature rankers as discussed in Section 8.7 can be seen in Table 8.12. All metrics rate features 8 (Equities) and 9 (Split & High Yielding Trusts) highly. This is in agreement with the expert-supplied rank. All measures apart from FR rank 9 first, with feature

TABLE 8.11 Features in the trust crisis dataset

Feature Number	Feature Name	Expert-Defined Ordering
0	Bank Debt	2
1	Bank Interest Rate	6
2	Zero Dividend Preference Shares (ZDPs)	6
3	Income Shares	6
4	Capital Shares	6
5	Ordinary Income & Residual Capital Shares	6
6	Other Share Classes	6
7	Gross Assets	6
8	Equities	1
9	Split & High Yielding Trusts	1
10	Fixed Interest & Convertible Stock	1
11	Cash (estimate)	1
12	Debtors less Creditors	6
13	Total Dividends	6
14	Management Fees	4
15	Charge to Capital	3
16	Admin Fees	5
17	SORP Compliance	6

TABLE 8.12 Feature ranker results for the trust crisis dataset

Feature	Expert	FR	Re	IG	GR	1R	χ^2
8	1	0.205	0.130	0.146	0.150	78.5	18.2
9	1	0.169	0.183	0.309	0.363	83.9	51.0
10	1	0.158	0.036	0.0	0.0	77.6	0.0
11	1	0.109	0.040	0.0	0.0	73.2	0.0
0	2	0.102	0.013	0.0	0.0	76.7	0.0
15	3	0.0	0.036	0.0	0.0	77.6	0.0
14	4	0.0	0.020	0.0	0.0	77.6	0.0
16	5	0.0	0.003	0.0	0.0	77.6	0.0
1	6	0.062	0.062	0.0	0.0	75.0	0.0
2	6	0.099	0.012	0.0	0.0	75.8	0.0
3	6	0.0	0.009	0.0	0.0	75.8	0.0
4	6	0.0	0.008	0.0	0.0	76.7	0.0
5	6	0.023	0.014	0.0	0.0	74.1	0.0
6	6	0.0	0.004	0.0	0.0	77.6	0.0
7	6	0.0	0.006	0.0	0.0	71.4	0.0
12	6	0.007	0.007	0.0	0.0	75.8	0.0
13	6	0.088	0.005	0.0	0.0	75.0	0.0
17	6	0.031	0.085	0.001	0.001	77.6	0.126

TABLE 8.13 Spearman's rank correlation results

	FR	Re	IG	GR	1R	χ^2
ρ	0.61249	0.58566	0.58772	0.58772	0.55624	0.58772

8 second. However, only the FR metric correctly rates features 10 (Fixed Interest & Convertible Stock) and 11 (Cash) highly. After these attributes, FR ranks Bank Debt next. This is in fact the only measure to detect the importance of this feature.

Spearman's rank correlation coefficient, denoted ρ, is often used to determine the nature of a relationship between two variables. This correlation coefficient is useful here to examine the correlation of each metric ranking with that of the expert's ranking. The value ρ is calculated in the following way:

$$\rho = 1 - \frac{6\sum d^2}{n(n^2 - 1)} \tag{8.15}$$

where d is the difference between the ranks of the corresponding values of the variables, and n is the number of pairs of values. The values of ρ for the rankings produced by each metric can be found in Table 8.13. From the rankings it can be seen that all metrics exhibit a positive correlation, with the fuzzy-rough metric resulting in the strongest correlation.

To determine the significance of these values, they must be compared with the critical values of ρ with $(n - 2)$ degrees of freedom. At the 0.01, 0.02, and 0.05 levels of significance the critical values are 0.625, 0.564, and 0.475, respectively, for these data. To illustrate what this means, consider the following example. If the value of ρ for a set of data containing the same number of samples is 0.75, then it can be observed that this is larger than the critical value of ρ at the 0.01 level of significance (0.625). It could then be concluded that the obtained value of ρ is likely to occur by chance less than one time in a hundred (i.e., it is highly significant). All the metrics are above the 0.05 level, but only the 1R metric falls below the 0.02 significance level. There is less confidence that the correlation has not occurred by chance for 1R. The remainder of the metrics fall between the 0.01 and 0.02 significance levels, with the FR metric closest to the 0.01 level.

These results show that the FR metric is a useful gauge of information, producing results very much in line with the expert ranking. Out of all the methods, FR appears to produce an ordering of the features closest to the expert's.

8.8 SUMMARY

In this chapter a method for feature selection based on fuzzy-rough sets was presented. An algorithm for finding feature subsets, based on the new fuzzy-rough dependency measure, was introduced and illustrated by means of two simple examples. The examples were chosen to demonstrate the fact that FRFS can be applied to datasets containing crisp or real-valued features, or a mixture of both. Implemented optimizations were briefly discussed that can significantly improve the runtime of FRFS. This was demonstrated by comparing the execution times of the different implementations on real-world data. One real-world and several artificial datasets were also used to evaluate the utility of the fuzzy-rough measure and provide comparisons with other leading feature importance measures. The results show that the new metric presented here is slightly better than the leading measures at locating the relevant features.

FRFS can be applied to any of the domains highlighted previously where feature selection has been employed. For the purposes of this book, three challenging domains of interest were chosen to illustrate its potential utility: Web content categorization, complex systems monitoring, and algae population estimation.

The problem of Web content categorization is a significant one due to the explosive increase of information available on the Web and its increasing popularity. Techniques for automated categorization of Web documents help in the building of catalogues and facilitate the retrieval of material. In order to deal with the large number of features involved in such classification, feature selectors are typically used [309]. The dimensionality of the problem datasets can be sufficiently reduced to enable more sophisticated learning algorithms to perform their tasks. The work presented here looks specifically at addressing the issues of bookmark/favorite classification and Web page classification. FRFS reduces the size of the datasets involved by several orders of magnitude, retaining most of the information present in the datasets and making the classification task manageable.

In systems monitoring it is important to reduce the number of features involved for several reasons. First, there is an associated cost with the measurement of a feature. It is desirable simply from an expense-saving point of view to reduce the number of monitored variables. Second, the resultant transparency of the monitoring process can be improved if fewer variables are involved. Third, it is often observed that the accuracy of the monitoring system can be significantly improved using fewer variables [322]. FRFS is here applied to the water treatment plant dataset [38] as an example of how this fuzzy-rough method can be used within the systems monitoring domain. Additionally the new feature grouping and ant colony optimization-based methods are applied to this domain to show their potential utility.

Algae population estimation is another important area that benefits greatly from the use of feature selection. FRFS can be used to significantly reduce computer time and space requirements. Also it decreases the cost of obtaining measurements and increases runtime efficiency, making the system more viable economically. Through the use of fuzzy-rough sets the FRFS-based method can handle real-valued decision features (algae populations), unlike many existing approaches. The system does not alter the domain semantics, making any distilled knowledge human-readable.

CHAPTER 9

NEW DEVELOPMENTS OF FRFS

Fuzzy-rough set-based feature selection has been shown to be highly useful at reducing data dimensionality, but it possesses several problems that render it ineffective for large datasets. This chapter presents three new approaches to fuzzy-rough feature selection based on fuzzy similarity relations. In particular, a fuzzy extension to crisp discernibility matrices is proposed and utilized. Initial experimentation shows that the methods greatly reduce dimensionality while preserving classification accuracy.

9.1 INTRODUCTION

FRFS has been shown to be a highly useful technique in reducing data dimensionality [168]. However, several problems exist with the method. First, the complexity of calculating the Cartesian product of fuzzy equivalence classes becomes prohibitively high for large feature subsets. If the number of fuzzy sets per attribute is n, $n^{|R|}$ equivalence classes must be considered per attribute for feature subset R. Optimizations that attempt to alleviate this problem are given in [33,168], but the complexity is still too high. In [34], a compact computational domain is proposed to reduce the computational effort required to calculate fuzzy lower approximations for large datasets, based on some of the properties of fuzzy connectives.

Second, it was shown in [361] that in some situations the fuzzy lower approximation might not be a subset of the fuzzy upper approximation. This is undesirable from a theoretical viewpoint as it is meaningless for a lower approximation

Computational Intelligence and Feature Selection: Rough and Fuzzy Approaches, by Richard Jensen and Qiang Shen

of a concept to be larger than its upper approximation, which suggests that there is more certainty in the upper than the lower. It is also shown that the Cartesian product of fuzzy equivalence classes might not result in a family of fuzzy equivalence classes. These issues motivate the development of the techniques proposed in this chapter.

9.2 NEW FUZZY-ROUGH FEATURE SELECTION

This section presents three new techniques for fuzzy-rough feature selection based on fuzzy similarity relations [169].

9.2.1 Fuzzy Lower Approximation Based FS

The previous method for fuzzy-rough feature selection used a fuzzy partitioning of the input space in order to determine fuzzy equivalence classes. Alternative definitions for the fuzzy lower and upper approximations can be found in [282], where a T-transitive fuzzy similarity relation is used to approximate a fuzzy concept X:

$$\mu_{\underline{R_P}X}(x) = \inf_{y \in \mathbb{U}} I(\mu_{R_P}(x, y), \mu_X(y)) \tag{9.1}$$

$$\mu_{\overline{R_P}X}(x) = \sup_{y \in \mathbb{U}} T(\mu_{R_P}(x, y), \mu_X(y)) \tag{9.2}$$

Here I is a fuzzy implicator and T a t-norm. R_P is the fuzzy similarity relation induced by the subset of features P:

$$\mu_{R_P}(x, y) = \bigcap_{a \in P} \{\mu_{R_a}(x, y)\} \tag{9.3}$$

$\mu_{R_a}(x, y)$ is the degree to which objects x and y are similar for feature a. Many fuzzy similarity relations can be constructed for this purpose, for example,

$$\mu_{R_a}(x, y) = 1 - \frac{|a(x) - a(y)|}{|a_{\max} - a_{\min}|} \tag{9.4}$$

$$\mu_{R_a}(x, y) = \exp\left(-\frac{(a(x) - a(y))^2}{2{\sigma_a}^2}\right) \tag{9.5}$$

$$\mu_{R_a}(x, y) = \max\left(\min\left(\frac{(a(y) - (a(x) - \sigma_a))}{(a(x) - (a(x) - \sigma_a))},\right.\right.$$
$$\left.\left. \times \frac{((a(x) + \sigma_a) - a(y))}{((a(x) + \sigma_a) - a(x))}, 0\right)\right) \tag{9.6}$$

where ${\sigma_a}^2$ is the variance of feature a. As these relations do not necessarily display T-transitivity, the fuzzy transitive closure must be computed for each attribute [74]. The combination of feature relations in equation (9.3) has been shown to preserve T-transitivity [369].

9.2.1.1 *Reduction* In a similar way to the original FRFS approach, the fuzzy positive region can be defined as

$$\mu_{POS_{R_P}(Q)}(x) = \sup_{X \in \mathbb{U}/Q} \mu_{\underline{R_P}X}(x) \tag{9.7}$$

The resulting degree of dependency is

$$\gamma'_P(Q) = \frac{\sum_{x \in \mathbb{U}} \mu_{POS_{R_P}(Q)}(x)}{|\mathbb{U}|} \tag{9.8}$$

A fuzzy-rough reduct R can be defined as a subset of features that preserves the dependency degree of the entire dataset, that is, $\gamma'_R(\mathbb{D}) = \gamma'_{\mathbb{C}}(\mathbb{D})$. Based on this concept, a new fuzzy-rough QUICKREDUCT algorithm can be constructed that operates in the same way as Figure 8.1 but uses equation (9.8) to gauge subset quality. A proof of the monotonicity of the dependency function can be found in Section 9.4. Core features may be determined by considering the change in dependency of the full set of conditional features when individual attributes are removed:

$$Core(\mathbb{C}) = \{a \in \mathbb{C} | \gamma'_{\mathbb{C}-\{a\}}(Q) \le \gamma'_{\mathbb{C}}(Q)\} \tag{9.9}$$

9.2.1.2 *Example* The fuzzy connectives chosen for this example (and all others in this section) are the Łukasiewicz t-norm ($\max(x + y - 1, 0)$) and the Łukasiewicz fuzzy implicator ($\min(1 - x + y, 1)$). As recommended in [74], the Łukasiewicz t-norm is used as this produces fuzzy T-equivalence relations dual to that of a pseudo-metric. The use of the Łukasiewicz fuzzy implicator is also recommended as it is both a residual and S-implicator.

Using the fuzzy similarity measure defined in (9.6), the resulting relations are as follows for each feature in the dataset:

$$R_a(x, y) = \begin{pmatrix} 1.0 & 1.0 & 0.699 & 0.0 & 0.0 & 0.0 \\ 1.0 & 1.0 & 0.699 & 0.0 & 0.0 & 0.0 \\ 0.699 & 0.699 & 1.0 & 0.0 & 0.0 & 0.0 \\ 0.0 & 0.0 & 0.0 & 1.0 & 0.699 & 0.699 \\ 0.0 & 0.0 & 0.0 & 0.699 & 1.0 & 1.0 \\ 0.0 & 0.0 & 0.0 & 0.699 & 1.0 & 1.0 \end{pmatrix}$$

$$R_b(x,y) = \begin{pmatrix} 1.0 & 0.0 & 0.568 & 1.0 & 1.0 & 0.0 \\ 0.0 & 1.0 & 0.0 & 0.0 & 0.0 & 0.137 \\ 0.568 & 0.0 & 1.0 & 0.568 & 0.568 & 0.0 \\ 1.0 & 0.0 & 0.568 & 1.0 & 1.0 & 0.0 \\ 1.0 & 0.0 & 0.568 & 1.0 & 1.0 & 0.0 \\ 0.0 & 0.137 & 0.0 & 0.0 & 0.0 & 1.0 \end{pmatrix}$$

$$R_c(x,y) = \begin{pmatrix} 1.0 & 0.0 & 0.036 & 0.0 & 0.0 & 0.0 \\ 0.0 & 1.0 & 0.036 & 0.518 & 0.518 & 0.518 \\ 0.036 & 0.036 & 1.0 & 0.0 & 0.0 & 0.0 \\ 0.0 & 0.518 & 0.0 & 1.0 & 1.0 & 1.0 \\ 0.0 & 0.518 & 0.0 & 1.0 & 1.0 & 1.0 \\ 0.0 & 0.518 & 0.0 & 1.0 & 1.0 & 1.0 \end{pmatrix}$$

Again, the first step is to compute the lower approximations of each concept for each feature. Consider feature a and the decision concept $\{1, 3, 6\}$ in the example dataset:

$$\mu_{\underline{R_a}\{1,3,6\}}(x) = \inf_{y \in \mathbb{U}} I(\mu_{R_a}(x, y), \mu_{\{1,3,6\}}(y))$$

For object 3, the lower approximation is

$$\begin{aligned} \mu_{\underline{R_a}\{1,3,6\}}(3) &= \inf_{y \in \mathbb{U}} I(\mu_{R_a}(3, y), \mu_{\{1,3,6\}}(y)) \\ &= \inf\{I(0.699, 1), I(0.699, 0), I(1, 1), I(0, 0), I(0, 0), I(0, 1)\} \\ &= 0.301 \end{aligned}$$

For the remaining objects, the lower approximations are

$$\mu_{\underline{R_a}\{1,3,6\}}(1) = 0.0$$
$$\mu_{\underline{R_a}\{1,3,6\}}(2) = 0.0$$
$$\mu_{\underline{R_a}\{1,3,6\}}(4) = 0.0$$
$$\mu_{\underline{R_a}\{1,3,6\}}(5) = 0.0$$
$$\mu_{\underline{R_a}\{1,3,6\}}(6) = 0.0$$

For concept $\{2, 4, 5\}$, the lower approximations are

$$\mu_{\underline{R_a}\{2,4,5\}}(1) = 0.0$$
$$\mu_{\underline{R_a}\{2,4,5\}}(2) = 0.0$$
$$\mu_{\underline{R_a}\{2,4,5\}}(3) = 0.0$$
$$\mu_{\underline{R_a}\{2,4,5\}}(4) = 0.301$$

$$\mu_{\underline{R_a}\{2,4,5\}}(5) = 0.0$$
$$\mu_{\underline{R_a}\{2,4,5\}}(6) = 0.0$$

Hence the positive regions for each object are

$$\mu_{POS_{R_a}(Q)}(1) = 0.0$$
$$\mu_{POS_{R_a}(Q)}(2) = 0.0$$
$$\mu_{POS_{R_a}(Q)}(3) = 0.301$$
$$\mu_{POS_{R_a}(Q)}(4) = 0.301$$
$$\mu_{POS_{R_a}(Q)}(5) = 0.0$$
$$\mu_{POS_{R_a}(Q)}(6) = 0.0$$

The resulting degree of dependency is therefore

$$\begin{aligned}\gamma'_{\{a\}}(Q) &= \frac{\sum_{x \in \mathbb{U}} \mu_{POS_{R_a}(Q)}(x)}{|\mathbb{U}|} \\ &= \frac{0.602}{6} \\ &= 0.1003\end{aligned}$$

Calculating the dependency degrees for the remaining features results in

$$\gamma'_{\{b\}}(Q) = 0.3597 \quad \gamma'_{\{c\}}(Q) = 0.4078$$

As feature c results in the largest increase in dependency degree, this feature is selected and added to the reduct candidate. The algorithm then evaluates the addition of all remaining features to this candidate. Fuzzy similarity relations are combined using (9.3). This produces the following evaluations:

$$\gamma'_{\{a,c\}}(Q) = 0.5501 \quad \gamma'_{\{b,c\}}(Q) = 1.0$$

Feature subset $\{b, c\}$ produces the maximum dependency value for this dataset, and the algorithm terminates. The dataset can now be reduced to these features only. The complexity of the algorithm is the same as that of FRFS in terms of the number of dependency evaluations. However, the explosive growth of the number of considered fuzzy equivalence classes is avoided through the use of fuzzy similarity relations and (9.3). This ensures that for one subset, only one fuzzy similarity relation is used to compute the fuzzy lower approximation.

9.2.2 Fuzzy Boundary Region Based FS

Most approaches to crisp rough set FS and all approaches to fuzzy-rough FS use only the lower approximation for the evaluation of feature subsets. The lower approximation contains information regarding the extent of certainty of object membership to a given concept. However, the upper approximation contains information regarding the degree of uncertainty of objects, and hence this information can be used to discriminate between subsets. For example, two subsets may result in the same lower approximation but one subset may produce a smaller upper approximation. This subset will be more useful as there is less uncertainty concerning objects within the boundary region (the difference between upper and lower approximations). The fuzzy-rough boundary region for a fuzzy concept X may thus be defined:

$$\mu_{BND_{R_P}(X)}(x) = \mu_{\overline{R_P}X}(x) - \mu_{\underline{R_P}X}(x) \tag{9.10}$$

The fuzzy-rough negative region for all decision concepts can be defined as follows:

$$\mu_{NEG_{R_P}}(x) = N(\sup_{X\in\mathbb{U}/Q} \mu_{\overline{R_P}X}(x)) \tag{9.11}$$

In classical rough set theory, the negative region is always empty for partitions [402]. It is interesting to note that the fuzzy-rough negative region is also always empty when the decisions are crisp. However, this is not necessarily the case when decisions are fuzzy. Further details can be found in Section 9.4.

9.2.2.1 *Reduction* As the search for an optimal subset progresses, the object memberships to the boundary region for each concept diminishes until a minimum is achieved. For crisp rough set FS, the boundary region will be zero for each concept when a reduct is found. This may not necessarily be the case for fuzzy-rough FS due to the additional uncertainty involved. The uncertainty for a concept X using features in P can be calculated as follows:

$$U_P(X) = \frac{\sum_{x\in\mathbb{U}} \mu_{BND_{R_P}(X)}(x)}{|\mathbb{U}|} \tag{9.12}$$

This is the average extent to which objects belong to the fuzzy boundary region for the concept X. The total uncertainty degree for all concepts, given a feature subset P is defined as

$$\lambda_P(Q) = \frac{\sum_{X\in\mathbb{U}/Q} U_P(X)}{|\mathbb{U}/Q|} \tag{9.13}$$

This is related to the conditional entropy measure, which considers a combination of conditional probabilities $H(Q|P)$ in order to gauge the uncertainty present using features in P. In the crisp case the minimization of this measure can be used to discover reducts: if the entropy for a feature subset P is zero, then the subset is a reduct [163].

In the method above, the overall uncertainty is evaluated via averaging the uncertainty of all decision concepts. The uncertainty for a concept is itself an average of the belongingness of objects to the fuzzy boundary region. A better way of measuring the uncertainty within the boundary region of a concept X is to calculate the fuzzy entropy:

$$U'_P(X) = \sum_{x \in \mathbb{U}} - \frac{\mu_{BND_{R_P}(X)}(x)}{|BND_{R_P}(X)|} \log_2 \frac{\mu_{BND_{R_P}(X)}(x)}{|BND_{R_P}(X)|} \tag{9.14}$$

$$\lambda'_P(\mathbb{D}) = \frac{\sum_{D_j \in R_{\mathbb{D}}} U'_P(D_j)}{\sum_{D_n \in R_{\mathbb{D}}} (|D_n|)^{-1}} \tag{9.15}$$

This will be minimized when all fuzzy boundary regions are empty; hence $\lambda'_P(\mathbb{D}) = \lambda_P(\mathbb{D}) = 0$ and therefore P must be a fuzzy-rough reduct.

Again, a QUICKREDUCT-style algorithm can be constructed for locating fuzzy-rough reducts based on this measure. Instead of maximizing the dependency degree, the task of the algorithm is to minimize the total uncertainty degree. When this reaches the minimum for the dataset, a fuzzy-rough reduct has been found. A proof of the monotonicity of the total uncertainty degree can be found in Section 9.4.

9.2.2.2 *Example* To determine the fuzzy boundary region, the lower and upper approximations of each concept for each feature must be calculated. Consider feature a and concept $\{1, 3, 6\}$:

$$\mu_{BND_{R_a}(\{1,3,6\})}(x) = \mu_{\overline{R_a}\{1,3,6\}}(x) - \mu_{\underline{R_a}\{1,3,6\}}(x)$$

For object 4, the boundary region is

$$\begin{aligned} \mu_{BND_{R_a}(\{1,3,6\})}(4) &= \sup_{y \in \mathbb{U}} T(\mu_{R_a}(4, y), \mu_{\{1,3,6\}}(y)) \\ &\quad - \inf_{y \in \mathbb{U}} I(\mu_{R_a}(4, y), \mu_{\{1,3,6\}}(y)) \\ &= 0.699 - 0.0 \\ &= 0.699 \end{aligned}$$

For the remaining objects, the boundary regions are

$$\mu_{BND_{R_a}(\{1,3,6\})}(1) = 1.0$$
$$\mu_{BND_{R_a}(\{1,3,6\})}(2) = 1.0$$

$$\mu_{BND_{R_a}(\{1,3,6\})}(3) = 0.699$$
$$\mu_{BND_{R_a}(\{1,3,6\})}(5) = 1.0$$
$$\mu_{BND_{R_a}(\{1,3,6\})}(6) = 1.0$$

Hence the uncertainty for concept {1, 3, 6} is

$$\begin{aligned} U_a(\{1,3,6\}) &= \frac{\sum_{x\in\mathbb{U}} \mu_{BND_{R_a}(\{1,3,6\})}(x)}{|\mathbb{U}|} \\ &= \frac{1.0+1.0+0.699+0.699+1.0+1.0}{6} \\ &= 0.899 \end{aligned}$$

For concept {2, 4, 5}, the uncertainty is

$$\begin{aligned} U_a(\{2,4,5\}) &= \frac{\sum_{x\in\mathbb{U}} \mu_{BND_{R_a}(\{2,4,5\})}(x)}{|\mathbb{U}|} \\ &= \frac{1.0+1.0+0.699+0.699+1.0+1.0}{6} \\ &= 0.899 \end{aligned}$$

From this, the total uncertainty for feature a is calculated as follows:

$$\begin{aligned} \lambda_a(Q) &= \frac{\sum_{X\in\mathbb{U}/Q} U_a(X)}{|\mathbb{U}/Q|} \\ &= \frac{0.899+0.899}{2} \\ &= 0.899 \end{aligned} \tag{9.16}$$

The values of the total uncertainty for the remaining features are

$$\lambda_{\{b\}}(Q) = 0.640 \ \lambda_{\{c\}}(Q) = 0.592$$

As feature c results in the smallest total uncertainty, it is chosen and added to the reduct candidate. The algorithm then considers the addition of the remaining features to the subset:

$$\lambda_{\{a,c\}}(Q) = 0.500 \quad \lambda_{\{b,c\}}(Q) = 0.0$$

The subset $\{b, c\}$ results in the minimal uncertainty for the dataset, and the algorithm terminates. This is the same subset as that chosen by the fuzzy lower approximation-based method above. Again, the complexity of the algorithm is the same as that of FRFS but avoids the Cartesian product of fuzzy equivalence classes. However, for each evaluation both the fuzzy lower and upper approximations are considered, and hence the calculation of the fuzzy boundary region is more costly than that of the fuzzy lower approximation alone.

9.2.3 Fuzzy-Rough Reduction with Fuzzy Entropy

Fuzzy entropy itself can be used to find fuzzy-rough reducts. A subset $P \subseteq \mathbb{C}$ induces a fuzzy similarity relation (R_P) with corresponding foresets $F_1, F_2, \ldots, F_n$. Similarly, the foresets induced by the (fuzzy) decision feature $\mathbb{D}$ are $D_1, D_2, \ldots, D_n$. The fuzzy entropy for a foreset F_i can be defined as

$$H(F_i) = \sum_{D_j \in R_{\mathbb{D}}} \frac{-p(D_j|F_i)\log_2 p(D_j|F_i)}{|D_j|} \tag{9.17}$$

where, $p(D_j|F_i)$ is the relative frequency of foreset F_i with respect to the decision D_j, and is defined

$$p(D_j|F_i) = \frac{|D_j \cap F_i|}{|F_i|} \tag{9.18}$$

Based on these definitions, the fuzzy entropy for an attribute subset P is defined as follows:

$$E(P) = \sum_{F_i \in R_P} \frac{|F_i|}{\sum_{Y_i \in R_P} |Y_i|} H(F_i) \tag{9.19}$$

This fuzzy entropy is monotonic and can be used to gauge the utility of feature subsets in a similar way to that of the fuzzy-rough measure. By dividing the entropy by $\log_2(\sum_{D_n \in R_{\mathbb{D}}}(|D_n|)^{-1})$, the measure will be normalized. This can be integrated into a QUICKREDUCT-style algorithm, employing a greedy hill-climbing approach. Again, as the measure monotonically decreases with addition of features, the search algorithm seeks to minimize this value in a manner similar to the boundary region minimization approach.

9.2.3.1 Example Returning to the example dataset, the fuzzy entropy measure is used to determine fuzzy-rough reducts. The algorithm begins with an empty subset, and considers the addition of individual features. The attribute that results in the greatest decrease in fuzzy entropy will ultimately be added to the reduct candidate. For attribute a, the fuzzy entropy is calculated as follows $(A = \{a\})$:

$$E(A) = \sum_{F_i \in R_A} \frac{|F_i|}{\sum_{Y_i \in R_A} |Y_i|} H(F_i)$$

Each foreset F_i corresponds to one row in the matrix R_A:

F_1	1.0	1.0	0.699	0.0	0.0	0.0
F_2	1.0	1.0	0.699	0.0	0.0	0.0
F_3	0.699	0.699	1.0	0.0	0.0	0.0
F_4	0.0	0.0	0.0	1.0	0.699	0.699
F_5	0.0	0.0	0.0	0.699	1.0	1.0
F_6	0.0	0.0	0.0	0.699	1.0	1.0

Considering F_1, H(F_1) must be calculated:

$$H(F_1) = \sum_{D_j \in R_{\mathbb{D}}} \frac{-p(D_j|F_1)\log_2 p(D_j|F_1)}{|D_j|}$$

Each foreset D_j corresponds to one row in the matrix $R_{\mathbb{D}}$:

D_1	1.0	0.0	1.0	0.0	0.0	1.0
D_2	0.0	1.0	0.0	1.0	1.0	0.0
D_3	1.0	0.0	1.0	0.0	0.0	1.0
D_4	0.0	1.0	0.0	1.0	1.0	0.0
D_5	0.0	1.0	0.0	1.0	1.0	0.0
D_6	1.0	0.0	1.0	0.0	0.0	1.0

For D_1,

$$\frac{-p(D_1|F_1)\log_2 p(D_1|F_1)}{|D_1|} = \frac{-(1.699/2.699)\log_2(1.699/2.699)}{3.0}$$

Calculating this for each D_j produces

$$H(F_1) = 0.140 + 0.177 + 0.140 + 0.177 + 0.177 + 0.140 = 0.951$$

The procedure is repeated for each remaining foreset:

$$H(F_2) = 0.951,\ H(F_3) = 0.871,\ H(F_4) = 0.871,$$
$$H(F_5) = 0.951,\ H(F_6) = 0.951$$

Hence the fuzzy entropy is

$$E(A) = \sum_{F_i \in R_A} \frac{|F_i|}{\sum_{Y_i \in R_A} |Y_i|} H(F_i)$$
$$= 0.926$$

Repeating this process for the remaining attributes gives

$$E(B) = 0.921$$
$$E(C) = 0.738$$

From this it can be seen that attribute c will cause the greatest decrease in fuzzy entropy. This attribute is chosen and added to the potential reduct, $R \leftarrow R \cup \{c\}$. The process iterates and the two fuzzy entropy values calculated are

$$E(\{a, c\}) = 0.669$$
$$E(\{b, c\}) = 0.0$$

Adding attribute b to the reduct candidate results in the minimum entropy for the data, and the search terminates, outputting the subset $\{b, c\}$. The dataset can now be reduced to only those attributes appearing in the reduct.

9.2.4 Fuzzy-Rough Reduction with Fuzzy Gain Ratio

The information gain (IG) [281] is the expected reduction in entropy resulting from partitioning the dataset objects according to a particular feature. For the fuzzy case this can be expressed as

$$IG(P \cup \{a\}) = E(P) - E(P \cup \{a\}) \tag{9.20}$$

One limitation of the IG measure is that it favors features with many values. The gain ratio (GR) seeks to avoid this bias by incorporating another term, split information, that is sensitive to how broadly and uniformly the attribute splits the considered data. Again, for the fuzzy case this can be expressed as

$$SP(Q) = \sum_{F_i \in R_Q} \frac{|F_i|}{\sum_{Y_i \in R_Q} |Y_i|} \log_2 \frac{|F_i|}{\sum_{Y_i \in R_Q} |Y_i|} \tag{9.21}$$

The gain ratio is then defined as follows:

$$GR(P \cup \{a\}) = \frac{IG(P \cup \{a\})}{SP(P \cup \{a\})} \tag{9.22}$$

When this is minimized, $P \cup \{a\}$ is a fuzzy-rough reduct due to the monotonicity of the fuzzy entropy measure.

9.2.5 Fuzzy Discernibility Matrix Based FS

As mentioned previously, there are two main branches of research in crisp rough set-based FS: those based on the dependency degree and those based on discernibility matrices. The developments given above are solely concerned with the extension of the dependency degree to the fuzzy-rough case. Hence methods constructed based on the crisp dependency degree can be employed for fuzzy-rough FS.

By extending the discernibility matrix to the fuzzy case, it is possible to employ approaches similar to those in crisp rough set FS to determine fuzzy-rough reducts. A first step toward this is presented in [361,378] where a crisp discernibility matrix is constructed for fuzzy-rough selection. A threshold is used, breaking the rough set ideology, which determines which features are to appear in the matrix entries. However, information is lost in this process as membership degrees are not considered. Search based on the crisp discernibility may result in reducts that are not true fuzzy-rough reducts.

9.2.5.1 *Fuzzy Discernibility* The approach presented here extends the crisp discernibility matrix by employing fuzzy clauses. Each entry in the fuzzy discernibility matrix is a fuzzy set, to which every feature belongs to a certain degree. The extent to which a feature a belongs to the fuzzy clause C_{ij} is determined by the fuzzy discernibility measure:

$$\mu_{C_{ij}}(a) = N(\mu_{R_a}(i, j)) \tag{9.23}$$

where N denotes fuzzy negation and $\mu_{R_a}(i, j)$ is the fuzzy similarity of objects i and j, and hence $\mu_{C_{ij}}(a)$ is a measure of the fuzzy discernibility. For the crisp case, if $\mu_{C_{ij}}(a) = 1$, then the two objects are distinct for this feature; if $\mu_{C_{ij}}(a) = 0$, the two objects are identical. For fuzzy cases where $\mu_{C_{ij}}(a) \in (0, 1)$, the objects are partly discernible. (The choice of fuzzy similarity relation must be identical to that of the fuzzy-rough dependency degree approach to find corresponding reducts.) Each entry in the fuzzy indiscernibility matrix is then a set of attributes and their corresponding memberships:

$$C_{ij} = \{a_x | a \in \mathbb{C}, x = N(\mu_{R_a}(i, j))\} i, j = 1, \ldots, |\mathbb{U}| \tag{9.24}$$

For example, an entry C_{ij} in the fuzzy discernibility matrix might be

$$C_{ij} : \{a_{0.4}, b_{0.8}, c_{0.2}, d_{0.0}\}$$

This denotes that $\mu_{C_{ij}}(a) = 0.4$, $\mu_{C_{ij}}(b) = 0.8$, and so forth. In crisp discernibility matrices these values are either 0 or 1 as the underlying relation is an equivalence relation. The example clause can be viewed as indicating the value of each feature—the extent to which the feature discriminates between the two objects i and j. The core of the dataset is defined as

$$Core(\mathbb{C}) = \{a \in \mathbb{C} | \exists C_{ij}, \mu_{C_{ij}}(a) > 0, \forall f \in \{\mathbb{C} - a\} \mu_{C_{ij}}(f) = 0\} \quad (9.25)$$

9.2.5.2 *Fuzzy Discernibility Function* As with the crisp approach, the entries in the matrix can be used to construct the fuzzy discernibility function:

$$f_D(a_1^*, \ldots, a_m^*) = \wedge\{\vee C_{ij}^* | 1 \leq j < i \leq |\mathbb{U}|\} \quad (9.26)$$

where $C_{ij}^* = \{a_x^* | a_x \in C_{ij}\}$. The function returns values in [0, 1], which can be seen to be a measure of the extent to which the function is satisfied for a given assignment of truth values to variables. To discover reducts from the fuzzy discernibility function, the task is to find the minimal assignment of the value 1 to the variables such that the formula is maximally satisfied. By setting all variables to 1, the maximal value for the function can be obtained as this provides the most discernibility between objects.

Crisp discernibility matrices can be simplified by removing duplicate entries and clauses that are supersets of others. A similar degree of simplification can be achieved for fuzzy discernibility matrices. Duplicate clauses can be removed as a subset that satisfies one clause to a certain degree will always satisfy the other to the same degree.

9.2.5.3 *Decision-Relative Fuzzy Discernibility Matrix* As with the crisp discernibility matrix, for a decision system the decision feature must be taken into account for achieving reductions; only those clauses with different decision values are included in the crisp discernibility matrix. For the fuzzy version, this is encoded as

$$f_D(a_1^*, \ldots, a_m^*) = \{\wedge\{\{\vee C_{ij}^*\} \leftarrow q_{N(\mu_{R_q}(i,j))}\} | 1 \leq j < i \leq |\mathbb{U}|\} \quad (9.27)$$

for decision feature q, where $\leftarrow$ denotes fuzzy implication. This construction allows the extent to which decision values differ to affect the overall satisfiability of the clause. If $\mu_{C_{ij}}(q) = 1$ then this clause provides maximum discernibility (i.e., the two objects are maximally different according to the fuzzy similarity measure). When the decision is crisp and crisp equivalence is used, $\mu_{C_{ij}}(q)$ becomes 0 or 1.

9.2.5.4 *Reduction* For the purposes of finding reducts, use of the fuzzy intersection of all clauses in the fuzzy discernibility function may not provide enough information for evaluating subsets. Here it may be more informative to consider the individual satisfaction of each clause for a given set of features. The degree of satisfaction of a clause C_{ij} for a subset of features P is defined as

$$SAT_P(C_{ij}) = \bigcup_{a \in P} \{\mu_{C_{ij}}(a)\} \tag{9.28}$$

Returning to the example, if the subset $P = \{a, c\}$ is chosen, the resulting degree of satisfaction of the clause is

$$SAT_P(C_{ij}) = \{0.4 \vee 0.2\} = 0.6$$

using the Łukasiewicz t-conorm, $\min(1, x + y)$.

For the decision-relative fuzzy indiscernibility matrix, the decision feature q must be taken into account also:

$$SAT_{P,q}(C_{ij}) = SAT_P(C_{ij}) \leftarrow \mu_{C_{ij}}(q) \tag{9.29}$$

For the example clause, if the corresponding decision values are crisp and are different, the degree of satisfaction of the clause is

$$\begin{aligned} SAT_{P,q}(C_{ij}) &= SAT_P(C_{ij}) \leftarrow 1 \\ &= 0.6 \leftarrow 1 \\ &= 0.6 \end{aligned}$$

For a subset P, the total satisfiability of all clauses can be calculated as

$$SAT(P) = \frac{\sum_{i,j \in \mathbb{U}, i \neq j} SAT_{P,q}(C_{ij})}{\sum_{i,j \in \mathbb{U}, i \neq j} SAT_{\mathbb{C},q}(C_{ij})} \tag{9.30}$$

where $\mathbb{C}$ is the full set of conditional attributes, and hence the denominator is a normalizing factor. If this value reaches 1 for a subset P, then the subset is a fuzzy-rough reduct. A proof of the monotonicity of the function $SAT(P)$ can be found in Section 9.4.

Many methods available from the literature for the purpose of finding reducts for crisp discernibility matrices are applicable here also. The Johnson reducer [253] is extended and used herein to illustrate the concepts involved. This is a simple greedy heuristic algorithm that is often applied to discernibility functions to find a single reduct. Subsets of features found by this process have no guarantee of minimality, but are generally of a size close to the minimal.

The algorithm begins by setting the current reduct candidate, P, to the empty set. Then each conditional feature appearing in the discernibility function is evaluated according to the heuristic measure used. For the standard Johnson algorithm this is typically a count of the number of appearances a feature makes within clauses; features that appear more frequently are considered to be more significant. The feature with the highest heuristic value is added to the reduct candidate, and all clauses in the discernibility function containing this feature are removed. As soon as all clauses have been removed, the algorithm terminates and returns the subset P. P is assured to be a fuzzy-rough reduct as all clauses contained within the discernibility function have been addressed. However, as with the other approaches, the subset may not necessarily have minimal cardinality.

The complexity of the algorithm is the same as that of FRFS in that $O((n^2 + n)/2)$ calculations of the evaluation function ($SAT(P)$) are performed in the worst case. Additionally this approach requires the construction of the fuzzy discernibility matrix, which has a complexity of $O(a * o^2)$ for a dataset containing a attributes and o objects.

9.2.5.5 *Example* For the example dataset, the fuzzy discernibility matrix needs to be constructed based on the fuzzy discernibility given in equation (9.23) using the standard negator, and fuzzy similarity in equation (9.6). For objects 2 and 3 the resulting fuzzy clause is

$$\{a_{0.301} \vee b_{1.0} \vee c_{0.964}\} \leftarrow q_{1.0}$$

where $\leftarrow$ denotes fuzzy implication. The fuzzy discernibility of objects 2 and 3 for attribute a is 0.301, indicating that the objects are partly discernible for this feature. The objects are fully discernible with respect to the decision feature, indicated by $q_{1.0}$. The full set of clauses is

$$\begin{aligned}
C_{12}: &\ \{a_{0.0} \vee b_{1.0} \vee c_{1.0}\} \leftarrow q_{1.0}\\
C_{13}: &\ \{a_{0.301} \vee b_{0.432} \vee c_{0.964}\} \leftarrow q_{0.0}\\
C_{14}: &\ \{a_{1.0} \vee b_{0.0} \vee c_{1.0}\} \leftarrow q_{1.0}\\
C_{15}: &\ \{a_{1.0} \vee b_{0.0} \vee c_{1.0}\} \leftarrow q_{1.0}\\
C_{16}: &\ \{a_{1.0} \vee b_{1.0} \vee c_{1.0}\} \leftarrow q_{0.0}\\
C_{23}: &\ \{a_{0.301} \vee b_{1.0} \vee c_{0.964}\} \leftarrow q_{1.0}\\
C_{24}: &\ \{a_{1.0} \vee b_{1.0} \vee c_{0.482}\} \leftarrow q_{0.0}\\
C_{25}: &\ \{a_{1.0} \vee b_{1.0} \vee c_{0.482}\} \leftarrow q_{0.0}\\
C_{26}: &\ \{a_{1.0} \vee b_{0.863} \vee c_{0.482}\} \leftarrow q_{1.0}\\
C_{34}: &\ \{a_{1.0} \vee b_{0.431} \vee c_{1.0}\} \leftarrow q_{1.0}\\
C_{35}: &\ \{a_{1.0} \vee b_{0.431} \vee c_{1.0}\} \leftarrow q_{1.0}\\
C_{36}: &\ \{a_{1.0} \vee b_{1.0} \vee c_{1.0}\} \leftarrow q_{0.0}\\
C_{45}: &\ \{a_{0.301} \vee b_{0.0} \vee c_{0.0}\} \leftarrow q_{0.0}\\
C_{46}: &\ \{a_{0.301} \vee b_{1.0} \vee c_{0.0}\} \leftarrow q_{1.0}\\
C_{56}: &\ \{a_{0.0} \vee b_{1.0} \vee c_{0.0}\} \leftarrow q_{1.0}
\end{aligned}$$

The feature selection algorithm then proceeds in the following way. Each individual feature is evaluated according to the measure defined in equation (9.30). For feature a, this is

$$\begin{aligned} SAT(\{a\}) &= \frac{\sum_{i,j\in\mathbb{U},i\neq j} SAT_{\{a\},q}(C_{ij})}{\sum_{i,j\in\mathbb{U},i\neq j} SAT_{\mathbb{C},q}(C_{ij})} \\ &= \frac{11.601}{15} \\ &= 0.773 \end{aligned}$$

Similarly for the remaining features,

$$SAT(\{b\}) = 0.782 \quad SAT(\{c\}) = 0.830$$

The feature that produces the largest increase in satisfiability is c. This feature is added to the reduct candidate, and the search continues:

$$SAT(\{a, c\}) = 0.887 \quad SAT(\{b, c\}) = 1.0$$

The subset $\{b, c\}$ is found to satisfy all clauses maximally, and the algorithm terminates. This subset is a fuzzy-rough reduct.

9.2.6 Vaguely Quantified Rough Sets (VQRS)

Equations (9.1) and (9.2) have been conceived with the purpose of conserving the traditional lower and upper approximations in mind. Indeed, when X and R_P are both crisp, it can be verified that the original crisp definitions are recovered. Note, in particular, how the inf and sup operations play the same role as the $\forall$ and $\exists$ quantifiers, and how a change in a single element can thus have a large impact on (9.1) and (9.2). This makes fuzzy-rough sets equally susceptible to noisy data (which is difficult to rule out in real-life applications) as their crisp counterparts.

To make up for this shortcoming, [64,65] proposed to soften the universal and existential quantifier by means of vague quantifiers like *most* and *some*. Mathematically they modeled such vague quantifiers in terms of Zadeh's [410] notion of a regularly increasing fuzzy quantifier Q: an increasing $[0, 1] \rightarrow [0, 1]$ mapping that satisfies the boundary conditions $Q(0) = 0$ and $Q(1) = 1$.

Examples of fuzzy quantifiers can be generated by means of the following parametrized formula, for $0 \leq \alpha < \beta \leq 1$, and x in $[0, 1]$,

$$Q_{(\alpha,\beta)}(x) = \begin{cases} 0, & x \leq \alpha \\ \frac{2(x-\alpha)^2}{(\beta-\alpha)^2}, & \alpha \leq x \leq \frac{\alpha+\beta}{2} \\ 1 - \frac{2(x-\beta)^2}{(\beta-\alpha)^2}, & \frac{\alpha+\beta}{2} \leq x \leq \beta \\ 1, & \beta \leq x \end{cases} \tag{9.31}$$

For instance, $Q_{(0.1,0.6)}$ and $Q_{(0.2,1)}$ might be used respectively to reflect the vague quantifiers *some* and *most* from natural language.

Once a couple (Q_l, Q_u) of fuzzy quantifiers is fixed, the Q_l-upper and Q_u-lower approximation of a fuzzy set A under a fuzzy relation R are defined by

$$\mu_{\underline{R_P X}}^{Q_u}(y) = Q_u\left(\frac{|R_P y \cap X|}{|R_P y|}\right) \tag{9.32}$$

$$\mu_{\overline{R_P X}}^{Q_l}(y) = Q_l\left(\frac{|R_P y \cap X|}{|R_P y|}\right) \tag{9.33}$$

for all y in $\mathbb{U}$. In other words, an element y belongs to the lower approximation of X if most of the elements related to y are included in X. Likewise an element belongs to the upper approximation of X if some of the elements related to y are included in X. Remark that when X and R_P are a crisp set and a crisp equivalence relation, respectively, the approximations may still be noncrisp. In this case, note also that when

$$Q_{\rangle x_l}(x) = \begin{cases} 0, & x \le x_l \\ 1, & x > x_l \end{cases} \qquad Q_{\ge x_u}(x) = \begin{cases} 0, & x < x_u \\ 1, & x \ge x_u \end{cases}$$

with $0 \le x_l < x_u \le 1$ are used as quantifiers, Ziarko's variable precision rough set (VPRS) model is recovered, and moreover when the following is used,

$$Q_{\exists}(x) = \begin{cases} 0, & x = 0 \\ 1, & x > 0 \end{cases} \qquad Q_{\forall}(x) = \begin{cases} 0, & x < 1 \\ 1, & x = 1 \end{cases}$$

Pawlak's standard rough set model is obtained as a particular case of the VQRS approach.

As such, the VQRS model puts dealing with noisy data into an interesting new perspective: it inherits both the flexibility of VPRSs for dealing with classification errors (by relaxing the membership conditions for the lower approximation, and tightening those for the upper approximation) and that of fuzzy sets for expressing partial constraint satisfaction (by distinguishing different levels of membership to the upper/lower approximation).

The VQRS-based definition of positive region is as follows:

$$POS_{R_P}^{Q_u}(y) = \sup_{X \in \mathbb{U}} \mu_{\underline{R_P X}}^{Q_u}(y) \tag{9.34}$$

Hence the VQRS degree of dependency of d on B, $\gamma_B^{Q_u}$, can be defined analogously as in equation (9.8) but is nonmonotonic due to the extra noise tolerance. The nonmonotonicity of the dependency degree, which occurs also in the VPRS approach, generates a number of complications that are both of theoretical and practical concern. First, it can occur that $POS_B^{Q_u} \not\subseteq POS_{\mathbb{C}}^{Q_u}$. For such a subset B, computing $\gamma_B^{Q_u}$ as in equation (9.8) results in a dependency degree that is strictly greater than 1. Note that this problem does not occur when the decision system is consistent. Nonmonotonicity also restricts the effectiveness of heuristic algorithms like (fuzzy-rough) QUICKREDUCT and REVERSEREDUCT, in a sense that neither of them is guaranteed to produce true (fuzzy) decision reducts. While from a theoretical point of view this is a fairly heavy price to pay, in practice the algorithms can still be used to produce sufficiently good attribute subsets.

Finally, the following proposition reveals an interesting relationship between the FRS- and the VQRS-based approach in case the second parameter of the fuzzy quantifier $Q_{(\alpha,\beta)}$ is equal to 1. Assume $(\mathbb{U}, \mathbb{C} \cup \{d\})$ is a consistent decision system, $B \subseteq \mathbb{C}$, $Q_u = Q_{(\alpha,1)}$, and I is an implicator that satisfies the confinement principle [345]. Then $\gamma_B^{Q_u} = 1$ if and only if $\gamma_B = 1$.

9.3 EXPERIMENTATION

This section presents the initial experimental evaluation of the selection methods for the task of pattern classification, over nine benchmark datasets from [38] and [168] with two classifiers.

9.3.1 Experimental Setup

FRFS uses a pre-categorization step that generates associated fuzzy sets for a dataset. For the new fuzzy-rough methods, the Łukasiewicz fuzzy connectives are used, with fuzzy similarity defined in (9.6). After feature selection the datasets are reduced according to the discovered reducts. These reduced datasets are then classified using the relevant classifier. (Obviously the feature selection step is not employed for the unreduced dataset.)

Two classifiers were employed for the purpose of evaluating the resulting subsets from the feature selection phase: JRip [59] and PART [381,382]. JRip learns propositional rules by repeatedly growing rules and pruning them. During the growth phase, features are added greedily until a termination condition is satisfied. Features are then pruned in the next phase subject to a pruning metric. Once the ruleset is generated, a further optimization is performed where classification rules are evaluated and deleted based on their performance on randomized data. PART generates rules by means of repeatedly creating partial decision trees from data. The algorithm adopts a divide-and-conquer strategy such that it removes instances covered by the current ruleset during processing. Essentially a classification rule is created by building a pruned tree for the current set of instances; the leaf with the highest coverage is promoted to a rule.

9.3.2 Experimental Results

Table 9.1 compares the reduct size and runtime data for FRFS, fuzzy boundary region-based FS (BFRFS), fuzzy lower approximation-based FS (LFRFS),

TABLE 9.1 Reduct size and time taken

Dataset	Objects	Attrs.	Reduct Size				Time Taken (s)				
			FRFS	BFRFS	LFRFS	FDM	FRFS	BFRFS	LFRFS	FDM setup	FDM
Cleveland	297	14	11	9	9	9	24.11	8.78	3.32	8.75	1.93
Glass	214	10	9	9	10	9	1.61	3.30	1.53	4.28	0.60
Heart	270	14	11	8	8	8	11.84	3.61	2.17	7.31	1.46
Ionosphere	230	35	11	9	9	8	61.80	8.53	3.77	14.09	3.45
Olitos	120	26	10	6	6	6	11.20	1.29	0.72	3.61	0.46
Water 2	390	39	11	7	7	7	96.58	21.37	12.12	43.44	18.48
Water 3	390	39	12	7	7	7	158.73	27.36	13.44	44.43	16.95
Web	149	2557	24	20	21	18	5642.65	949.69	541.85	357.11	1425.58
Wine	178	14	10	6	6	6	1.42	1.69	0.97	4.16	0.44

and fuzzy discernibility matrix-based FS (FDM). It can be seen that the new fuzzy-rough methods find smaller subsets than FRFS in general. The fuzzy boundary region-based method finds smaller or equally sized subsets than the LFRFS. This is to be expected as BFRFS includes fuzzy upper approximation information in addition to that of the fuzzy lower approximation. Of all the methods the fuzzy discernibility matrix-based approach finds the smallest fuzzy-rough reducts. It is often seen in crisp rough set FS that discernibility matrix-based approaches find smaller subsets on average than those that rely solely on dependency degree information. This comes at the expense of setup time as can be seen in the table. Fuzzy clauses must be generated for every pair of objects in the dataset. The new fuzzy-rough methods are also quicker in computing reducts than FRFS, due mainly to the computation of the Cartesian product of fuzzy equivalence classes that FRFS must perform.

It is interesting to observe from the results that some datasets greatly benefit from all feature selection methods, such as the Cleveland dataset. In this case noisy or misleading features must be present. This is to be expected as the data originates from patient measurements (cholesterol level, blood pressure, etc), not all of which can be expected to influence the class attribute. It is also interesting to note that for some datasets, such as Glass, the effect of feature selection is always detrimental. For data such as this, all features are necessary to be able to distinguish between classes.

FRFS has been experimentally evaluated with other leading FS methods (e.g., relief-F and entropy-based approaches [163,168]) and has been shown to outperform these in terms of resulting classification performance. Hence only comparisons to FRFS are given here. Table 9.2 shows the average classification accuracy as a percentage obtained using 10-fold cross-validation. The classification was initially performed on the unreduced dataset, followed by the reduced datasets that were obtained using the feature selection techniques. All techniques perform similarly, with classification accuracy improving or remaining the same for most datasets. FRFS performs equally well; however, this is at the cost of extra features and extra time required to find reducts. The performance of the FDM method is generally slightly worse than the other methods. This can be attributed partly to the fact that the method produces smaller subsets for data reduction.

9.3.3 Fuzzy Entropy Experimentation

This section presents the initial experimental evaluation of the fuzzy entropy-based selection methods.

Table 9.3 compares the reduct size for fuzzy entropy-based FS (E), fuzzy boundary region-based FS (B), fuzzy lower approximation-based FS (L), fuzzy boundary/entropy FS (BE), and fuzzy gain ratio FS (GR). It can be seen that the new entropy-based fuzzy-rough methods find smaller subsets in general. The entropy-based methods perform similarly, with the fuzzy gain ratio measure finding the smallest subsets in general. This demonstrates the utility of considering the split information when evaluating subset quality.

TABLE 9.2 Resulting classification accuracies (%)

Dataset	JRip					PART				
	Unred.	FRFS	BFRFS	LFRFS	FDM	Unred.	FRFS	BFRFS	LFRFS	FDM
Cleveland	52.19	54.55	54.55	54.55	54.55	50.17	52.19	53.20	53.20	53.20
Glass	71.50	69.63	65.89	71.50	65.89	67.76	68.22	70.56	70.56	67.76
Heart	77.41	78.89	78.52	78.52	78.52	73.33	78.52	76.30	76.30	76.30
Ionosphere	86.52	87.83	88.26	88.26	86.96	88.26	91.30	86.09	86.09	85.23
Olitos	70.83	70.83	71.67	64.17	63.33	57.50	62.50	67.50	58.33	64.17
Water 2	83.85	84.36	85.64	85.64	82.82	83.08	82.31	84.62	84.62	78.97
Water 3	82.82	82.82	79.74	81.28	80.00	83.33	80.51	80.26	79.23	79.74
Web	58.39	58.39	43.62	55.03	44.97	42.95	63.09	52.35	57.72	44.97
Wine	92.70	89.33	95.50	95.50	88.20	93.82	93.82	94.38	94.38	94.38

TABLE 9.3 Reduct size

Dataset	Objects	Features	Reduct size				
			E	B	L	BE	GR
Cleveland	297	14	10	9	9	10	10
Glass	214	10	9	9	10	10	9
Heart	270	14	9	8	8	8	9
Ionosphere	230	35	8	9	9	10	8
Olitos	120	26	6	6	6	6	6
Water 2	390	39	7	7	7	7	7
Water 3	390	39	7	7	7	7	7
Web	149	2557	23	20	21	20	18
Wine	178	14	6	6	6	6	6

Table 9.4 shows the average classification accuracy as a percentage obtained using 10-fold cross-validation. The classification accuracies are also presented in Figures 9.1 and 9.2 for each of the nine datasets. The classification was initially performed on the unreduced dataset, followed by the reduced datasets which were obtained using the feature selection techniques.

9.4 PROOFS

Theorem 1 LFRFS monotonicity. Suppose that $P \subseteq \mathbb{C}$, a is an arbitrary conditional feature that belongs to the dataset, and Q is the set of decision features. Then $\gamma'_{P\cup\{a\}}(Q) \geq \gamma'_P(Q)$.

Proof 1 The fuzzy lower approximation of a concept X is

$$\mu_{\underline{R_{P\cup\{a\}}}X}(x) = \inf_{y\in\mathbb{U}} I(\mu_{R_{P\cup\{a\}}}(x, y), \mu_X(y))$$

From (9.3) it can be seen that

$$\mu_{R_{P\cup\{a\}}}(x, y) = \mu_{R_a}(x, y) \wedge \mu_{R_P}(x, y)$$

From the properties of t-norms it can be seen that $\mu_{R_{P\cup\{a\}}}(x, y) \leq \mu_{R_P}(x, y)$. Thus $I(\mu_{R_{P\cup\{a\}}}(x, y), \mu_X(y)) \geq I(\mu_{R_P}(x, y), \mu_X(y))$, $\forall x, y \in \mathbb{U}$, $X \in \mathbb{U}/Q$, and hence $\mu_{\underline{R_{P\cup\{a\}}}X}(x) \geq \mu_{\underline{R_P}X}(x)$. The fuzzy positive region of X is

$$\mu_{POS_{R_{P\cup\{a\}}}(Q)}(x) = \sup_{X\in\mathbb{U}/Q} \mu_{\underline{R_{P\cup\{a\}}}X}(x)$$

TABLE 9.4 Resulting classification accuracies (%)

Dataset	JRip						PART					
	Unred.	E	B	L	BE	GR	Unred.	E	B	L	BE	GR
Cleveland	52.19	53.53	54.55	54.55	53.20	53.53	50.17	56.22	53.20	53.20	57.23	56.22
Glass	71.50	65.89	65.89	71.50	71.50	65.89	67.76	70.56	70.56	67.76	67.76	70.56
Heart	77.41	80.37	78.52	78.52	78.15	80.37	73.33	78.51	76.30	76.30	76.30	78.51
Ionosphere	86.52	84.37	88.26	88.26	89.15	84.37	88.26	86.95	86.09	86.09	88.26	86.95
Olitos	70.83	67.50	71.67	64.17	65.83	67.50	57.50	61.67	67.50	58.33	69.16	56.67
Water 2	83.85	82.30	85.64	85.64	84.36	83.59	83.08	83.59	84.62	84.62	84.10	82.31
Water 3	82.82	81.29	82.56	81.03	84.10	81.29	83.33	80.76	81.03	80.77	85.39	80.76
Web	58.39	53.02	46.97	55.03	50.37	52.34	42.95	55.70	55.03	57.72	52.34	53.69
Wine	92.70	94.94	95.50	95.50	93.82	91.57	93.82	94.94	94.38	94.38	94.94	93.82

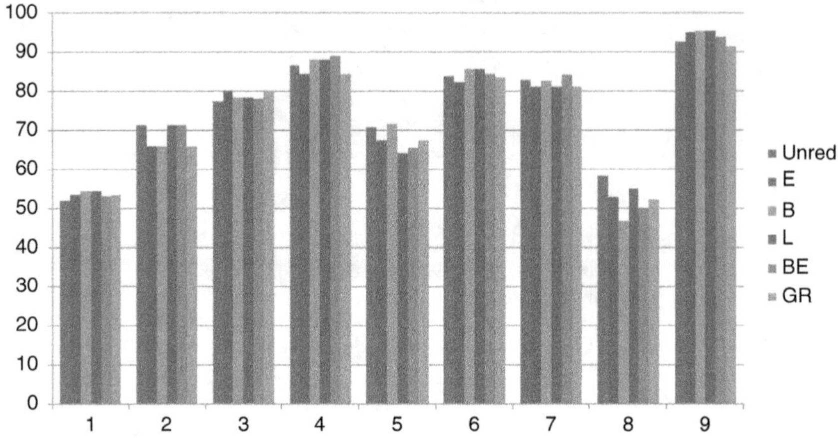

Figure 9.1 Performance: JRip

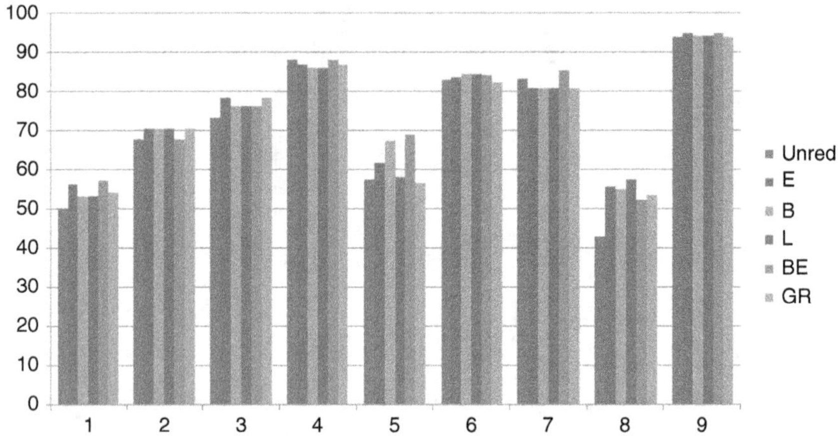

Figure 9.2 Performance: PART

so

$$\mu_{POS_{R_{P\cup\{a\}}}(Q)}(x) \geq \mu_{POS_{R_P}(Q)}(x)$$

and therefore

$$\gamma'_{P\cup\{a\}}(Q) \geq \gamma'_P(Q).$$

Theorem 2 BFRFS monotonicity. Suppose that $P \subseteq \mathbb{C}$, a is an arbitrary conditional feature that belongs to the dataset, and Q is the set of decision features. Then $\lambda_{P\cup\{a\}}(Q) \leq \lambda_P(Q)$.

Proof 2 The fuzzy boundary region of a concept X for an object x and set of features $P \cup \{a\}$ is defined as

$$\mu_{BND_{R_{P\cup\{a\}}}(X)}(x) = \mu_{\overline{R_{P\cup\{a\}}}X}(x) - \mu_{\underline{R_{P\cup\{a\}}}X}(x)$$

For the fuzzy upper approximation component of the fuzzy boundary region,

$$\mu_{\overline{R_{P\cup\{a\}}}X}(x) = \sup_{y\in\mathbb{U}} T(\mu_{R_{P\cup\{a\}}}(x, y), \mu_X(y))$$

It is known from Theorem 1 that $\mu_{R_{P\cup\{a\}}}(x, y) \leq \mu_{R_P}(x, y)$, so $\mu_{\overline{R_{P\cup\{a\}}}X}(x) \leq \mu_{\overline{R_{P\cup\{a\}}}X}(x)$. As $\mu_{\underline{R_{P\cup\{a\}}}X}(x) \geq \mu_{\underline{R_P}X}(x)$, then $\mu_{BND_{R_{P\cup\{a\}}}(X)}(x) \leq \mu_{BND_{R_P}(X)}(x)$. Thus $U_{P\cup\{a\}}(Q) \leq U_P(Q)$, and therefore $\lambda_{P\cup\{a\}}(Q) \leq \lambda_P(Q)$.

Theorem 3 FDM monotonicity. Suppose that $P \subseteq \mathbb{C}$, a is an arbitrary conditional feature that belongs to the dataset, and Q is the set of decision features. Then $SAT(P \cup \{a\}) \geq SAT(P)$.

Proof 3 For a clause C_{ij}, the degree of satisfaction for a given set of features $P \cup \{a\}$ is

$$\begin{aligned} SAT_{P\cup\{a\}}(C_{ij}) &= \bigcup_{z\in P\cup\{a\}} \{\mu_{C_{ij}}(z)\} \\ &= SAT_P(C_{ij}) \cup \mu_{C_{ij}}(a) \end{aligned}$$

derived from the properties of the t-conorm. Thus $SAT_{P\cup\{a\}}(C_{ij}) \geq SAT_P(C_{ij})$ for all clauses. Hence $SAT_{P\cup\{a\},q}(C_{ij}) \geq SAT_{P,q}(C_{ij})$. The overall degree of satisfaction for subset $P \cup \{a\}$ is

$$SAT(P \cup \{a\}) = \frac{\sum_{i,j\in\mathbb{U},i\neq j} SAT_{P\cup\{a\},q}(C_{ij})}{\sum_{i,j\in\mathbb{U},i\neq j} SAT_{\mathbb{C},q}(C_{ij})}$$

The denominator is a normalizing factor and can be ignored. As $SAT_{P\cup\{a\},q}(C_{ij}) \geq SAT_{P,q}(C_{ij})$ for all clauses, then $\sum_{i,j\in\mathbb{U},i\neq j} SAT_{P\cup\{a\},q}(C_{ij}) \geq \sum_{i,j\in\mathbb{U},i\neq j} SAT_{P,q}(C_{ij})$. Therefore $SAT(P \cup \{a\}) \geq SAT(P)$.

Theorem 4 FDM reducts are fuzzy-rough reducts. Suppose that $P \subseteq \mathbb{C}$, a is an arbitrary conditional feature that belongs to the dataset and Q is the set of decision features. If P maximally satisfies the fuzzy discernibility function, then P is a fuzzy-rough reduct.

Proof 4 The fuzzy positive region for a subset P is

$$\mu_{POS_{R_P}(Q)}(x) = \sup_{X\in\mathbb{U}/Q} \inf_{y\in\mathbb{U}}\{\mu_{R_P}(x, y) \rightarrow \mu_X(y)\}$$

The dependency function is maximized when each x belongs maximally to the fuzzy positive region. Hence

$$\inf_{x\in\mathbb{U}} \sup_{X\in\mathbb{U}/Q} \inf_{y\in\mathbb{U}}\{\mu_{R_P}(x, y) \rightarrow \mu_X(y)\}$$

is maximized only when P is a fuzzy-rough reduct. This can be rewritten as the following:

$$\inf_{x,y\in\mathbb{U}}\{\mu_{R_P}(x, y) \rightarrow \mu_{R_q}(x, y)\}$$

when using a fuzzy similarity relation in the place of crisp decision concepts, as $\mu_{[x]_R} = \mu_R(x, y)$ [86]. Each $\mu_{R_P}(x, y)$ is constructed from the t-norm of its constituent relations:

$$\inf_{x,y\in\mathbb{U}}\{T_{a\in P}(\mu_{R_a}(x, y)) \rightarrow \mu_{R_q}(x, y)\}$$

This may be reformulated as

$$\inf_{x,y\in\mathbb{U}}\{S_{a\in P}(\mu_{R_a}(x, y) \rightarrow \mu_{R_q}(x, y))\} \tag{9.35}$$

Considering the fuzzy discernibility matrix approach, the fuzzy discernibility function is maximally satisfied when

$$\{\wedge\{\{\vee C^*_{xy}\} \leftarrow q_{N(\mu_{R_q}(x,y))}\}|1 \le y < x \le |\mathbb{U}|\}$$

is maximized. This can be rewritten as

$$T_{x,y\in\mathbb{U}}(S_{a\in P}(N(\mu_{R_a}(x, y))) \leftarrow N(\mu_{R_q}(x, y)))$$

because each clause C_{xy} is generated by considering the fuzzy similarity of values of each pair of objects x, y. Through the properties of the fuzzy connectives, this may be rewritten as

$$T_{x,y\in\mathbb{U}}(S_{a\in P}(\mu_{R_a}(x, y) \rightarrow \mu_{R_q}(x, y))) \tag{9.36}$$

When this is maximized, (9.35) is maximized and so the subset P must be a fuzzy-rough reduct.

Theorem 5 Fuzzy-rough negative region is always empty for crisp decisions. Suppose that $P \subseteq \mathbb{C}$ and Q is the set of decision features. If Q is crisp, then the fuzzy-rough negative region is empty.

Proof 5 The fuzzy-rough negative region for subset P is

$$\mu_{NEG_{R_P}}(x) = N\left(\sup_{X \in \mathbb{U}/Q} \mu_{\overline{R_P}X}(x)\right)$$

For the negative region to be empty, all object memberships must be zero. Hence

$$\forall x, \sup_{X \in \mathbb{U}/Q} \mu_{\overline{R_P}X}(x) = N(0) = 1$$

Expanding this gives

$$\forall x, \sup_{X \in \mathbb{U}/Q} \sup_{y \in \mathbb{U}} T(\mu_{R_P}(x, y), \mu_X(y)) = 1$$

For this to be maximized, there must be a suitable X and y such that

$$\forall x, \exists X, \exists y, T(\mu_{R_P}(x, y), \mu_X(y)) = 1$$

Setting $y = x$, the formula above holds as the decisions are crisp, so each x must belong fully to one decision X, $\mu_X(x) = 1$. Therefore the fuzzy-rough negative region is always empty for crisp decisions. When the decisions are fuzzy and $\sup_{x \in X} \mu_X(x) < 1$ then the fuzzy-rough negative region will be nonempty.

Theorem 6 EFRFS reducts are fuzzy-rough reducts. Suppose that $P \subseteq \mathbb{C}$. If $E(P) = 0$, then P is a fuzzy-rough reduct.

Proof 6 Equation (9.7) can be rewritten as [168],

$$\mu_{POS_{R_P}(\mathbb{D})}(x) = \sup_{D_j} \sup_{F_i} \min\left(\inf_{y \in \mathbb{U}} I(\mu_{F_i}(y), \mu_{D_j}(y))\right)$$

If P is a fuzzy-rough reduct, then it must be the case that $F_i \subseteq D_j$ or $F_i \cap D_j = \emptyset$ for all F_i, D_j. If $F_i \subseteq D_j$, then $p(D_j|F_i) = 1$, and if $F_i \cap D_j = \emptyset$, then $p(D_j|F_i) = 0$ $\forall F_i, D_j$. Therefore each $H(F_i) = 0$, and $E(P) = 0$.

9.5 SUMMARY

This chapter has presented three new techniques for fuzzy-rough feature selection based on the use of fuzzy T-transitive similarity relations, that alleviate problems encountered with FRFS. The first development, based on fuzzy lower approximations, uses the similarity relations to construct approximations of decision concepts and evaluates these through a new measure of feature dependency. The second development employs the information in the fuzzy boundary region to guide the feature selection search process. When this is minimized, a fuzzy-rough reduct has been obtained. The third development extends the concept of the discernibility matrix to the fuzzy case, allowing features to belong to entries to a certain degree. An example FS algorithm is given to illustrate how reductions may be achieved. Note that no user-defined thresholds are required for any of the methods, although a choice must be made regarding fuzzy similarity relations and connectives.

Further work in this area will include a more in-depth experimental investigation of the proposed methods and the impact of the choice of relations and connectives. Additionally the development of fuzzy discernibility matrices here allows the extension of many existing crisp techniques for the purposes of finding fuzzy-rough reducts. In particular, by reformulating the reduction task in a propositional satisfiability (SAT) framework, SAT solution techniques may be applied that should be able to discover such subsets, guaranteeing their minimality. The performance may also be improved through simplifying the fuzzy discernibility function further. This could be achieved by considering the properties of the fuzzy connectives and removing clauses that are redundant in the presence of others.

CHAPTER 10

FURTHER ADVANCED FS METHODS

This chapter discusses several further FS developments. First, a new area in feature selection, fuzzy set-based feature grouping, is presented. Although the method given here is used within fuzzy-rough feature selection, it can be used with any feature selector that employs a suitable evaluation function. Indeed it may also be applied to the standard crisp rough set-based method, RSAR. Also in this chapter a novel framework for selecting features via ant colony optimization [40,82] is detailed. This is applied to finding optimal fuzzy-rough subsets but can replace the search mechanism for most feature selection methods.

10.1 FEATURE GROUPING

By its definition the degree of dependency measure (whether using crisp or fuzzy-rough sets) always lies in the range [0, 1], with 0 indicating no dependency and 1 indicating total dependency. For example, two subsets of the conditional attributes in a dataset may have the following dependency degrees:

$$\gamma'_{\{a,b,c\}}(\mathbb{D}) = 0.54, \quad \gamma'_{\{a,c,d\}}(\mathbb{D}) = 0.52$$

In traditional rough sets, it would be said that the attribute set $\{a, b, c\}$ has a higher dependency value than $\{a, c, d\}$ and so would make the better candidate to produce a minimal reduct. This may not be the case when considering real datasets that contain noise and other discrepancies. In fact it is possible

---

Computational Intelligence and Feature Selection: Rough and Fuzzy Approaches, by Richard Jensen and Qiang Shen

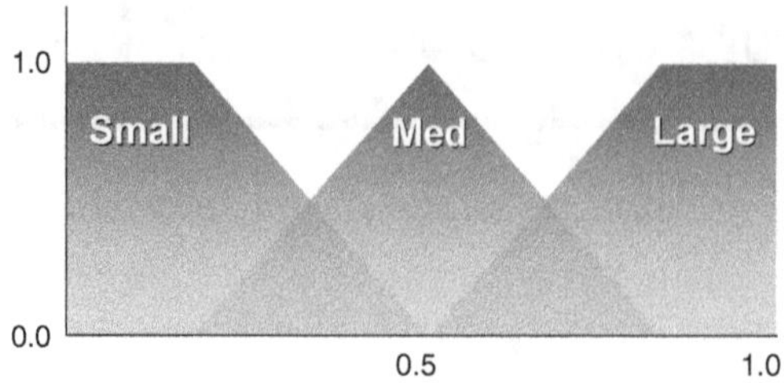

Figure 10.1 Possible fuzzification of dependency

that $\{a, c, d\}$ is the best candidate for this and other unseen related datasets. By fuzzifying the output values of the dependency function, this problem may be successfully tackled. In addition attributes may be *grouped* at stages in the selection process depending on their dependency label, speeding up the reduct search.

10.1.1 Fuzzy Dependency

In order to achieve this, several fuzzy sets must be defined over the dependency range (e.g., Figure 10.1). This leads to the next problem area: How are these sets to be defined? There is also the problem of how many fuzzy sets should be used to produce the most useful results. Initially these may be defined beforehand by an expert and refined through experimentation. However, to fit in with the rough set ideology, it would be interesting to investigate how to automatically generate these sets purely from the dataset itself. This could be achieved through fuzzy clustering techniques such as fuzzy c-means [30,88]. For the time being it is assumed that these fuzzy sets have already been defined.

The goal of FRFS and RSAR is to find a (possibly minimal) subset of the conditional attributes for which the dependency metric is at a maximum (ideally the value 1). In the case of fuzzy equivalence classes, where an element of uncertainty is introduced, the maximum degree of dependency may be substantially less than this. In fact the maximum dependency for different datasets may be quite different due to differing levels of uncertainty. The maximum for dataset A may be 0.9, whereas for dataset B the maximum may be only 0.2. Given a feature subset resulting in a dependency degree of 0.19, for dataset A this is quite a small value, but for dataset B this is quite large, so some way of scaling the dependency value depending on the dataset is required.

10.1.2 Scaled Dependency

The following is one potential way of achieving this for a subset P of all conditional attributes $\mathbb{C}$:

$$\gamma''_P(\mathbb{D}) = \frac{\gamma'_P(\mathbb{D})}{\gamma'_{\mathbb{C}}(\mathbb{D})}$$

In the example above, the scaled dependency degree for dataset A is now 0.21 (which fuzzifies to *Small*) and for dataset B is 0.95 (which fuzzifies to *Large*). However, a further problem is encountered as the search for a reduct nears its conclusion. In this situation almost all of the dependency values are mapped to *Large* because of their underlying closeness in value. This means that too large a group of attributes will be selected every time. Additionally, if the data is noisy, it may be the case that $\gamma''_P(\mathbb{D}) > 1$ as the dependency degree of the full set of conditional attributes may be greater than that of a particular attribute subset. An alternative scaling approach to combat both of these problems is to use the extreme values at each level of search. As soon as the reduct candidates have been evaluated, the highest and lowest dependencies ($\gamma'_{high}(\mathbb{D})$ and $\gamma'_{low}(\mathbb{D})$) are used as follows to scale the dependency degree of subset P:

$$\gamma''_P(\mathbb{D}) = \frac{\gamma'_P(\mathbb{D}) - \gamma'_{low}(\mathbb{D})}{\gamma'_{high}(\mathbb{D}) - \gamma'_{low}(\mathbb{D})}$$

This type of expression is also called an *index* [29]. By this method the attribute subset with the highest dependency value will have a scaled dependency ($\gamma''_P(\mathbb{D})$) of 1. The subset with the lowest will have a scaled dependency of 0. In so doing, the definition of the fuzzy sets need not be changed for different datasets; one definition should be applicable to all.

The next question to address is how to handle those scaled dependencies that fall at the boundaries. For example, a value may partially belong to both *Small* and *Medium*. A simple strategy is to choose the single fuzzy label with the highest membership value. However, this loses the potentially useful information of dual fuzzy set membership. Another strategy is to take both labels as valid, considering both possibilities within the feature selection process. If, for example, a dependency value lies within the labels *Small* and *Medium* then it is considered to belong to both groups.

10.1.3 The Feature Grouping Algorithm

The new fuzzy-rough QUICKREDUCT algorithm (FQR) which employs scaling and fuzzy dependencies can be seen in Figure 10.2. In this algorithm *Cands* contains sets of attributes and their corresponding dependency degrees when added to the current reduct candidate. Once each remaining attribute has been evaluated, the dependencies are scaled according to γ'_{high} and γ'_{low}. Next the decision is made on which feature(s) to add to the current reduct. In the previous fuzzy-rough QUICKREDUCT algorithm this would amount to selecting the feature providing the highest gain in dependency degree. Here other strategies may be employed; for example, attributes may be selected individually or in groups. This is discussed in more detail below.

Note that in addition to being applied to fuzzy-rough feature selection, this method may also be applied to crisp RSAR. Given a dataset containing crisp values, the dependency values may be fuzzified similarly (with scaling) so that

```
FuzzyQuickReduct(ℂ,𝔻)
Input: ℂ, the set of all conditional features; 𝔻, the set of
decision features
Output: R, the feature subset
R ← {}; γ'best
while γ'best ≠ γ'prev
   Cands ← {}
   γ'prev = γ'best
   γ'high = 0; γ'low = 1
   foreach x ∈ (ℂ − R)
      T ← R ∪ {x}
      Cands ← Cands ∪ (x, γ'T(𝔻))
      if γ'T(𝔻) > γ'high
         γ'high = γ'T(𝔻)
      else if γ'T(𝔻) < γ'low
         γ'low = γ'T(𝔻)
   Cands ← scale(Cands, γ'high, γ'low)
   R ← R ∪ selectFeatures(Cands)
   γ'best = γ'R(𝔻)
return R
```

Figure 10.2 Fuzzy dependency-based QuickReduct

groups of attributes may be selected at one time. The algorithm for this is exactly the same as the one given in Figure 10.2, except the dependency function used is now based on crisp rough sets. Additionally this may be applied to any feature selector where the result of subset evaluation is a value in the range [0, 1].

10.1.4 Selection Strategies

When using fuzzy degrees of dependency, it is possible to change strategy at any stage of the attribute selection process. The main distinction to make in the set of possible strategies is whether features are chosen individually or in groups.

10.1.4.1 Individuals In this subset of strategies, attributes are chosen one at a time in a similar fashion to that of FRFS. However, the choice of attribute depends on its corresponding linguistic label(s) and membership obtained from the dependency degree. In the example fuzzification of dependency given in Figure 10.1, attributes are grouped into the categories $small$, $medium$, and $large$. A representative attribute of the required label can be chosen randomly or based on the extent to which the attribute belongs to the fuzzy set itself. Those individual attributes lying on set boundaries are assigned both fuzzy labels. Other issues include which particular group of attributes to consider. Intuitively it would seem most appropriate to consider those belonging to the $Large$ group only; however it may be worthwhile investigating *Small* and *Medium*-grouped attributes at different stages of the search process.

10.1.4.2 Grouping To speed up the reduct search process, many attributes may be added to a reduct candidate at once, according to their label. For instance, selecting only those attributes considered to be *Large* would appear to be a suitable strategy. It may also be beneficial to add different groups of attributes at various stages of the search. To include diversity, cross-group selection is a method that picks representative attributes from each fuzzy label and adds them to the reduct candidate. Again, strategies may be changed during search; for example, it might be worthwhile to use the cross-group strategy first, followed by selecting *Large*-grouped attributes later.

One problem encountered in grouping attributes in this way is that in later stages there are sometimes too many attributes in the required label. Therefore it is usually best to revert to individual selection when this becomes a problem, making the search more accurate. Alternatively, taking suitable α-cuts will limit the total number selected.

10.1.5 Algorithmic Complexity

The complexity of this algorithm is the same as that of QUICKREDUCT, $O((n^2 + n)/2)$. This is because, in the worst case, the algorithm will add individual features (and not groups), following the same route as the standard algorithm. However, when groups of features are selected, certain areas of the QUICKREDUCT search space are avoided, decreasing the time taken to find a reduct.

10.2 ANT COLONY OPTIMIZATION-BASED SELECTION

Swarm intelligence (SI) is the property of a system whereby the collective behaviors of simple agents interacting locally with their environment cause coherent functional global patterns to emerge [40]. SI provides a basis with which it is possible to explore collective (or distributed) problem solving without centralized control or the provision of a global model. For example, ants are capable of finding the shortest route between a food source and their nest without the use of visual information. Hence, despite possessing no global world model, they adapt to changes in the environment. SI techniques based on the behavior of real ant colonies used to solve discrete optimization problems are classed as ant colony optimization (ACO) techniques [40]. ACO techniques have been successfully applied to a large number of difficult combinatorial problems like quadratic assignment [221] and the traveling salesman [82] problem, to routing in telecommunications networks, scheduling, and others.

ACO is particularly attractive for feature selection as there seems to be no heuristic that can guide search to the optimal minimal subset every time. Additionally it can be the case that ants discover the best feature combinations as they proceed throughout the search space. This section discusses how ant colony optimization may be applied to the difficult problem of finding optimal feature subsets.

10.2.1 Ant Colony Optimization

The ability of real ants to find shortest routes is mainly due to their depositing of pheromone as they travel; each ant probabilistically prefers to follow a direction rich in this chemical. The pheromone decays over time, resulting in much less pheromone on less popular paths. Given that over time the shortest route will have the highest rate of ant traversal, this path will be reinforced and the others diminished until all ants follow the same, shortest path (the "system" has converged to a single solution). It is also possible that there are many equally short paths - this situation can be handled by ACO as well. In this situation the rates of ant traversal over the short paths will be roughly the same, resulting in these paths being maintained while others are ignored. Additionally, if a sudden change to the environment occurs (e.g., a large obstacle appears on the shortest path), the system responds to this and will eventually converge to a new solution. Further details on ACO algorithms and their evaluation can be found in [40,82]. In general, an ACO algorithm can be applied to any combinatorial problem as far as it is possible to define:

- *Appropriate problem representation.* A description of the problem as a graph with a set of nodes and edges between nodes.
- *Heuristic desirability (η) of edges.* A suitable heuristic measure of the "goodness" of paths from one node to every other connected node in the graph.
- *Construction of feasible solutions.* A mechanism to efficiently create possible solutions.
- *Pheromone updating rule.* A suitable method of updating the pheromone levels on edges with a corresponding evaporation rule. Typical methods involve selecting the n best ants and updating the paths they chose.
- *Probabilistic transition rule.* A mechanism for the determination of the probability of an ant traversing from one node in the graph to the next.

Each ant in the artificial colony maintains a memory of its history—remembering the path it has chosen so far in constructing a solution. This history can be used in the evaluation of the resulting created solution and may also contribute to the decision process at each stage of solution construction.

Two types of information are available to ants during their graph traversal, local and global, controlled by the parameters β and α, respectively. Local information is obtained through a problem-specific heuristic measure. The extent to which this influences an ant's decision to traverse an edge is controlled by the parameter β. This parameter will guide ants toward paths that are likely to result in good solutions. Global knowledge is also available to ants through the deposition of artificial pheromone on the graph edges by their predecessors over time. The impact of this knowledge on an ant's traversal decision is determined by the parameter α. Good paths discovered by past ants will have a higher amount of associated pheromone. How much pheromone is deposited, and when, is dependent on the characteristics of the problem. No other local or global knowledge is available to the ants in the standard ACO model. However, the inclusion of such information by extending the ACO framework has been investigated [40] with some success.

10.2.2 Traveling Salesman Problem

To illustrate how ACO may be applied to artificial systems, its application to the traveling salesman problem (TSP) [198] is presented here. The TSP is a combinatorial optimization problem where, given N cities and a distance function d between cities, a minimal tour that goes through every city exactly once is to be found.

The TSP is represented as a graph, with nodes representing cities and edges representing journies between cities. The heuristic desirability of edge (i, j) is the inverse of the distance between those cities $(1/d(i, j))$, where $i \neq j$. Pheromone is increased proportional to the inverse of the tour length. These two measures can be thought of as providing different information about the problem: the heuristic measure provides local information and pheromone global information. The information is combined to form the so-called probabilistic transition rule, denoting the probability of an ant in city i choosing to travel to city j at time t:

$$p_{ij}^k(t) = \frac{[\tau_{ij}(t)]^\alpha.[\eta_{ij}]^\beta}{\sum_{l \in J_i^k}[\tau_{il}(t)]^\alpha.[\eta_{il}]^\beta} \tag{10.1}$$

where k is the number of ants, J_i^k the set of k's possible next cities, η_{ij} the heuristic desirability of choosing city j when at city i, and $\tau_{ij}(t)$ the amount of virtual pheromone on edge (i, j). The choice of α and β is determined experimentally. Typically several experiments are performed by varying each parameter and choosing the values that produce the best results.

With this representation a number of ants may be placed at nodes (cities) in the graph and traverse edges to form a tour. The pheromone of the edges corresponding to the best tours are reinforced, and a new set of ants make their way through the graph. This process continues until an optimum has been discovered or a certain number of generations of ants has been tried.

10.2.3 Ant-Based Feature Selection

By following similar principles, the feature selection task may be reformulated into an ACO-suitable problem [160]. ACO requires a problem to be represented as a graph—here nodes represent features, with the edges between them denoting

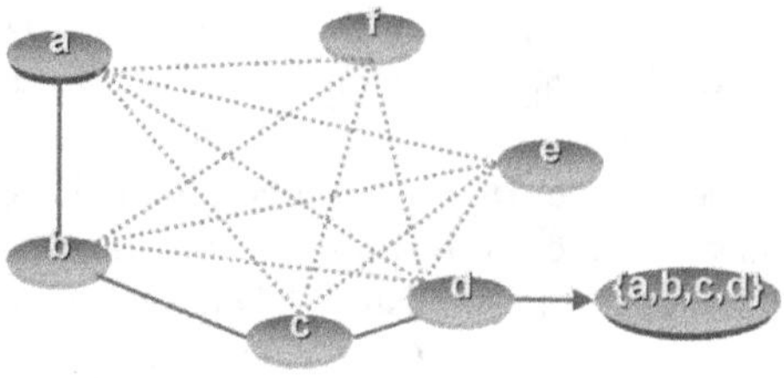

Figure 10.3 ACO problem representation for FS

the choice of the next feature. The search for the optimal feature subset is then an ant traversal through the graph where a minimum number of nodes are visited that satisfies the traversal stopping criterion. Figure 10.3 illustrates this setup—the ant is currently at node a and has a choice of which feature to add next to its path (dotted lines). It chooses feature b next based on the transition rule, then c and then d. Upon arrival at d, the current subset $\{a, b, c, d\}$ is determined to satisfy the traversal stopping criterion (e.g., a suitably high classification accuracy has been achieved with this subset). The ant terminates its traversal and outputs this feature subset as a candidate for data reduction.

A suitable heuristic desirability of traversing between features could be any subset evaluation function, for example, an entropy-based measure [281] or the fuzzy-rough set dependency measure [164]. Depending on how optimality is defined for the particular application, the pheromone may be updated accordingly. For instance, subset minimality and "goodness" are two key factors, so the pheromone update should be proportional to "goodness" and inversely proportional to size. The transition rule is the same as that given in equation (10.1), but J_i^k in this case is the set of ant k's unvisited features. There is also the possibility of allowing the removal of features here. If feature h has been selected already, an alternative transition rule may be applied to determine the probability of removing this attribute. However, this is an extension of the approach and is not necessary to perform feature selection.

10.2.3.1 Selection Process The overall process of ACO feature selection can be seen in Figure 10.4. The process begins by generating a number of ants, k; the ants are then placed randomly on the graph (i.e., each ant starts with one random feature). Alternatively, the number of ants to place on the graph may be set equal to the number of features within the data; each ant starts path construction at a different feature. From these initial positions, they traverse edges probabilistically until a traversal stopping criterion is satisfied. The resulting subsets are gathered and then evaluated. If an optimal subset has been found or the algorithm has executed a certain number of times, then the process halts and outputs the best feature subset encountered. If neither condition holds, then the pheromone is updated, a new set of ants are created, and the process iterates once more.

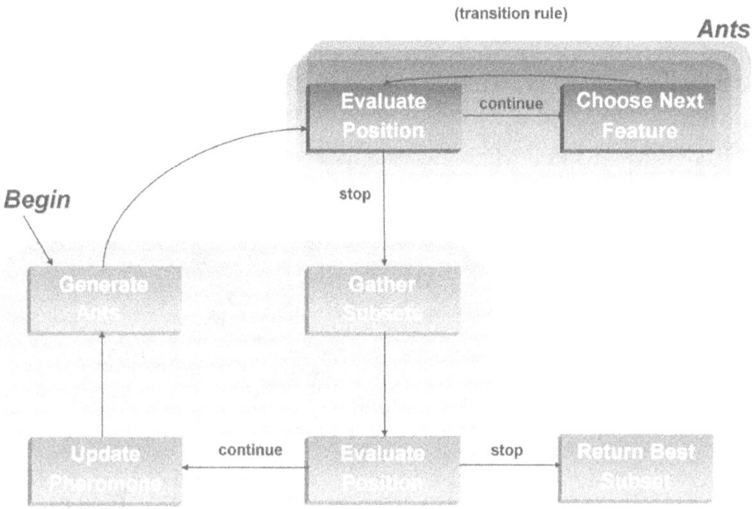

Figure 10.4 Overview of a general ACO-based FS system

10.2.3.2 Complexity Analysis The time complexity of the ant-based approach to feature selection is $O(IAk)$, where I is the number of iterations, A the number of original features, and k the number of ants. This can be seen from Figure 10.4. In the worst case each ant selects all the features. As the heuristic is evaluated after each feature is added to the reduct candidate, this will result in A evaluations per ant. After one iteration in this scenario, Ak evaluations will have been performed. After I iterations the heuristic will be evaluated IAk times.

This method is attractive for feature selection as there seems to be no heuristic that can guide search to the optimal minimal subset every time. Additionally it should be the case that ants will discover best feature combinations as they traverse the graph. This information will be reflected in the pheromone matrix used by other ants to guide their own local search.

10.2.3.3 Pheromone Update Depending on how optimality is defined for the particular application, the pheromone may be updated accordingly. For instance, subset minimality and "goodness" are two key factors so the pheromone update must be proportional to "goodness" and inversely proportional to size.

To tailor this mechanism to find fuzzy-rough set reducts, it is necessary to use the dependency measure given in equation (8.2) as the stopping criterion. This means that an ant will stop building its feature subset when the dependency of the subset reaches the maximum for the dataset (the value 1 for consistent datasets). The dependency function may also be chosen as the heuristic desirability measure, but this is not necessary. It may in fact be of more useful to employ a nonrough set related heuristic for this purpose to avoid the pitfalls of a QuickReduct style search. An alternative measure such as an entropy-based heuristic [281] may allow the method to avoid feature combinations that may mislead the fuzzy-rough set-based heuristic. Again, the time complexity of this fuzzy-rough ant-based method will be the same as that mentioned earlier, $O(IAk)$.

The pheromone on each edge is updated according to the following formula:

$$\tau_{ij}(t+1) = (1-\rho)\tau_{ij}(t) + \Delta\tau_{ij}(t) \tag{10.2}$$

where

$$\Delta\tau_{ij}(t) = \sum_{k=1}^{n} \left(\frac{\gamma'(S^k)}{|S^k|} \right) \tag{10.3}$$

This is the case if the edge (i, j) has been traversed; $\Delta\tau_{ij}(t)$ is 0 otherwise. The value ρ is a decay constant used to simulate the evaporation of the pheromone, and S^k is the feature subset found by ant k. The pheromone is updated according to both the fuzzy-rough measure of the "goodness" of the ant's feature subset (γ') and the size of the subset itself. By this definition, all ants update the pheromone. Alternative strategies may be used for this, such as allowing only the ants with the best feature subsets to proportionally increase the pheromone.

10.2.3.4 Approach Comparison Upon inspection, a number of similarities and differences can be observed between the approaches described previously. In the genetic and ant-based FS a population of potential solutions are considered on each algorithmic cycle. In the genetic case, this population is evaluated and then used to produce the next population. However, in the ant-based case, each population is erased after generation; the only lasting effect the population has is in the updated pheromone levels that are used in the next generation cycle. Simulated annealing-based FS (discussed in Section 4.2.4) considers an individual solution candidate only. New candidates are produced by mutating the current solution in a similar manner to genetic FS. The extent of mutation decreases with time, allowing convergence to a single, hopefully optimal, solution. All of these approaches employ a certain degree of randomness in their search for optimal subsets.

10.3 SUMMARY

This chapter has presented a new direction in feature selection, namely feature grouping. It was shown how fuzzifying particular evaluation functions, the fuzzy-rough or crisp rough set dependency degree, can lead to group and individual selection based on linguistic labels—more closely resembling human reasoning. In fact such grouping can be applied to most FS algorithms that use an evaluation function that returns values in [0, 1]. Choosing grouped features instead of individuals also decreases the time taken to reach potentially optimal subsets.

Conventional hill-climbing approaches to feature selection often fail to find minimal data reductions. Some guiding heuristics are better than others for this, but as no perfect heuristic exists, there can be no guarantee of optimality. When minimal data reductions are required, other search mechanisms must be employed. Although these methods also cannot ensure optimality, they provide a means by which the best feature subsets might be found. This chapter has presented a new method for feature selection based on ant colony optimization for this purpose. Unlike semantics-destroying approaches, this approach maintains the underlying semantics of the feature set; thereby it ensures that the resulting models are interpretable and the inference explainable. This is evident from the application of FRFS to the problem domains. In particular, for complex systems monitoring (Chapter 12) where fuzzy rule induction is performed, the resulting ruleset is highly comprehensible.

CHAPTER 11

APPLICATIONS II: WEB CONTENT CATEGORIZATION

Because of the explosive growth of electronically stored information, automatic methods must be developed to aid users in maintaining and using the abundance of information effectively. Sorting through even a fraction of the available data by hand can be very difficult. In particular, the sheer volume of redundancy present must be dealt with, leaving only the information-rich data to be processed. Information filtering and retrieval systems are therefore acquiring increasing prominence as automated aids in quickly sifting through Web-based information. This chapter focuses on the application of FRFS to this task of automatic Web data classification. Two tasks are considered: bookmark categorization and Web page categorization.

11.1 TEXT CATEGORIZATION

Text categorization has often been defined as the content-based assignment of one or more predefined categories to text. As stated in [239], it has become important from two points of view. First, it is important from the information retrieval (IR) viewpoint because of the rapid growth of textual information sources, which requires a greater amount of information processing. Text categorization can be used as a means to efficiently classify this textual data, or to provide support to further IR processes by performing such tasks as document filtering or information extraction. Second, it is important from the machine learning (ML) viewpoint as text categorization provides ML with an application field. The approach that

Computational Intelligence and Feature Selection: Rough and Fuzzy Approaches, by Richard Jensen and Qiang Shen

ML takes in automatic categorization is to generate a means of classification by the use of induction over examples that have been categorized previously (a form of supervised learning).

The categorization of textual documents involves two main phases: training and classification. The training phase involves examining sample documents and retrieving those keywords deemed important. These sets of keywords are large, rendering most text classifiers intractable, so a feature selection step is performed. Induction is carried out, and a means of classifying future data is the output. The classification phase uses the classification means from the training process to classify new documents. Several methods exist for this purpose. An outline of these is given below; details can be found in [162,365].

11.1.1 Rule-Based Classification

For rule-based approaches to classification, a set of rules and an appropriate classifier are required. Each individual rule has a set of preconditions and an associated decision. If a document matches the preconditions, then it is classified according to the decision value.

A simple form of this is the Boolean Exact Model [301], which employs exact Boolean existential rules. The problem with this is that it is inflexible—only documents for which the rule returns true match the rule. The Boolean Inexact Model BIM [301,302] bypasses this problem by providing a scoring mechanism so that the rule with the highest score classifies the document. If a term exists in a document and is also present in the corresponding rule, then the score for that rule is increased. If there is more than one rule that fires, then all rules must agree on the classification. If there is a conflict, then the classification is undecidable.

The main advantage of BIM is that it is fast (the computations involved are simple). BIM can also be quite accurate. A drawback is that words can have many meanings—something that the BIM cannot differentiate.

Fuzzy classifiers are another rule-based technique for classification. These follow the usual approach for the construction of fuzzy rule-based systems [267]. All precondition memberships are evaluated, and the necessary logical conjunctions integrated using the conventional minimum operator to derive classifications of new data.

11.1.2 Vector-Based Classification

The vector space model (VSM) [365,300] considers document representatives as binary vectors embedded in an n-dimensional Euclidean space (n is the total number of keywords). As there tends to be a large number of keywords involved, the dimensionality is also very high.

Each document and query is represented as a point in the space. To obtain the document vector, the keywords contained in the document are obtained and ranked according to their weight in the document. They are then converted to a vector. Any missing keywords (according to the universal keyword set) are marked as absent.

The procedure can be divided into three stages. The first stage is document indexing, where content-bearing terms are extracted from the document text. The second stage is the weighting of the indexed terms to enhance retrieval of documents relevant to the user. The last stage ranks the document with respect to the query according to the similarity measure. The similarity measure used here is the cosine coefficient, which measures the angle between the rule vector **x** and the query vector **q**, and is defined as

$$Sim(\mathbf{x}, \mathbf{q}) = \frac{\sum_{i=1}^{|\mathbb{C}|} \mathbf{x}_i \cdot \mathbf{q}_i}{\sqrt{\sum_{i=1}^{|\mathbb{C}|} (\mathbf{x}_i)^2 \cdot \sum_{i=1}^{|\mathbb{C}|} (\mathbf{q}_i)^2}} \tag{11.1}$$

There are a number of disadvantages with the VSM. There is the lack of justification for some of the vector operations (e.g., the choice of similarity function and the choice of term weights). It is barely a retrieval model as it does not explicitly model relevance. There is also the assumption that a query and a document can be treated the same. However, the simplicity of the model is attractive—and this is probably why it is the most popular retrieval model today.

11.1.3 Latent Semantic Indexing

Latent semantic indexing is a variant of the vector retrieval model outlined above that takes into account the dependencies between terms [87,75]. Unlike other models, LSI treats words as if they are not independent of each other; it attempts to automatically derive and model interrelationships between them.

The term-document matrix can be considered to be a "bag of documents" and is split into a set of k orthogonal factors. Similarity is computed in the following way:

1. Choose k (not greater than the number of terms or documents).
2. Add the weighted vectors for each term—multiply each vector by term weight—sum each element separately.
3. Repeat for query or second document.
4. Compute inner product—multiply corresponding elements and add.

An advantage that LSI has is that it reduces the original number of dimensions. Vectors that are similar are assigned to similar terms. This composite term is then mapped to a single dimension. However, as with most retrieval models, words with similar meanings confound LSI. Also the computations required are expensive, but a lot of this is carried out in advance.

11.1.4 Probabilistic

Another classic retrieval and classification method is the probabilistic retrieval technique [106], where the probability that a specific document will be judged relevant to a specific query, is based on the assumption that the terms are distributed differently in relevant and nonrelevant documents. The probability formula is usually derived from Bayes's theorem. Given a particular document x and a set of categories, the probability of x belonging to each category in the category set is calculated. The document is classified into the category that produced the highest probability.

11.1.5 Term Reduction

In text categorization the high dimensionality of the term space can be problematic, so some form of dimensionality reduction (DR) is employed. This can also be beneficial as it tends to reduce *overfitting*, where a classifier is tuned also to the contingent, rather than just the necessary characteristics of the training data [309]. It is often the case that the training data undergoes *several* stages of feature reduction, the final one being feature subset selection. The following measures and techniques are the main DR preprocessors to subset selection:

- *Stop word removal*. This is a simple technique for removing very information-poor terms. Stop words are connectives such as articles ("the," "a," "an," etc.) and contribute very little (if anything) to the classification of documents. Care must be taken when adopting this approach as it is feasible that some information-rich words might be mistakenly viewed as stop words.
- *Document frequency*. Again, this is a simple and effective reduction strategy [7]. It has been shown that the most informative terms are those with low to medium document frequency. By removing those attributes that occur the most (and to some extent those that occur rarely), the dimensionality of the document is reduced with no or little loss in information. To make this effective, stop words should be removed beforehand; otherwise, only topic-neutral words may remain after reduction.
- *Word stemming*. Word suffixes are removed, leaving only the root of the word. This is an attempt to reduce words with similar meanings to the same root; for example, *retailing* and *retailer* both contain the same root, namely *retail*. Word stemming is not guaranteed to work, however, as many words with different meanings share a common root.
- *Other functions*. There has been much research into using sophisticated information-theoretic term selection functions, such as chi-square [328] and correlation coefficient [248]. These functions have been shown to produce better results than document frequency.

Most text classification systems use one or more of the DR techniques above. For example, in the studies presented here, document frequency and stop word removal were implemented in performing a degree of initial data reduction. However, the use of this alone is insufficient to provide the extent of reduction required for efficient keyword search. Further reduction must take place by the use of feature selectors in order to remove the many remaining information-poor features.

11.2 SYSTEM OVERVIEW

The World Wide Web (WWW) is an information resource whose full potential may not be realized unless its content is adequately organized and described. This not only applies to the vast network of Web pages but also to users' personal repositories of Web page bookmarks. However, due to the immense size and dynamicity of the Web, manual categorization is not a practical solution to this problem. There is a clear need for automated classification of Web content.

To demonstrate the applicability of the described fuzzy-rough methods, two relevant domains of interest regarding the Web were selected, namely bookmark classification and Web site classification. However, these domains are quite distinct, possessing features and problems that must be independently addressed. This section will initially present those features that are common to both before focusing on domain-dependent issues.

Both applications employ a similar general system architecture in order to reduce dimensionality and perform categorization. A key issue in the design of the system was that of modularity; it should be modeled in such a way as to enable the straightforward replacement of existing techniques with new methods. The current implementations allow this flexibility by dividing the overall process into several independent submodules (see Figure 11.1):

- *Splitting of training and testing datasets.* Datasets were generated from large textual corpora and separated randomly into training and testing sets. Each dataset is a collection of documents, either bookmarks or Web pages depending on the application.
- *Keyword acquisition.* Given the output from the previous module, keywords are extracted and weighted according to their perceived importance in the document, resulting in a new dataset of weight-term pairs. Note that in this work, no sophisticated keyword acquisition techniques are used as the current focus of attention is on the evaluation of attribute reduction. However, the use of more effective keyword acquisition techniques recently built in the area of information retrieval would help improve the system's overall classification performance further.
- *Keyword selection.* As the newly generated datasets can be too large, mainly due to keyword redundancy, to perform classification at this stage, a dimensionality reduction step is carried out using FRFS. If this step is preceded by a discretization phase, RSAR may also be applied to the data.
- *Keyword filtering.* Used only in testing, this simple module filters the keywords obtained during acquisition, using the reduct generated in the keyword selection module.
- *Classification.* This final module uses the reduced dataset to perform the actual categorization of the test data. More efficient and effective classifiers can be employed for this step, but for simplicity only conventional classifiers are adopted here to show the power of attribute reduction. Better

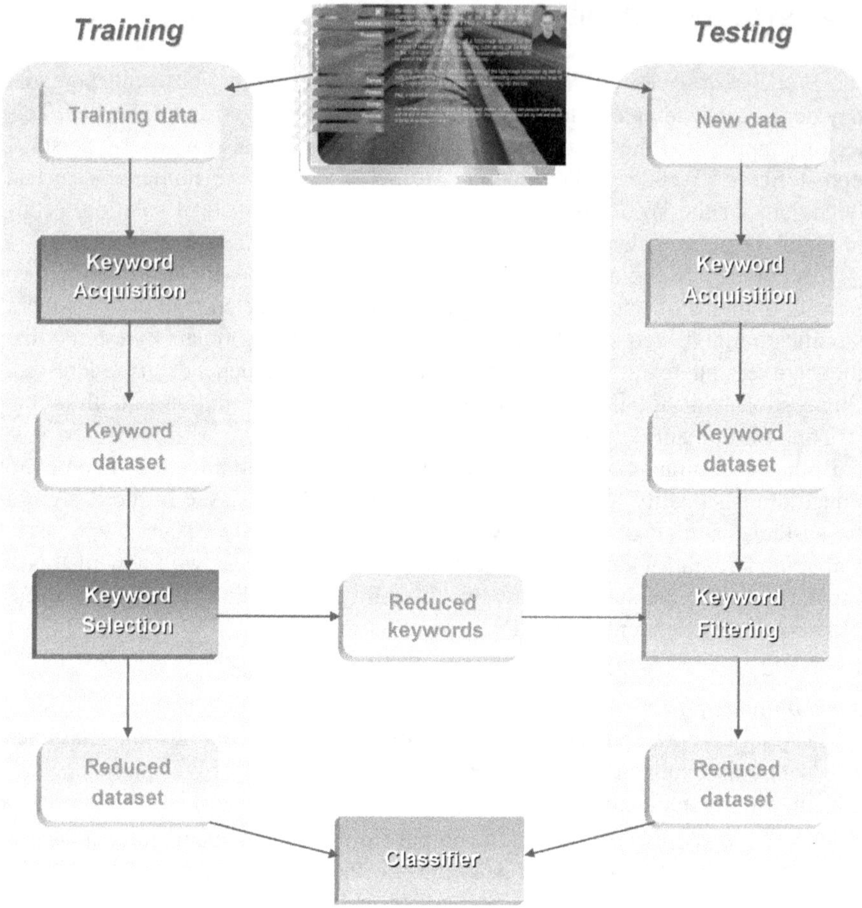

Figure 11.1 Modular decomposition of the classification system

classifiers are expected to produce more accurate results, although not necessarily to enhance the comparisons between classifiers that use reduced or unreduced datasets.

11.3 BOOKMARK CLASSIFICATION

As the use of the World Wide Web becomes more prevalent and the size of personal repositories grows, adequately organizing and managing bookmarks becomes crucial. Several years ago, in recognition of this problem, web browsers included support for tree-like folder structures for organizing bookmarks. These enable the user to browse through their repository to find the necessary information. However manual uniform resource locater (URL) classification and organization can be difficult and tedious when there are more than a few dozen bookmarks to classify—something that goes against the whole grain of the bookmarking concept.

Many usability studies (e.g., [195]) indicate that a deep hierarchy results in less efficient information retrieval as many traversal steps are required, so users are more likely to make mistakes. Users also do not have the time and patience to arrange their collection into a well-ordered hierarchy. The present work can help extract information from this relatively information-poor domain in order to classify bookmarks automatically.

11.3.1 Existing Systems

As this area of research is relatively new, there has been very little research carried out. Most bookmark utilities provide no means of automatic classification; the systems outlined here are the only ones known with such functionality.

11.3.1.1 Bookmark Organizer Bookmark Organizer (BO) [217] is an application that attempts to provide automatic assistance in organizing bookmark repositories by their conceptual categories. It can operate in two modes:

1. *Fully automatic*. If the users do not know which category the bookmark belongs to, they can request BO to insert it in the relevant folder by applying an automatic clustering technique to the document to which it refers.
2. *Semi-automatic*. The user specifies the node into which to insert the new bookmark. The user can now request the bookmarks to be re-organized automatically.

Bookmarks are organized by applying the hierarchical agglomerative clustering (HAC) technique to the text contained within the document the bookmark refers to. This is outlined below:

Step 1. **Start** with a set of singleton clusters, each contains one object.

Step 2. **Repeat** the following steps iteratively **until** there is only one cluster.

- **Identify** the two clusters that are the most similar.
- **Merge** them together into a single cluster.

This system constructs folders based on the text in the document, which should give a clear indication as to the category to which it belongs, and also enables automatic and semi-automatic classification (a useful feature). However, the automatically generated folder titles are quite unclear and often meaningless to the user.

11.3.1.2 PowerBookmarks This semistructured database application (as reported in [202]) aims to provide personalized organization and management of bookmarks in a multi-user environment.

PowerBookmarks automatically classifies documents by content; it parses metadata from bookmarked URLs to index and classify them. The metadata in this case are automatically generated from the document that the bookmark refers to, essentially a summary of the information the document contains. It avoids the verbose labeling problem encountered in BO by using existing classifiers with manually selected labels (e.g., "Sports/Football"). Significant keywords are extracted from documents according to the word frequency analysis. Queries are then issued to an external classifier, the results of which are processed to ensure consistency with existing classifications.

11.3.2 Overview

A large set of bookmarks was used as the training dataset. This database was generated by collating various online bookmark lists into one uniform collection. Each bookmark is pre-classified into a relevant category (e.g., "Sports" or "Computing/Java"). An additional testing dataset of "unseen" bookmarks was also compiled from online resources. To clarify the operation of the system, an example is included. The following bookmark is one of many contained in the database of bookmarks under the category *Programming/Java*:

```
<A HREF = "http://java.sun.com/Performance/"> Ways to Increase
Java Performance</A>
```

The two previously outlined methods require information in addition to that stored in the bookmark database. Both require further processing of the documents referred to by the bookmarks. The sorting system developed here relies solely on the information contained within the bookmark database—no other information is used to determine the feature reductions or classifications. The system is modular in structure, allowing various subcomponents to be replaced with alternative implementations if the need arises. The main modules are *keyword acquisition*, *feature selection*, and *classification*.

11.3.2.1 Keyword Acquisition To retain as much information as possible, all fields residing within the bookmark database are considered. For every bookmark the URL is divided into its slash-separated parts, with each part regarded as a keyword. In a similar fashion the bookmark title is split into terms with stop words removed.

In order to compare the similarity of bookmarks, a suitable representation must be chosen. Each bookmark is considered to be a vector where the ith element is the weight of term i according to some weighting method (a metric). The size of the vector is equal to the total number of keywords determined from the training documents.

This module produces weight-term pairs given a dataset. Each encountered word in a URL or title field is assigned a weight according to the metric used. Several metrics were implemented for this purpose:

- *Boolean existential metric*. All keywords that exist in the document are given a weight of 1, those that are absent are assigned 0 [301].
- *Frequency count metric*. The normalized frequency of the keywords in the document is used as the weight [302].
- *TF-IDF*. The term frequency-inverse document frequency metric [303] assigns higher weights to those keywords that occur frequently in the current document but not in most others. It is calculated using the formula $w(t, i) = F_i(t) \times \log \frac{N}{N_t}$, where $F_i(t)$ is the frequency of term t in document i, N is the number of documents in the collection, and N_t is the total number of documents that contain t.

For the example bookmark, the keywords {*java, sun, com, performance*} are obtained from the URL, and the keywords {*ways, increase, java, performance*} from the title field. By way of the simple Boolean existential metric, the vector elements relating to these keywords will each contain the value 1, the remainder 0.

The resulting sets of weight-term pairs, no matter which keyword acquisition metric is adopted, are large in size and need to be greatly reduced to be of any practical use for classification. Hence the next step is dimensionality reduction.

11.3.2.2 Dimensionality Reduction Given the weight-term sets, this module aims to significantly reduce their size while retaining their information content and preserving the semantics of those remaining keywords. FRFS is applied here to achieve this goal. Once a reduct has been calculated, the dataset can then be reduced by deleting those attributes that are absent from the reduct. The reduced dataset is now in a form that can be used by the classification module.

Returning to the example, it may be decided by this module that the term "com" provides little or no useful information. The column relating to this term is removed from the main dataset. This process is repeated for all keywords deemed by FRFS to be information poor.

11.3.2.3 Classification This module attempts to classify a given bookmark or bookmarks using the reduced keyword datasets obtained by the feature selection stage. Each bookmark has been transformed into a weight-term vector by the keyword acquisition process. For investigation purposes, two different inference techniques were implemented to perform classification: the Boolean inexact model and the vector space model.

To illustrate the operation of the classification methods, consider the training and testing vectors presented in Figure 11.2. The training data consists of two objects, one classified to the "Sport" category and the other classified to "News." Values in the vector represent the frequency of occurrence of terms in the training item.

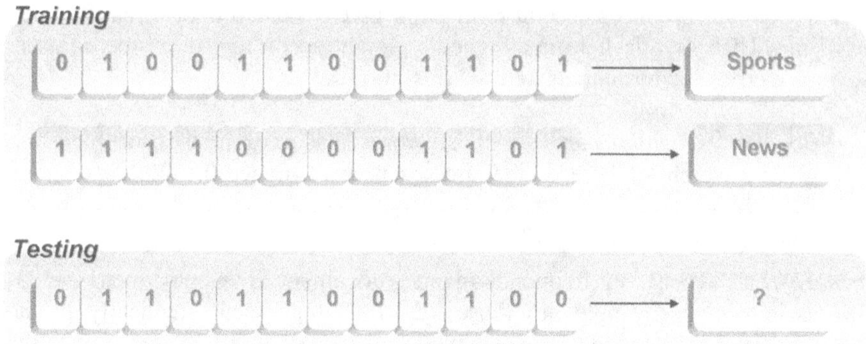

Figure 11.2 Training and testing vectors

For BIM classification, the training data (viewed as rules) is used to classify the new object by comparing term values in the vectors and incrementing a score if they are the same. Classifying using the first object results in a score of 10/12, as 10 of the term values match this training object. With the second object a score of 7/12 is produced. Hence, for this example, the test object is classified to the "Sport" category.

In VSM, the similarity measure defined in equation (11.1) is used to determine the closeness of the test object to the training examples. For the first training item, the computed similarity is $(5/\sqrt{6 \cdot 6}) = 0.83$. For the second item, the similarity is calculated to be $(4/\sqrt{7 \cdot 6}) = 0.62$. Again, the test object is determined to belong to the "Sport" category.

11.3.3 Results

The experiments presented here attempt to test whether FRFS is a useful tool for reducing data while retaining the information content. For this domain the FRFS approach produces exactly the same results as the crisp RSAR approach as all equivalence classes are crisp. Random-reduct (RR) generation (i.e., generating reducts randomly) was used to compare the results. This method deletes random attributes from the dataset, but it is constrained to leave the same number of attributes present as the FRFS method. The results of these approaches can be found in Table 11.2.

It can be seen from Table 11.1 that using the present work, the amount of attributes was reduced to around 35% of the original. This demonstrates the large amount of redundancy present in the original datasets. In light of the fact that bookmarks contain very little useful information, the results are a little better than anticipated.

A comparison of the performance of the dimensionality reduction techniques is presented in Table 11.2. The table shows that the overall accuracy is poor (obviously the random reduction gives worst results). The main point to make

TABLE 11.1 Comparison of the original number of features (unreduced) with those resulting from FRFS reduction

Dataset	Attributes (URL)	Attributes (Title)
Unreduced	1397	1283
FRFS-reduced	514	424

TABLE 11.2 Comparison of reduction strategies with the unreduced dataset

Dataset	VSM	BIM
Unreduced	55.6%	49.7%
FRFS-reduced	49.1%	47.3%
RR-reduced	37.3%	34.9%

here is that the ability of the system to classify new data depends entirely on the quality (and to a certain extent the quantity) of the training data. It cannot, in general, be expected that the FRFS-reduced experiments should perform much better than the original unreduced dataset, which only allows a rather low classification rate.

In light of the fact that bookmarks contain very little useful information, the results are unsurprising and perhaps a little better than anticipated. As stated earlier, the goal is to investigate how useful fuzzy-rough feature selection is in reducing the training dataset.

It is interesting to compare how the use of reduced datasets may affect the classification accuracy as opposed to that of the unreduced dataset. Importantly, the performance of the reduced dataset is almost as good as the original. Although a very small amount of important information might have been lost in the attribute reduction, this information loss is not significant enough to reduce classification accuracy significantly, while the reduction of dimensionality is substantial.

Results clearly show that FRFS (and RSAR) can be used to significantly reduce the dimensionality of the training dataset without much loss in information content. The measured drop in classification accuracy was 2.4% (using BIM) and 6.5% (using VSM) for the training dataset, which is within acceptable bounds.

TABLE 11.3 Performance: training data (using VSM)

Method	Attributes	Average Precision	Average Accuracy
Original	2557	—	—
Crisp	29	73.7%	76.2%
Fuzzy	23	78.1%	83.3%

11.4 WEB SITE CLASSIFICATION

There are an estimated 1 billion Web pages available on the WWW with around 1.5 million Web pages being added every day [148]. The difficult task is to find a particular Web page that satisfies a user's requirements by traversing hyperlinks. To aid this process, many Web directories have been developed—some rely on manual categorization while others make decisions automatically. However, as Web page content is vast and dynamic, manual categorization is becoming increasingly impractical. Automatic Web site categorization is therefore required to deal with these problems.

11.4.1 Existing Systems

Very recently research was carried out into this complex text classification area. In fact it has given rise to specific techniques for problems such as indexing and term selection as Web pages can be considered to be a special kind of document. Web pages contain text much like any other document, but the fact that they contain pointers to *other* documents makes them an interesting area for research.

An indexing technique specific to these documents has been proposed by [11]. A Web page tends to have many other pages pointing toward it. The authors reason that these documents can be combined forming an artificial document that can be considered to be a compilation of "short reviews" of the Web page. It is this compilation that is used in classification, not the original document.

Another indexing technique has been put forward [107] where a document representation is obtained by the combined indexing of both the original document and its children (the documents that it points to). This is of interest for Web site classification as often the children of a document contain relevant information for the site as a whole.

The topic of indexing is not the only one to have been investigated, and different methods of classifier induction have been proposed. In [48] a Web page categorization approach was developed based on the hypothesis that for each document-classification pair (d, c), two values can be associated: the authority $a(d, c)$ and the hub value $h(d, c)$. The authority value is a measure of the "authoritativeness" of d on c in terms of the number of c-related pages pointing to it. The hub value measures the "informativeness" of d on c in terms of how many c-related documents it points *to*. Therefore the purpose of this system is to identify the n most authoritative and the n most informative documents that are associated with the document, given a classification c. By issuing a query c to ALTAVISTA, an initial set of relevant documents is obtained, giving a starting point for the induction algorithm.

A method for the induction of classifiers for hierarchical category sets was developed in [295], using neural networks to achieve this. The system is composed of two networks: *gating* and *expert*. A gating network for a category c_i is a neural network that decides if a document d_j might be a plausible candidate for categorization under any child categories of c_i. Document d_j is propagated to all child nodes of c_i if the decision is positive; otherwise, no propagation takes place. An expert network for c_i simply decides whether document d_j should be classified to c_j. A classification tree consisting of gating or expert networks is used for document classification; leaf nodes are expert networks and internal nodes may be gating or expert. This allows a document to be classified under internal nodes and leaf nodes simultaneously.

11.4.2 Overview

There is usually much more information contained in a web document than a bookmark. Additionally information can be structured within a Web page that may indicate a relatively higher or lower importance of the contained text. For example, terms appearing within a <TITLE> tag would be expected to be more informative than the majority of those appearing within the document body at large. Because of this, keywords are weighted not only according to their statistical occurrence but also to their location within the document itself. These weights are almost always real valued, which can be a problem for most feature selectors unless data discretization takes place (a source of information loss). This motivates the application of FRFS to this domain.

The training and testing datasets were generated using Yahoo [393]. Five classification categories were used, namely Art & Humanity, Entertainment, Computers & Internet, Health, and Business & Economy. A total of 280 Web sites were collected from Yahoo categories and classified into these categories, 56 sites per category resulting in a balanced dataset. From this collection of data, the keywords, weights and corresponding classifications were collated into a single dataset (see Figure 11.3).

11.4.3 Results

11.4.3.1 Comparison with RSAR For this set of experiments FRFS is compared with the crisp RSAR approach. As the unreduced dataset exhibits high

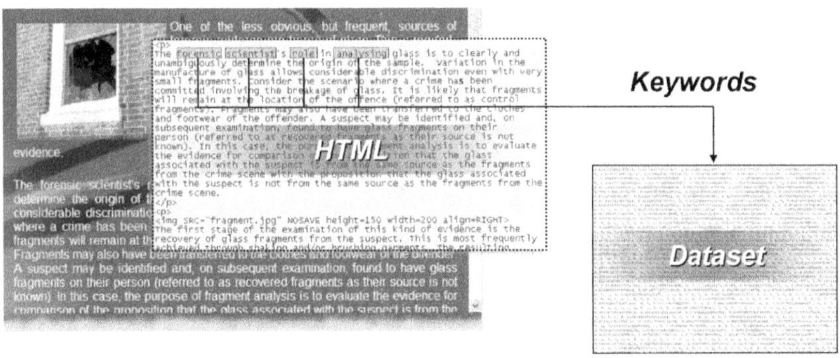

Figure 11.3 Keyword extraction

dimensionality (2557 attributes), it is too large to evaluate (hence the need for keyword selection). Using crisp RSAR the original attribute set was reduced to 29 (1.13% of the full set of attributes). However, using FRFS the number of selected attributes was only 23 (0.90% of the full attribute set). It is interesting that the FRFS reduct and crisp RSAR reduct share four attributes in common. With such a large reduction in attributes, it must be shown that classification accuracy does not suffer in the FRFS-reduced system.

In addition to the classification accuracy the precision of the system is presented. Precision is defined here to be the ratio of the number of correctly classified documents to the total number of correctly and incorrectly classified documents (expressed as a percentage). This differs from the classification accuracy in that the accuracy also considers documents that have not been able to be classified by the system.

To see the effect of dimensionality reduction, the system was tested on the original training data first; the results are summarized in Table 11.3. The results are averaged over all the classification categories. Clearly, the fuzzy method exhibits better precision and accuracy rates. This performance was achieved using fewer attributes than the crisp approach.

Table 11.4 contains the results for experimentation on 140 previously unseen Web sites. For the crisp case the average precision and accuracy are both rather low. With FRFS there is a significant improvement in both the precision and classification accuracy. Again, this more accurate performance is achieved while using fewer attributes.

It must be pointed out that although the testing accuracy is rather low, this is largely to do with the poor performance of the simple classifiers used. The fact that VSM-based results are much better than those using BIM-based classifiers shows that when a more accurate classification system is employed, the accuracy can be considerably improved with the involvement of the same attributes. Nevertheless, the purpose of the present experimental studies is to compare the performance of the two attribute reduction techniques based on the common use of any given classifier. Thus only the relative accuracies are important. Also the FRFS approach requires a reasonable fuzzification of the input data, while the fuzzy sets are herein generated by simple statistical analysis of the dataset with no attempt made at optimizing these sets. A fine-tuned fuzzification will certainly improve the performance of FRFS-based systems [224].

TABLE 11.4 Performance: unseen data

Method	% Original Attributes	Classifier	Average Precision	Average Accuracy
Crisp	1.13	BIM	23.3%	11.7%
		VSM	45.2%	50.3%
Fuzzy	0.90	BIM	16.7%	20.0%
		VSM	74.9%	64.1%

Finally, it is worth noting that the classifications were checked automatically. Many Web sites can be classified to more than one category; however, only the designated category is considered to be correct here.

11.4.3.2 Comparison with ACO-Based FRFS

Table 11.5 shows the resulting degree of dimensionality reduction, performed via selecting informative keywords, by the standard fuzzy-rough method (FRFS) and the ACO-based approach (AntFRFS). AntFRFS is run several times, and the results averaged both for classification accuracy and number of features selected. It can be seen that both methods drastically reduce the number of original features. AntFRFS performs the highest degree of reduction, with an average of 14.1 features occurring in the reducts it locates.

To see the effect of dimensionality reduction on classification accuracy, the system was tested on the original training data and a test dataset. The results are summarized in Table 11.6. Clearly, the fuzzy-rough methods exhibit better resultant accuracies for the test data than the unreduced method for all classifiers. This demonstrates that feature selection using either FRFS or AntFRFS can greatly aid classification tasks. It is of additional benefit to rule inducers as the induction time is decreased and the generated rules involve significantly fewer features. AntFRFS improves on FRFS in terms of the size of subsets found and resulting testing accuracy for QSBA and PART, but not for C4.5 and JRip.

The challenging nature of this particular task can be seen in the overall low accuracies produced by the classifiers (perhaps due to overfitting), though improved somewhat after feature selection. Both fuzzy-rough approaches require a reasonable fuzzification of the input data, while the fuzzy sets are herein generated by simple statistical analysis of the dataset with no attempt made at optimizing these sets. A fine-tuned fuzzification will certainly improve the performance of FRFS-based systems. Finally, it is worth noting that the classifications were checked automatically. Many Web pages can be classified to more than one category; however, only the designated category is considered to be correct here.

TABLE 11.5 Extent of feature reduction

Original	FRFS	AntFRFS
2557	17	14.10

TABLE 11.6 Classification performance

	Original		FRFS		AntFRFS	
Classifier	Train	Test	Train	Test	Train	Test
C4.5	95.89	44.74	86.30	57.89	81.27	48.39
QSBA	100.0	39.47	82.19	46.05	69.86	50.44
JRip	72.60	56.58	78.08	60.53	64.84	51.75
PART	95.89	42.11	86.30	48.68	82.65	48.83

11.5 SUMMARY

This chapter has presented a fuzzy-rough method to aid the classification of Web content, with promising results. In many text categorization problems, feature weights within datasets are real valued, posing a problem for many feature selection methods. FRFS can handle this type of data without the need for a discretizing step beforehand. In particular, while retaining fewer attributes than the conventional crisp rough set-based technique, the work resulted in an overall higher classification accuracy.

CHAPTER 12

APPLICATIONS III: COMPLEX SYSTEMS MONITORING

The ever-increasing demand for dependable, trustworthy intelligent diagnostic and monitoring systems, as well as knowledge-based systems in general, has focused much of the attention of researchers on the knowledge-acquisition bottleneck. The task of gathering information and extracting general knowledge from it is known to be the most difficult part of creating a knowledge-based system. Complex application problems, such as reliable monitoring and diagnosis of industrial plants, are likely to present large numbers of features, many of which will be redundant for the task at hand [268,310]. Additionally inaccurate and/or uncertain values cannot be ruled out. Such applications typically require convincing explanations about the inference performed. Therefore a method to allow automated generation of knowledge models of clear semantics is highly desirable.

The most common approach to developing expressive and human-readable representations of knowledge is the use of *if-then* production rules [192]. Yet real-life problem domains usually lack generic and systematic expert rules for mapping feature patterns onto their underlying classes. The present work aims to induce low-dimensionality rulesets from historical descriptions of domain features that are often of high dimensionality. In particular, a recent fuzzy rule induction algorithm (RIA), as first reported in [51], is taken to act as the starting point for this. The premise attributes of the induced rules are represented by fuzzy variables, facilitating the modeling of the inherent uncertainty of the knowledge domain. It should be noted, however, that the flexibility of the system discussed here allows the incorporation of almost any rule induction algorithm that uses

Computational Intelligence and Feature Selection: Rough and Fuzzy Approaches, by Richard Jensen and Qiang Shen

descriptive set representation of features. The choice of the current RIA is largely due to its recency and the simplicity in implementation. Provided with sets of continuous feature values, the RIA is able to induce classification rules to partition the feature patterns into underlying categories.

There exists a number of approaches relevant to the rule induction task at hand, both from the point of view of applications and that of computational methods. For example, the FAPACS (fuzzy automatic pattern analysis and classification system) algorithm documented in [12,50] is able to discover fuzzy association rules in relational databases. It works by locating pairs of features that satisfy an "interestingness" measure that is defined in terms of an adjusted difference between the observed and expected values of relations. This algorithm is capable of expressing linguistically both the regularities and the exceptions discovered within the data. Modifications to the Fuzzy ID3 (itself an augmentation of Quinlan's original ID3 [281]) rule induction algorithm have been documented [126] to better support fuzzy learning. In a similar attempt [157] has proposed modifications to decision trees to combine traditional symbolic decision trees with approximate reasoning, offered by fuzzy representation. This approach redefines the methodology for knowledge inference, resulting in a method best suited to relatively stationary problems.

A common disadvantage of these techniques is their sensitivity to high dimensionality. This may be remedied using conventional work such as principal components analysis (PCA) [77, 100]. As indicated previously, although efficient, PCA irreversibly destroys the underlying semantics of the feature set. Further reasoning about the derivation from transformed principal features is almost always humanly impossible.

In order to speed up the RIA and reduce rule complexity, a preprocessing step is required. This is particularly important for tasks where learned rulesets need regular updating to reflect the changes in the description of domain features. This step reduces the dimensionality of potentially very large feature sets while minimizing the loss of information needed for rule induction. It has an advantageous side effect in that it removes redundancy from the historical data. This also helps simplify the design and implementation of the actual pattern classifier, by determining what features should be made available to the system. In addition the reduced input dimensionality increases the processing speed of the classifier, leading to better response times. Most significant, however, is the fact that fuzzy-rough feature selection (FRFS) preserves the semantics of the surviving features after removing any redundant ones. This is essential in satisfying the requirement of user readability of the generated knowledge model, as well as ensuring the understandability of the pattern classification process.

This chapter is structured as follows: The problem domain is described, with the key factors that motivate the use of feature selection detailed. Additionally key areas that need to be addressed within the monitoring system are discussed. Experimental analysis, including comparisons with other feature selection methods and RIAs, is presented next. The application of feature grouping and ACO-based fuzzy-rough feature selection to this domain is also investigated.

12.1 THE APPLICATION

In order to evaluate further the utility of the FRFS approach and to illustrate its domain independence, a challenging test dataset was chosen, namely the Water Treatment Plant Database [38] (in addition to the experimental evaluation carried out in the last chapter).

12.1.1 Problem Case

The dataset itself is a set of historical data charted over 521 days, with 38 different input features measured daily. Each day is classified into one of 13 categories depending on the operational status of the plant. However, these can be collapsed into just two or three categories (i.e., *Normal* and *Faulty*, or *OK*, *Good*, and *Faulty*) for plant monitoring purposes as many classifications reflect similar performance. The collapsed form of classification is also performed to balance the class distributions present in the data. Because of the efficiency of the actual plant the measurements were taken from, all faults appear for short periods (usually single days) and are dealt with immediately. This does not allow for a lot of training examples of faults, which is a clear drawback if a monitoring system is to be produced. Note that this dataset has been utilized in many previous studies, including that reported in [321] (to illustrate the effectiveness of applying crisp RSAR as a preprocessing step to rule induction, where a different RIA is adopted from here).

The 38 conditional features account for the following five aspects of the water treatment plant's operation (see Figure 12.1):

1. Input to plant (9 features)
2. Input to primary settler (6 features)
3. Input to secondary settler (7 features)
4. Output from plant (7 features)
5. Overall plant performance (9 features)

The original dataset was split into 75% training and 25% testing data, maintaining the proportion of classifications present. It is likely that not all of the 38 input features are required to determine the status of the plant, hence the dimensionality reduction step. However, choosing the most informative features is a difficult task as there will be many dependencies between subsets of features. There is also a monetary cost involved in monitoring these inputs, so it is desirable to reduce this number.

12.1.2 Monitoring System

This work follows the original approach for complex systems monitoring developed in [321]. The original monitoring system consisted of several modules as

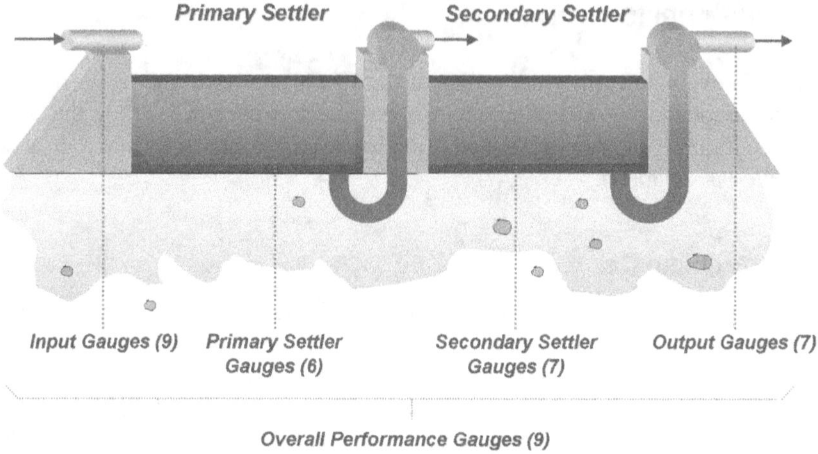

Figure 12.1 Water treatment plant, with number of measurements shown at different points in the system

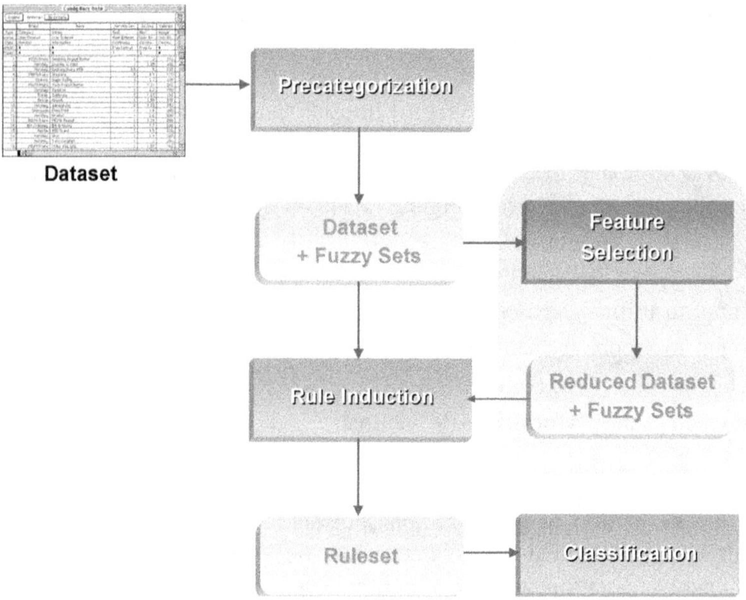

Figure 12.2 Modular decomposition of the implemented system

shown in Figure 12.2. It is this modular structure that allows the new FRFS technique to replace the existing crisp method.

Originally a precategorization step preceded feature selection where feature values were quantized. To reduce potential loss of information, the original use of just the dominant symbolic labels of the discretized fuzzy terms is now replaced by a fuzzification procedure. This leaves the underlying feature values unchanged but generates a series of fuzzy sets for each feature. These sets are generated entirely from the data while exploiting the statistical data attached to the dataset (in keeping with the rough set ideology in that the dependence of learning upon information provided outside of the training dataset is minimized). This module may be replaced by alternative fuzzifiers, or expert-defined fuzzification if available. Based on these fuzzy sets and the original real-valued dataset, FRFS calculates a reduct and reduces the dataset accordingly. Finally, fuzzy rule induction is performed on the reduced dataset using the modeling algorithm given in Section 3.2.2.1. Note that this algorithm is not optimal, nor is the fuzzification. Yet the comparisons given below are fair due to their common background. Alternative fuzzy modeling techniques can be employed if available.

12.2 EXPERIMENTAL RESULTS

This section first provides the results for the FRFS-based approach compared with the unreduced approach. Next a comparative experimental study is carried out between various dimensionality reduction methods; namely FRFS, entropy-based feature selection, PCA, and a random reduction technique.

The experiments were carried out over a tolerance range (with regard to the employment of the RIA) in steps of 0.01. As mentioned earlier, a suitable value for the threshold α must be chosen before rule induction can take place. However, the selection of α tends to be an application-specific task; a good choice for this threshold that provides a balance between a resultant ruleset's complexity and accuracy can be found by experiment. It should be noted that due to the fuzzy rule induction method chosen, all approaches generate exactly the same number of rules (as the number of classes of interest), but the arities in different rulesets differ. This way a potential complexity factor in the comparative studies can be avoided as otherwise the sizes of learned rulesets would need to be considered. Only the complexity in each learned rule is therefore examined,

12.2.1 Comparison with Unreduced Features

First of all, it is important to show that at least the use of features selected does not significantly reduce the classification accuracy as compared to the use of the full set of original features. For the 2-class problem the fuzzy-rough set-based feature selector returns 10 features out of the original 38.

Figure 12.3 compares the classification accuracies of the reduced and unreduced datasets on both the training and testing data. As can be seen, the FRFS results are almost always better than the unreduced accuracies over the tolerance range. The best results for FRFS were obtained when α is in the range 0.86 to 0.90, producing a classification accuracy of 83.3% on the training set and 83.9% for the test data. Compare this FRFS result with the optimum for the unreduced approach, which gave an accuracy of 78.5% for the training data and 83.9% for the test data.

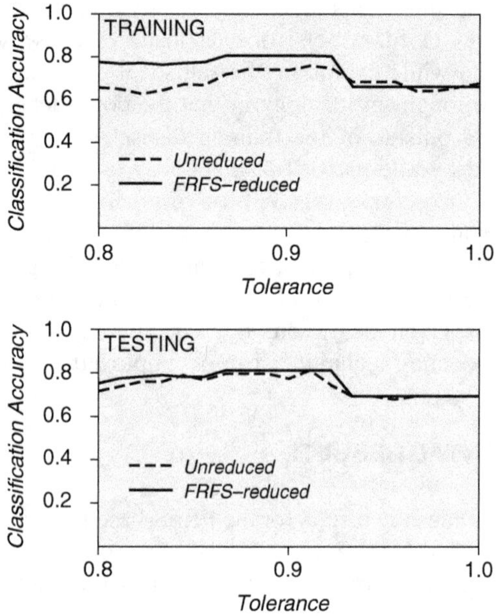

Figure 12.3 Training and testing accuracies for the 2-class dataset over the tolerance range

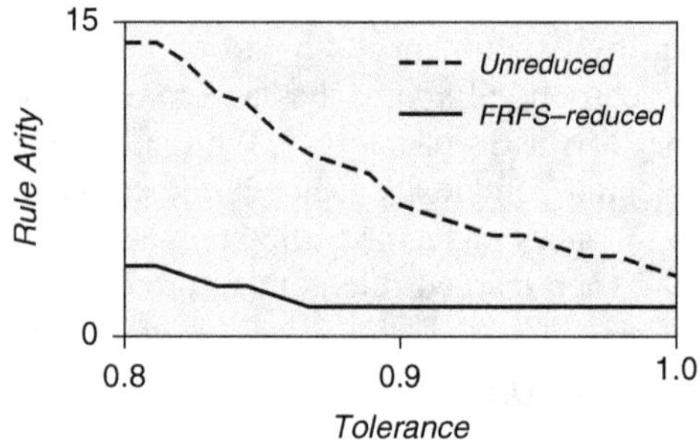

Figure 12.4 Average rule arities for the 2-class dataset

Use of the FRFS-based approach allows rule complexity to be greatly reduced. Figure 12.4 charts the average rule complexity over the tolerance range for the two approaches. Over the range of α values, FRFS produces significantly less complex rules while having a higher resultant classification accuracy. The average rule arity of the FRFS optimum is 1.5 ($\alpha \in (0.86, 0.9)$), which is less than that of the unreduced optimum, 6.0.

The 3-class dataset is a more challenging problem, as is reflected in the overall lower classification accuracies produced. The fuzzy-rough method chooses 11

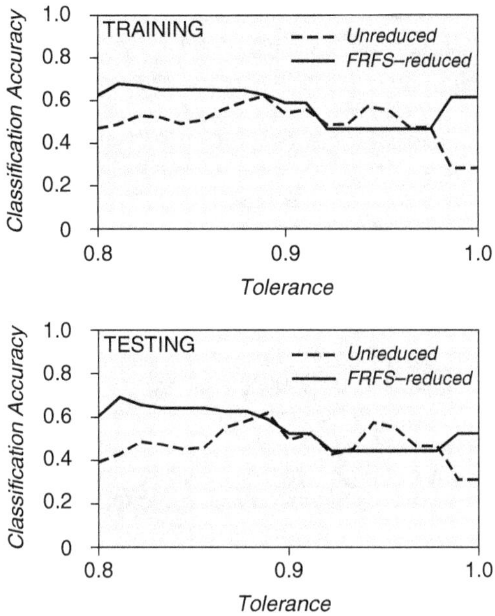

Figure 12.5 Training and testing accuracies for the three-class dataset over the tolerance range

out of the original 38 features. The results of both approaches are presented in Figure 12.5. Again, it can be seen that FRFS outperforms the unreduced approach on the whole. The best classification accuracy obtained for FRFS was 70.0% using the training data, 71.8% for the test data ($\alpha = 0.81$). For the unreduced approach, the best accuracy obtained was 64.4% using the training data, 64.1% for the test data ($\alpha = 0.88$).

Figure 12.6 compares the resulting rule complexity of the two approaches. It is evident that rules induced using FRFS as a preprocessor are simpler, with

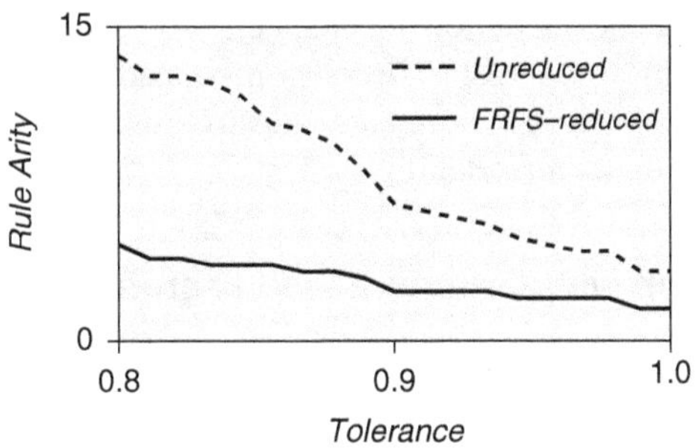

Figure 12.6 Average rule arities for the 3-class dataset

Rules from FRFS-reduced data

IF SED-S IS Medium, THEN Situation IS Normal

IF PH-E IS NOT High AND SSV-E IS Low AND SSV-P IS NOT Medium AND PH-D IS NOT High AND DQO-D IS NOT Medium AND PH-S IS NOT High, THEN Situation IS Good

IF PH-E IS NOT High AND SSV-E IS Low AND SSV-P IS Low AND DQO-D IS NOT High AND SED-S IS Medium, THEN Situation IS Faulty

Rules from unreduced data

IF ZN-E IS NOT High AND SS-E IS NOT High AND SED-E IS NOT High AND SSV-D IS NOT High AND DBO-S IS Low AND SS-S IS NOT High AND SED-S IS Low, THEN Situation IS Normal

IF ZN-E IS Low AND PH-E IS NOT High AND SSV-E IS NOT High AND PH-P IS NOT High AND SSV-P IS NOT High AND PH-D IS NOT High AND DBO-D IS NOT Medium AND SSV-D IS NOT High AND SS-S IS NOT High, THEN Situation IS Good

IF SSV-E IS NOT High AND SSV-P IS Low AND DQO-D IS NOT High AND SSV-D IS NOT High AND SED-D IS NOT High AND DBO-S IS Low AND SS-S IS NOT High AND SSV-S IS NOT High AND SED-S IS Low, THEN Situation IS Faulty

Figure 12.7 A selection of generated rulesets

little loss in classification accuracy. In fact the simple rules produced regularly outperform the more complex ones generated by the unreduced approach. The average rule arity of the FRFS optimum is 4.0, which is less than that of the unreduced optimum, 8.33.

These results show that FRFS is useful not only in removing redundant feature measures but also in dealing with the noise associated with such measurements. To demonstrate that the resulting rules are comprehensible, two sets of rules produced by the induction mechanism are given in Figure 12.7, with feature descriptions in Table 12.1. The rules produced are reasonably short and understandable. However, when semantics-destroying dimensionality reduction techniques are applied, such readability is lost.

12.2.2 Comparison with Entropy-Based Feature Selection

To support the study of the performance of FRFS for use as a preprocessor to rule induction, a conventional entropy-based technique is now applied for comparison. This approach utilizes the entropy heuristic employed by machine learning techniques such as C4.5 [281] and EBR (described in Section 4.2.1.5).

TABLE 12.1 Features and their meaning within the treatment plant

Feature	Description
DBO-D	(input biological demand of oxygen to secondary settler)
DBO-S	(output biological demand of oxygen)
DQO-D	(input chemical demand of oxygen to secondary settler)
PH-D	(input pH to secondary settler)
PH-E	(input pH to plant)
PH-P	(input pH to primary settler)
PH-S	(output pH)
SED-D	(input sediments to secondary settler)
SED-E	(input sediments to plant)
SED-S	(output sediments)
SS-E	(input suspended solids to plant)
SS-S	(output suspended solids)
SSV-D	(input volatile suspended solids to secondary settler)
SSV-E	(input volatile suspended solids to plant)
SSV-P	(input volatile suspended solids to primary settler)
SSV-S	(output volatile suspended solids)
ZN-E	(input zinc to plant)

TABLE 12.2 Comparison of FRFS and entropy-based feature selection

Approach	Number of Classes	Selected Features	Number of Features	Training Accuracy	Testing Accuracy
FRFS	2	{0,2,6,10,12,15,22,24,26,37}	10	83.3%	83.9%
Entropy	2	{1,5,6,7,9,12,15,16,20,22,29,30,33}	13	80.7%	83.9%
FRFS	3	{2,3,6,10,12,15,17,22,27,29,37}	11	70.0%	71.8%
Entropy	3	{6,8,10,12,17,21,23,25,26,27,29,30,34,36}	14	70.0%	72.5%

The features that provide the most gain in information are selected. A summary of the results of this comparison can be seen in Table 12.2.

For both the 2-class and 3-class datasets, FRFS selects three fewer features than the entropy-based method. FRFS has a higher training accuracy and the same testing accuracy for the 2-class data using less features. However, for the 3-class data the entropy-based method produces a very slightly higher testing accuracy. Again, it should be noted that this is obtained with three additional features over the FRFS approach.

12.2.3 Comparison with PCA and Random Reduction

The previous comparisons ensured that little information loss is incurred due to FRFS. The question now is whether any other feature sets of a dimensionality 10 (for the 2-class dataset) and 11 (for the 3-class dataset) would perform similarly. To avoid a biased answer to this, without resorting to exhaustive computation,

70 sets of random reducts were chosen of size 10 for the 2-class dataset, and a further 70 of size 11 for the 3-class dataset to see what classification results might be achieved. The classification accuracies for each tolerance value are averaged.

The effect of using a different dimensionality reduction technique, namely PCA, is also investigated. To ensure that the comparisons are fair, only the first 10 principal components are chosen for the 2-class dataset (likewise, the first 11 for the 3-class dataset). As PCA irreversibly destroys the underlying dataset semantics, the resulting rules are not human-comprehensible but may still provide useful automatic classifications of new data.

The results of FRFS, PCA, and random approaches can be seen in Figure 12.8. for the 2-class dataset. On the whole, FRFS produces a higher classification accuracy than both PCA-based and random-based methods over the tolerance range. In fact FRFS results in the highest individual classification accuracy for training and testing data (see Table 12.3).

For the 3-class dataset the results of FRFS, PCA and random selection are shown in Figure 12.9. The individual best accuracies can be seen in Table 12.4. Again, FRFS produces the highest classification accuracy (71.8%), and it is almost always the best over the tolerance range. Although PCA produces a comparatively reasonable accuracy of 70.2%, this is at the expense of incomprehensible rules.

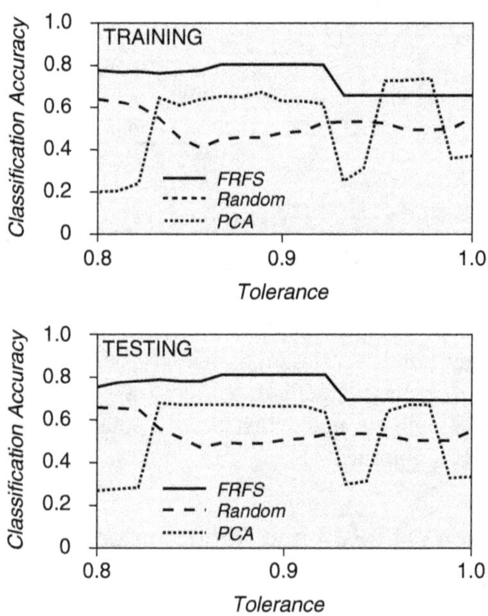

Figure 12.8 Training and testing accuracies for the 2-class dataset: comparison with PCA and random-reduction methods

TABLE 12.3 Best individual classification accuracies (2-class dataset) for FRFS, PCA, and random approaches

Approach	Training Accuracy	Testing Accuracy
FRFS	83.3%	83.9%
Random	66.4%	68.1%
PCA	76.7%	70.3%

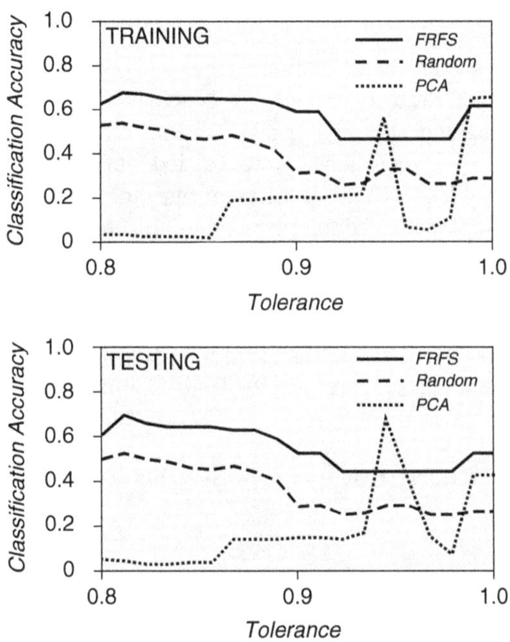

Figure 12.9 Training and testing accuracies for the 3-class dataset: Comparison with PCA and random-reduction methods.

TABLE 12.4 Best resultant classification accuracies (3-class dataset) for FRFS, PCA, and random approaches

Approach	Training Accuracy	Testing Accuracy
FRFS	70.0%	71.8%
Random	55.7%	54.3%
PCA	67.7%	70.2%

12.2.4 Alternative Fuzzy Rule Inducer

As stated previously, many fuzzy rule induction algorithms exist and can be used to replace the RIA adopted in the present monitoring system. In this section an example is given using Lozowski's algorithm as presented in [216]. This method extracts linguistically expressed fuzzy rules from real-valued features as with the subsethood-based RIA. Provided with training data, it induces approximate relationships between the characteristics of the conditional features and their underlying classes. However, as with many RIAs, this algorithm exhibits high computational complexity due to its generate-and-test nature. The effects of this become evident where high-dimensional data needs to be processed. Indeed, for this particular domain, feature selection is essential as running the RIA on all conditional features would be computationally prohibitive.

The results presented here compare the use of fuzzy-rough set based feature selection with the crisp rough set based method. For RSAR the data are discretized using the supplied fuzzy sets and reduction performed on the resulting dataset. The experiments were carried out over a tolerance range, required by the fuzzy RIA. This is a different threshold from that required in the subsethood-based approach. The tolerance here indicates the minimal confidence gap in the decision between a candidate rule and other competing contradictory rules. Again, the threshold is incremented in steps of 0.01.

As can be seen from Table 12.5, FRFS selects fewer attributes than the crisp method for the 2-class dataset and results in a higher classification accuracy over the entire tolerance range (Figure 12.10). Both results show that there is a lot

TABLE 12.5 Extent of dimensionality reduction

Method	Attributes 2-Class	Attributes 3-Class
FRFS	10	11
RSAR	11	11

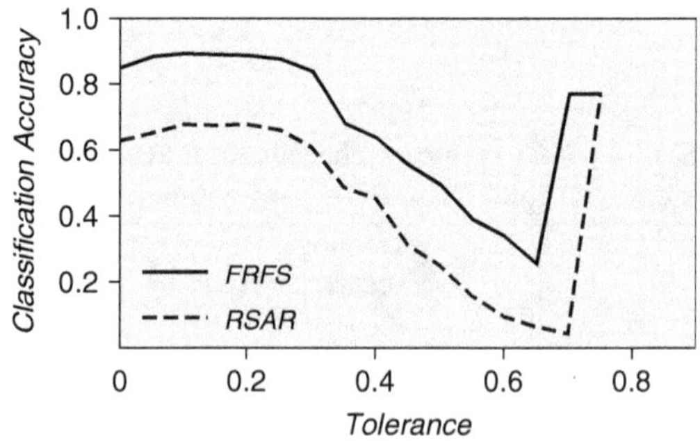

Figure 12.10 Classification accuracies for the 2-class dataset

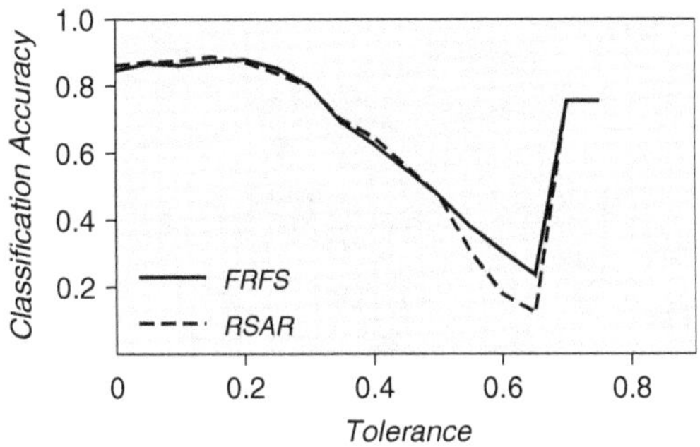

Figure 12.11 Classification accuracies for the 3-class dataset

of redundancy in the dataset that can be removed with little loss in classification accuracy. For the 3-class dataset the approaches perform similarly, with the FRFS method generally outperforming the other two, using the same number of attributes (but not *identical* attributes). The classification results can be seen in Figure 12.11.

12.2.5 Results with Feature Grouping

The experiments were carried out over a tolerance range (with regard to the employment of the RIA). It is worth reiterating here that due to the fuzzy rule induction method chosen (see Section 3.2.2.1), all approaches generate exactly the same number of rules (as the number of classes of interest), but the arities in different rulesets will differ. In all experiments here, FQR employs the strategy where all attributes belonging to *Large* (as defined in Section 10.1) are selected at each stage.

For the 2-class problem the FRFS feature selector returns 10 features out of the original 38, whereas FQR returns 12. Figure 12.12 compares the classification accuracies of the reduced and unreduced datasets on both the training and testing data. As can be seen, both the FRFS and FQR results are almost always better than the unreduced accuracies over the tolerance range. The best results for FQR were obtained in the range 0.85 to 0.88, producing a classification accuracy of 82.6% on the training set and 83.9% for the test data. For FRFS the best accuracies were 83.3% (training) and 83.9% (testing). Compare this with the optimum for the unreduced approach, which gave an accuracy of 78.5% for the training data and 83.9% for the test data.

The 3-class dataset is a more challenging problem, as is reflected in the overall lower classification accuracies produced. Both fuzzy-rough methods, FQR and FRFS, choose 11 out of the original 38 features (but not the *same* features). The results of these approaches can be seen in Figure 12.13. Again, the results show that both FQR and FRFS outperform the unreduced approach on the whole. The best classification accuracy obtained for FQR was 72.1% using the training data,

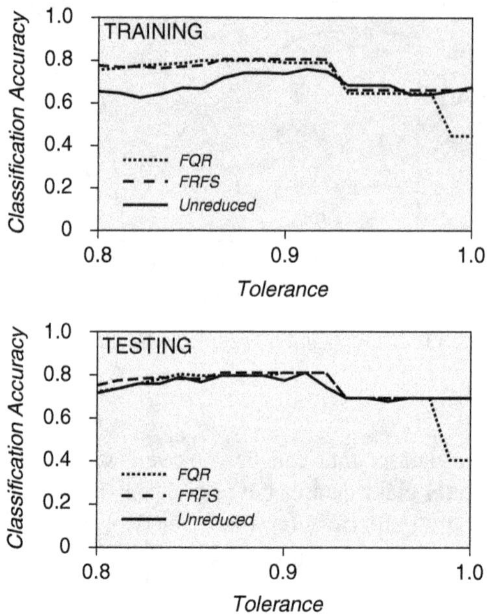

Figure 12.12 Classification accuracies for the 2-class dataset

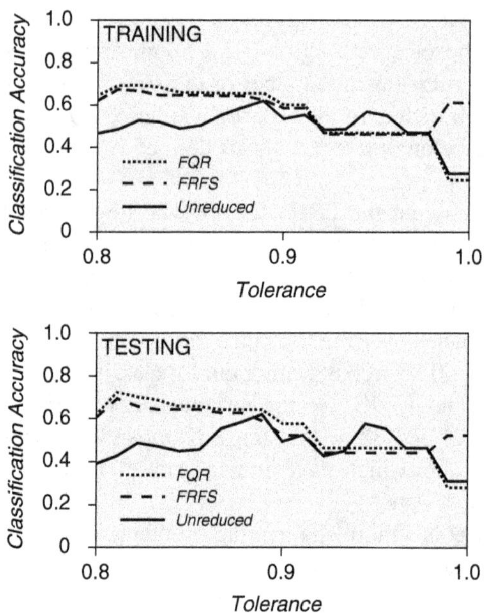

Figure 12.13 Classification accuracies for the 3-class dataset

74.8% for the test data. For FRFS the best results were 70.0% (training) and 71.8% (testing). In this case FQR has found a better reduction of the data. For the unreduced approach the best accuracy obtained was 64.4% using the training data, 64.1% for the test data.

12.2.6 Results with Ant-Based FRFS

The results from various comparative studies regarding ant-based FRFS are presented here. The first set of experiments compares the hill-climbing and ant-based fuzzy-rough methods. An investigation into another feature selector based on the entropy measure is then presented. This is followed by comparisons with PCA and support vector classifiers [272].

12.2.6.1 Comparison of Fuzzy-Rough Methods Three sets of experiments were carried out on both the (collapsed) 2-class and 3-class datasets. The first bypasses the feature selection part of the system, using the original water treatment dataset as input to C4.5, with all 38 conditional attributes. The second method employs FRFS to perform the feature selection before induction is carried out. The third uses the ant-based method described in Section 10.2.3 (antFRFS) to perform feature selection over a number of runs, and the results are averaged.

The results for the 2-class dataset can be seen in Table 12.6. Both FRFS and antFRFS significantly reduce the number of original attributes with antFRFS producing the greatest data reduction on average. As well as generating smaller reducts, antFRFS finds reducts of a higher quality according to the fuzzy-rough dependency measure. This higher quality is reflected in the resulting classification accuracies for both the training and testing datasets, with antFRFS outperforming FRFS.

Table 12.7 shows the results for the 3-class dataset experimentation. The hill-climbing fuzzy-rough method chooses 11 out of the original 38 features. The ant-based method chooses fewer attributes on average; however, this is at the cost of a lower dependency measure for the generated reducts. Again, the effect

TABLE 12.6 Results for the 2-class dataset

Method	Attributes	γ' Value	Training Accuracy	Testing Accuracy
Unreduced	38	—	98.5%	80.9%
FRFS	10	0.58783	89.2%	74.8%
antFRFS	9.55	0.58899	93.5%	77.9%

TABLE 12.7 Results for the 3-class dataset

Method	Attributes	γ' Value	Training Accuracy	Testing Accuracy
Unreduced	38	—	97.9%	83.2%
FRFS	11	0.59479	97.2%	80.9%
antFRFS	9.09	0.58931	94.8%	80.2%

can be seen in the classification accuracies, with FRFS performing slightly better than antFRFS. For both fuzzy methods the small drop in accuracy as a result of feature selection is acceptable.

By selecting a good feature subset from data, it is usually expected that the applied learning method should benefit, producing an improvement in results. However, when the original training (and test) data are very noisy, selected features may not necessarily be able to reflect all the information contained within the original entire feature set. As a result of removing less informative features, partial useful information may be lost. The goal of selection methods in this situation is to minimize this loss, while reducing the number of features to the greatest extent. Therefore it is not surprising that the classification performance for this challenging dataset can decrease upon data reduction, as shown in Table 12.7. However, the impact of feature selection can have different effects on different classifiers. With the use of an alternative classifier in Section 12.2.6.4, performance can be seen to improve for the test data.

The results here also show a marked drop in classification accuracy for the test data. This could be due to the problems encountered when dealing with datasets of small sample size. Overfitting can occur, where a learning algorithm adapts so well to a training set that the random disturbances present are included in the model as being meaningful. Consequently, as these disturbances do not reflect the underlying distribution, the performance on the test data will suffer. Although such techniques as cross-validation and bootstrapping have been proposed as a way of countering modeling inaccuracies, these still often exhibit high variance in error estimation [42].

12.2.6.2 Comparison with Entropy-Based Feature Selection As with the experimental studies carried out in Chapter 11, and in Section 12.2.2, to support the investigation of the performance of the fuzzy-rough methods, a conventional entropy-based technique is herein used again for comparison. A summary of the results of this comparison can be seen in Table 12.8.

For both the 2-class and 3-class datasets, FRFS and antFRFS select at least three fewer features than the entropy-based method. However, the entropy-based method outperforms the other two feature selectors with the resulting C4.5 classification accuracies. This is probably because C4.5 uses exactly the same entropy

TABLE 12.8 Results for the three selection methods

Approach	Number of Classes	Number of Features	Training Accuracy	Testing Accuracy
FRFS	2	10	89.2%	74.8%
antFRFS	2	9.55	93.5%	77.9%
Entropy	2	13	97.7%	80.2%
FRFS	3	11	97.2%	80.9%
antFRFS	3	9.09	94.8%	80.2%
Entropy	3	14	98.2%	80.9%

TABLE 12.9 Results for the 2-class and 3-class datasets using PCA

Accuracy		Number of Features								
	Class	5	6	7	8	9	**10**	11	12	13
Training (%)	2	80.0	80.0	80.0	80.0	80.3	**80.3**	80.3	80.8	82.1
Testing (%)	2	72.5	72.5	72.5	72.5	73.3	**73.3**	73.3	35.1	34.4
Training (%)	3	73.6	73.6	73.6	73.6	73.6	**75.9**	75.9	75.9	76.4
Testing (%)	3	80.9	80.9	80.9	80.9	80.9	**80.9**	80.9	80.9	80.2

measure in generating decision trees. The entropy-based measure will favor those attributes that are the most influential in the decision tree generation process.

12.2.6.3 Comparison with the Use of PCA

The effect of using PCA [77] is also investigated. Again, PCA is applied to the dataset, and the first n principal components are used. A range of values is chosen for n to investigate how the performance varies with dimensionality. The results are summarized in Table 12.9.

Both antFRFS and FRFS significantly outperform PCA on the 2-class dataset. Of particular interest is when 10 principal components are used, as this is roughly the same number chosen by antFRFS and FRFS. The resulting accuracy for PCA is 80.3% for the training data and 73.3% for the test data. For antFRFS, the accuracies were 93.5% (training) and 77.9% (testing), and for FRFS, 89.2% (training) and 74.8% (testing). In the 3-class dataset experimentation both fuzzy-rough methods produce much higher classification accuracies than PCA for the training data. For the test data, the performance is about the same, with PCA producing a slightly higher accuracy than antFRFS on the whole.

12.2.6.4 Comparison with the Use of a Support Vector Classifier

A possible limitation of employing C4.5 is that it performs some feature selection during the induction process. The resulting decision trees do not necessarily contain all the features present in the original training data. As a result it is beneficial to evaluate the use of an alternative classifier that uses all the given features. A support vector classifier [307] is used for this purpose, and trained by the sequential minimal optimization (SMO) algorithm [272]. The results of the application of this classifier can be found in Table 12.10.

TABLE 12.10 Results for the 2-class and 3-class datasets using SMO

Approach	Number of Classes	Number of Features	Training Accuracy	Testing Accuracy
Unreduced	2	38	80.0%	71.8%
FRFS	2	10	80.0%	72.5%
antFRFS	2	9.55	80.0%	72.5%
Unreduced	3	38	74.6%	80.9%
FRFS	3	11	73.6%	80.2%
antFRFS	3	9.09	73.6%	80.9%

For the 2-class dataset the training accuracy for both FRFS and antFRFS is the same as that of the unreduced approach. However, this is with significantly fewer attributes. Additionally the resulting testing accuracy *is* increased with these feature selection methods. With the more challenging 3-class problem, the training accuracies are slightly worse (as seen with the C4.5 analysis). The antFRFS method performs better than FRFS for the test data and is equal to the unreduced method, again using fewer features.

12.3 SUMMARY

Automated generation of feature pattern-based *if-then* rules is essential to the success of many intelligent pattern classifiers, especially when their inference results are expected to be directly human-comprehensible. This chapter has evaluated such an approach that integrates a recent fuzzy rule induction algorithm with a fuzzy-rough method for feature selection. Unlike semantics-destroying approaches such as PCA, this approach maintains the underlying semantics of the feature set, thereby ensuring that the resulting models are interpretable and the inference explainable. The method alleviates important problems encountered by traditional RSAR such as dealing with noise and real-valued features.

Overall, the experimental results presented in this chapter have shown the utility of employing FRFS, ant-based FRFS, and FQR as preprocessors to fuzzy rule induction. The work has demonstrated the potential benefits of using fuzzy dependencies and attribute grouping in the search for reducts. Not only are the runtimes of the induction and classification processes improved by this step (which for some systems are important factors), but the resulting rules are less complex.

In all experimental studies there has been no attempt to optimize the fuzzifications or the classifiers employed. It can be expected that the results obtained with optimization would be even better than those already observed. The generality of this approach should enable it to be applied to other domains. The ruleset generated by the RIA was not processed by any postprocessing tools so as to allow its behavior and capabilities to be revealed fully. By enhancing the induced ruleset through postprocessing, performance should improve.

CHAPTER 13

APPLICATIONS IV: ALGAE POPULATION ESTIMATION

Environmental issues have garnered a lot of attention in the last decade. Toxic and nontoxic waste production from a large variety of industrial plants and manufacturing processes is one of the most important areas. The future of humanity's food and water supply are influenced by this directly. Hence extreme care has to be taken in order to maintain the balance. It has also become clear that changes in farming and sewage water treatment can affect the ecology and chemistry of rivers and lakes.

Of particular interest are the communities of algae that flourish in such conditions. These communities are detrimental to water clarity and can endanger complex water life because of the resulting change in oxygen content. Human activities can also be affected by the toxic effects present in relation to algae growth. For example, cyanobacterial toxins are the naturally produced poisons stored in the cells of certain species of blue-green algae (*cyanobacteria*). These toxins fall into various categories. Some are known to attack the liver (hepatotoxins) or the nervous system (neurotoxins); others simply irritate the skin. These toxins cannot be treated by boiling water or typical forms of home water treatment.

Each of the many different species of alga have their own characteristics, responding very rapidly to changes in their environment. Ecologies where algae are present are thus heavily dependent on adequate chemical and physical balance in the environment. This has led to much active research interest concerning the impact that manufacturing, farming, and waste disposal have on nutrient content in rivers and how this can be combated.

Computational Intelligence and Feature Selection: Rough and Fuzzy Approaches, by Richard Jensen and Qiang Shen

An intelligent, automated tool for this task would be highly desirable, locating the parameters that control fluctuations in algae population and using this information to estimate changes [324]. Such a system could aid in a number of areas, including simulating hypothetical scenarios and predicting trends in algae communities, in addition to its intended estimation task. The collection of chemical concentration measurements in the water would be greatly simplified as a result of this tool. Only those tests that are absolutely necessary need to be conducted, which simplifies the testing process. An additional advantage of such a system is that the entire process would be decentralized, enabling individual testers to take samples and obtain results in situ. This would in turn reduce the cost associated with these measurements, and minimize response times.

Both the complexity of the estimation task and the interrelated nature of the chemicals diluted in the water suggest that knowledge acquisition could be an obstacle for the system. Eliciting knowledge from any source of data is notorious for its difficulty. Whether the source of information is a human expert or a dataset of experimental measurements, extracting general knowledge from it is a serious bottleneck in the development of a knowledge-based system in general.

Domain complexity typically adds a lot of difficulty to the knowledge acquisition process. Real-life application domains often yield complex data possessing a large number of attributes—many of which will be superfluous for the task at hand. In fact their presence has detrimental effects on data-driven knowledge acquisition and machine learning software. Such effects range from a drastic reduction in training speed to rendering knowledge acquisition intractable for the domain. This also affects the runtime of the system. Having additional quantities to measure tends to be a slow, error-prone, and cost-ineffective situation. In addition unknown, unavailable, or inaccurate values cannot be ruled out, while instruments inevitably develop faults of their own, and the humans reading them are not infallible. The absence of experts to interpret and check the data often proves to be another problem. These factors motivate the inclusion of a feature selection step [67,215]. Feature selection has been shown to be highly useful in reducing dimensionality while preserving the underlying meaning of the features involved [69,214].

13.1 APPLICATION DOMAIN

13.1.1 Domain Description

As stated previously, the aim of developing the system is to estimate the concentration of various different types of river alga, based on a set of chemical concentrations and other parameters [94]. To build the knowledge base, training samples were taken from different European rivers over the period of one year. These samples were analyzed to quantify the presence of several chemicals, including nitrates, nitrites and ammonia, phosphate, oxygen, and chloride. The pH of the water was also measured. In addition the algae population distributions for each of the species involved were determined in the samples. A number of additional factors were taken into account, such as the season, river size, and flow rate.

It is relatively easy to locate relations between one or two of these quantities and a species of algae. However, the process involves expertise in chemistry and biology and requires well-trained personnel and microscopic examination that cannot be automated given the state of the art. Thus the process can become expensive and slow if too many microscopic examinations are required. There are complex relations at work between the attributes of this application domain, be they conditional or decision: algae may influence one another, as well as be influenced by the concentration of chemicals. As such, there is expected to be some redundancy in the data; an important reason for the present utilization of approximation-based feature selection. Note that the work proposed here is not to eliminate the need for trained personnel but to minimize the need for potentially costly measurements and the subsequent visual examinations.

The algae estimation task had previously been attempted in [320]. The system, FuREAP, was reported to be a fuzzy-rough technique. However, this was not a true hybridization of fuzzy and rough sets but merely a combination of separate processes that performed fuzzy rule induction and crisp rough set reduction. Additionally the real-valued decision features were discretized, transforming the problem into a classification task rather than the more challenging problem of prediction, tackled in this chapter.

The application domain requires the system to be able to estimate the populations of 7 different species of alga based on 11 attributes of the river sample:

- Time of year the sample was taken, given as a season.
- Size of the river.
- Flow rate of the water.
- Eight chemical concentrations, including nitrogen in the form of nitrates, nitrites, ammonia, phosphate, the pH of the water, oxygen, and chloride

The dataset available for training includes 200 instances. The first three attributes of each instance (season, river size, and flow rate) are represented as fuzzy linguistic variables. Chemical concentrations and algae population estimates are given as continuous quantities. The dataset includes a few samples with missing values, where a measurement have not been obtainable for the sample. Of the 200 instances, two exhibiting mostly unknown values were removed from the data because of their low quality.

It is assumed that the river's water is perfectly homogeneous and that any sample of the water, no matter how small, is statistically representative. A few drops of each sample are examined visually via microscope and the number of algae are counted. Errors can occur in the collected data, and must somehow be handled sufficiently by future processing techniques. The use of fuzzy methods allows for errors in determining the population, and reflects the fact that a number of drops of water from a sample of a river are not necessarily statistically representative of the entire river.

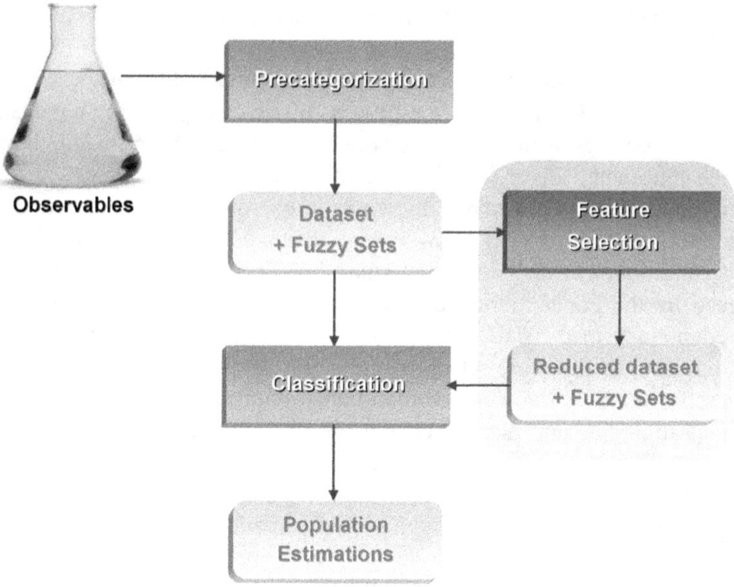

Figure 13.1 Modular decomposition of the implemented system.

A block diagram of the system is presented in Figure 13.1, showing both the training and runtime stages of the system.

During the training stage, 7 unreduced datasets (one per alga species) of 11 conditional attributes each are obtained from water samples. The datasets are reduced with FRFS to obtain 7 datasets with (on average) 7 conditional attributes each. These are then provided to the classifier, which induces 7 models (one for each species of alga).

During runtime, the water samples are analyzed to obtain only 7 (on average) of the original 11 conditional attributes, as per the reduct set chosen by FRFS. This simplifies, speeds up, and reduces the costs associated with the data-gathering stage. In running experimental simulations, these new 7-attribute datasets are used by classifiers to provide the system's user with estimations of the 7 algae populations.

13.1.2 Predictors

To facilitate a careful experiment-based analysis of the present work, five predictors were used to estimate the algae populations [382]: standard linear regression, a backpropagation neural network (BPNN), M5Prime, Pace regression, and a support vector-based system called SMOreg. The following briefly introduces these methods with details omitted (readers can refer to the respective references given).

The linear regression model [92] is applicable for numeric classification and prediction provided that the relationship between the input attributes and the output attribute is almost linear. The relation is then assumed to be a linear function of some parameters, the task being to estimate these parameters given training data. The estimates are often accomplished by the method of least squares, which consists of finding the values that minimize the sum of squares of the residuals. Once the parameters are established, the function can be used to estimate the output values for unseen data.

BPNNs [36] consist of a network of nodes arranged in several layers - the input, hidden, and output layers. Input and output layers buffer the input/output for the model, respectively. The hidden layer(s) provide a means for representing input relations. The network is trained by repeatedly presenting it with (labeled) training data and backpropagating any resulting errors in classification through it, adjusting weights between nodes in the process. This weight modification is achieved via the gradient of error curve.

M5Prime is a rational reconstruction of Quinlan's M5 model tree inducer [375]. While decision trees were designed for assigning nominal categories, this representation can be extended to numeric prediction by modifying the leaf nodes of the tree to contain a numeric value that is the average of all the dataset's values that the leaf applies to.

Projection adjustment by contribution estimation (Pace) regression [376] is a recent approach to fitting linear models, based on competing models. Pace regression improves on classical ordinary least squares regression by evaluating the effect of each variable and using a clustering analysis to improve the statistical basis for estimating their contribution to the overall regression.

SMOreg is a sequential minimal optimization algorithm for training a support vector regression using polynomial or radial basis function kernels [272, 346]. It reduces support vector machine training down to a series of smaller quadratic programming subproblems that have an analytical solution. This has been shown to be very efficient for prediction problems using linear support vector machines and/or sparse data sets.

13.2 EXPERIMENTATION

For each of the 7 algae datasets, 10-fold cross-validation [353] was used to estimate the predictor's performance. The experimental results are given as two types of graph: root mean squared error (RMSE) and mean absolute error (MAE). The mean absolute error is computed by summing the absolute difference between the actual and predicted target value for each instance and then taking the average. The root mean squared error is determined by summing the squared differences between actual and predicted values, and taking the square root of the average. Both quantities are given for each predictor over the 7 datasets.

13.2.1 Impact of Feature Selection

To investigate the impact of feature selection on predictor performance, the experimentation was carried out both with and without FRFS. The unreduced data for

TABLE 13.1 Features selected: FRFS

Species	Subset
1	{season, size, flow, 1, 2, 3, 7}
2	{season, size, flow, 1, 2, 3, 7, 8}
3	{season, size, flow, 1, 3, 4, 5, 7}
4	{season, size, flow, 1, 2, 5}
5	{season, size, flow, 1, 2, 4, 7, 8}
6	{season, size, flow, 1, 2, 4, 5, 8}
7	{season, size, flow, 1, 7, 8}

each species of alga were supplied to each predictor and used in evaluation via cross-validation. Then the same data were processed by FRFS to reduce dimensionality and evaluated in an identical fashion. This resulted in, on average, a 7-attribute dataset selected from the original, 11-attribute one.

The exact selected attributes were different for each alga species (as can be seen in Table 13.1), although certain attributes were present in all 7 reduct sets, namely the season, size of the river, flow rate of the water, and concentration 1. The obtained reducts could not be verified based on empirical evidence because the dataset documentation mentions the names of the concentration attributes but not their ordering in the data; hence it is needed to refer to the chemical concentrations by number rather than name. However, based on previous experience with FRFS [163], it is expected that the selected feature subsets would overall make sense to an expert. It must also be noted, however, that it is difficult to verify directly the quality of selected attributes, in default of a suitable quality metric. The most accessible way is therefore to use the reduced and unreduced data to train a learning system, and compare the results. This gives an indirect measure of subset quality.

The results of experimentation using linear regression can be found in Figure 13.2. It can be seen that both approaches perform similarly in terms of RMSE and MAE, with FRFS-based predictions somewhat more accurate in general. This trend is reflected in the results for M5Prime (presented in Figure 13.3) and Pace (Figure 13.5). For SMOreg (Figure 13.6) the results for both methods are very similar, which is to be expected as SVM methods are not sensitive to feature selection. It is worth reiterating that the task of the system is to reduce the number of measurements that must be obtained while maintaining prediction performance. This is clearly the case in these experiments.

Figure 13.4 shows the results for the BPNN-based predictor. Here a small difference in performance can be seen between the two approaches. The method that incorporates FRFS produces some improvement in accuracy for each algae estimation problem.

Again, note that the improvement in accuracies are obtained with fewer measured variables, which is important for dynamic systems where observables are often restricted, or where the cost of obtaining more measurements is high. In the river algae domain, for instance, providing different measurements has different

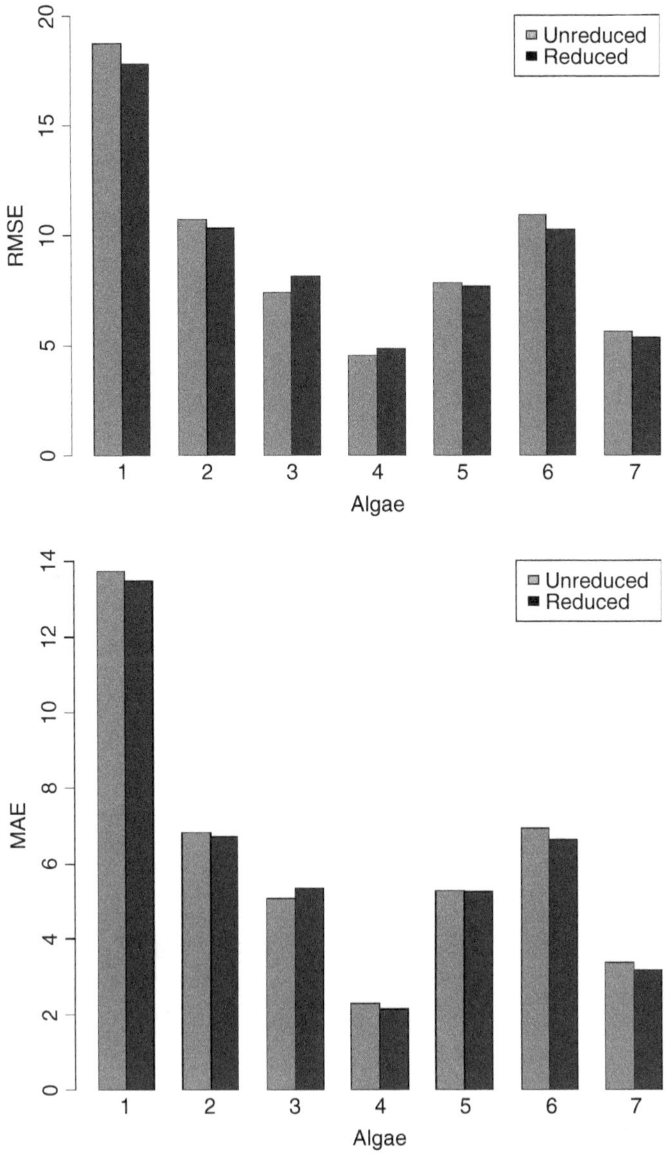

Figure 13.2 Unreduced and reduced data RMSEs and MAEs with linear regression

costs attached. It is trivial to give the time of year and size of river, but flow rate may need extra equipment. Additionally each of the measurements of concentration of chemicals may need its own process, requiring time, well-trained personnel, and money. Reducing the number of measurements to be made significantly enhances the potential of the estimator system.

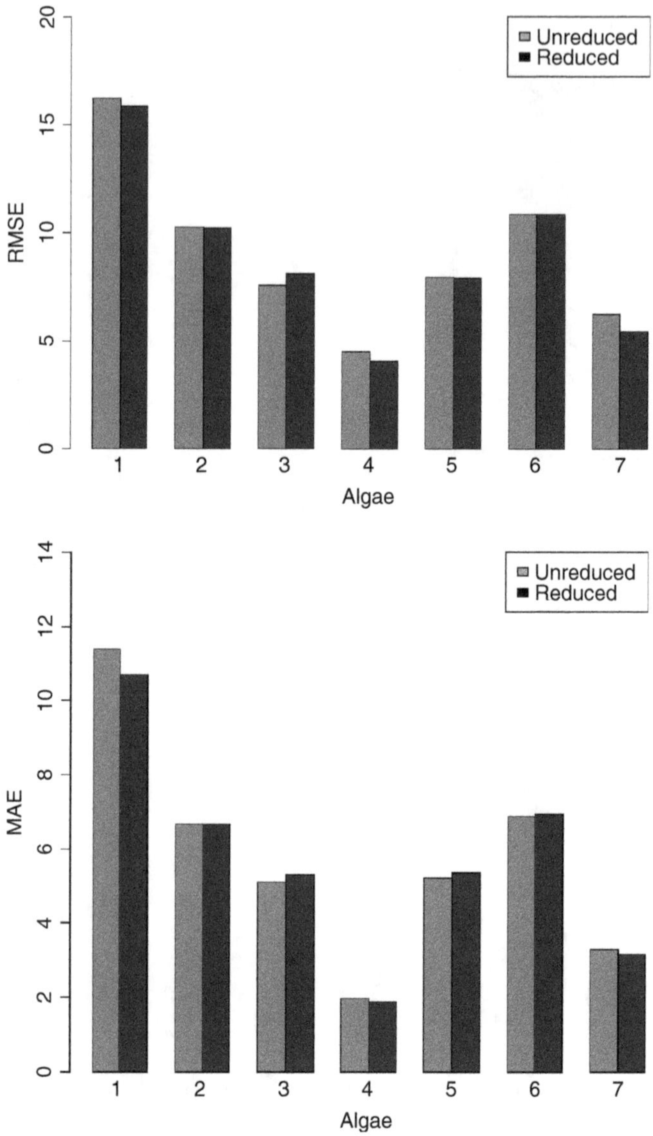

Figure 13.3 Unreduced and reduced data RMSEs and MAEs with M5Prime

13.2.2 Comparison with Relief

In order to further show the utility of feature selection, and in particular, the benefits of using FRFS, a well-established FS algorithm was chosen for experimental comparisons: Relief (see Section 4.2.1.1). Unlike most FS methods, both FRFS and Relief can handle continuous decision features. For the experimentation

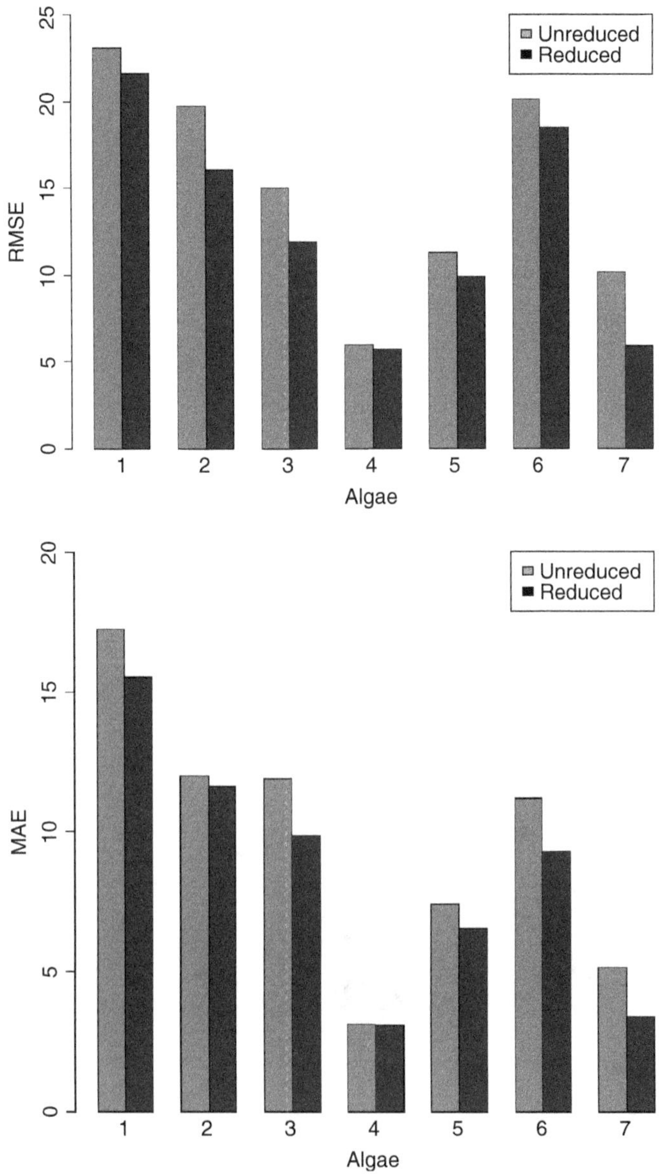

Figure 13.4 Unreduced and reduced data RMSEs and MAEs with BPNN

presented here, only those features that result in a final positive weight are selected (see Table 13.2).

Figures 13.7 to 13.13 show the results for the unreduced, FRFS-reduced, and RELIEF-reduced data for algae species 1 to 7. It is clear that estimators trained using data reduced by FRFS generally outperform those trained using

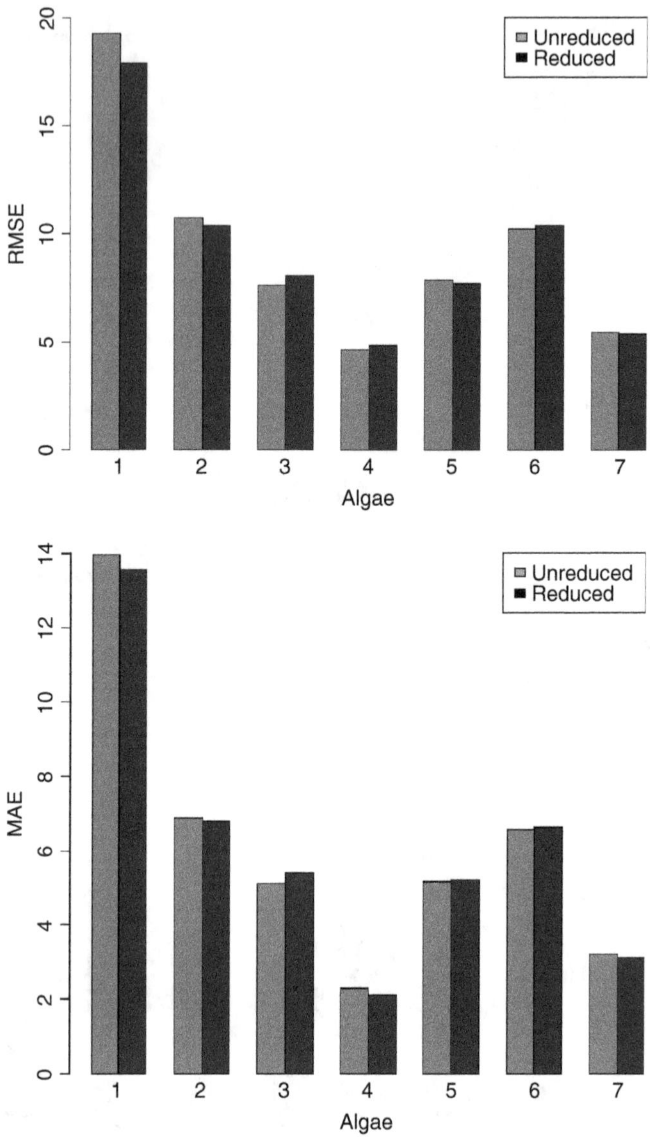

Figure 13.5 Unreduced and reduced data RMSEs and MAEs with Pace

RELIEF-reduced data for algae species 1, 5, and 7. RELIEF performs generally better than FRFS for species 3 and 4. For the remaining algae species, both methods perform equivalently. Note that for all species in general, population estimators that employ feature selection perform significantly better than those without. This suggests that the data contains features that are redundant, noisy, or irrelevant to the task at hand.

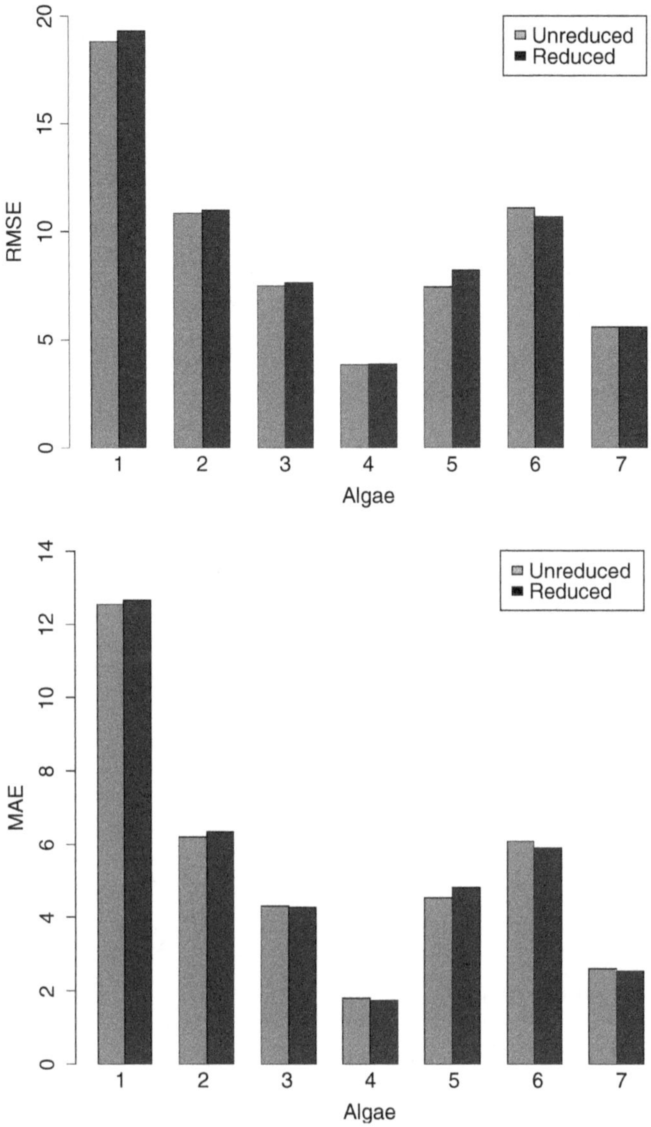

Figure 13.6 Unreduced and reduced data RMSEs and MAEs with SMOreg

The average RMSEs over all classifiers and algae species are 10.46 for the unreduced method, 9.93 for FRFS, and 10.17 for RELIEF. Similarly the average MAEs are 6.67 for the unreduced method, 6.38 for FRFS, and 6.48 for RELIEF. The FRFS-reduced method exhibits a lower average RMSE and MAE overall. The P-values for each classifier for FRFS and RELIEF can be found in Table 13.3, where comparisons are against the unreduced data performance. It can be seen

TABLE 13.2 Features selected: RELIEF

Species	Subset
1	{1}
2	{size, flow, 2, 3, 4, 5, 6, 7, 8}
3	{size, flow, 2, 3}
4	{size, 1, 2, 3, 4, 5, 6, 7}
5	{size, 1, 3, 4, 5, 6, 7, 8}
6	{size, 1, 3, 4, 5, 6, 7, 8}
7	{season, 1, 4, 5, 6, 7, 8}

that for both methods, the only statistically significant improvement is for neural net-based prediction.

13.2.3 Comparison with Existing Work

Work has been previously carried out on these data in [49], which uses a neural network-based approach to predict the populations. A multi-layer perceptron feed forward network with one hidden layer was employed, with direct connections between every input and output neuron.

The dataset used is exactly the same as the one investigated here, however, in [49], two power transformation techniques were applied. First, the variables were normalized taking into account skewed distributions or overly peaked or flat distributions. Second, the relationships between interval targets and interval variables were linearized. The transformed data were then split into training (80%), validation (10%), and testing (10%). The neural net attempts to find the best weights using the training data set. The validation set was used to assess the adequacy of the model and was also used for model fine-tuning. The test set is a holdout data set that is not used in training or validation. Its sole purpose is for obtaining a final, unbiased estimate of the generalization error of the trained network.

The results from this work are presented in Table 13.4 alongside the best RMSEs from the estimators considered in the present work, for both unreduced and reduced algae data. For algae species 1 and 4 the approaches proposed in this chapter result in a more accurate estimation of population. The remaining species result in a slightly worse performance. It is surprising that the results obtained here are very similar given that the data used in [49] has been substantially modified to better suit it for the requirements of the learning algorithm. The data used in the present research have not been altered in any way. Note that no optimization of fuzzifications has taken place.

13.3 SUMMARY

The control and limitation of waste production is a very important issue in preserving the fragile balance of river ecologies. River algae are very sensitive

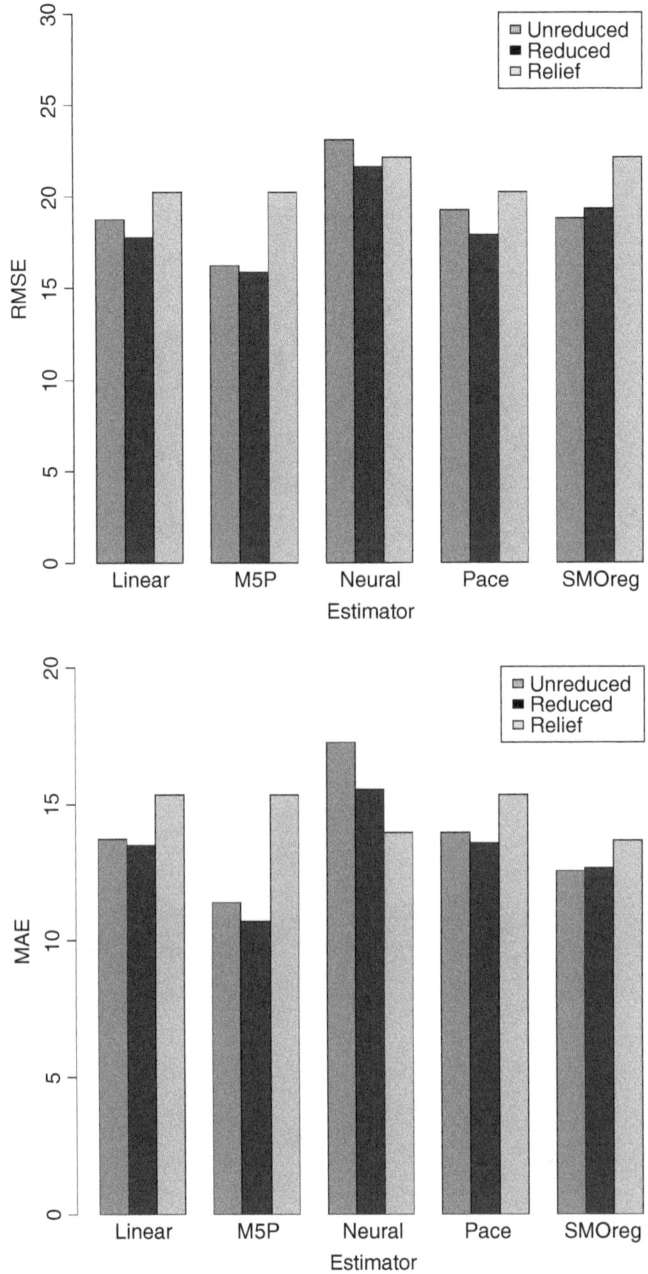

Figure 13.7 Unreduced, reduced, and RELIEF-reduced RMSEs and MAEs for species 1

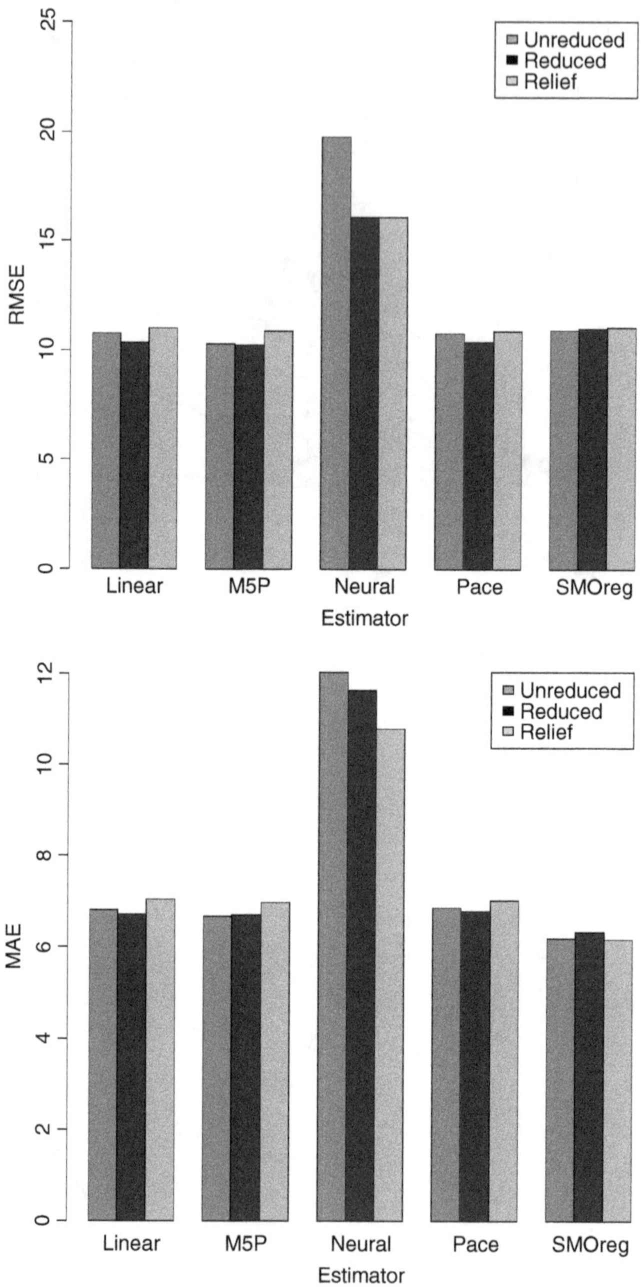

Figure 13.8 Unreduced, reduced, and RELIEF-reduced RMSEs and MAEs for species 2

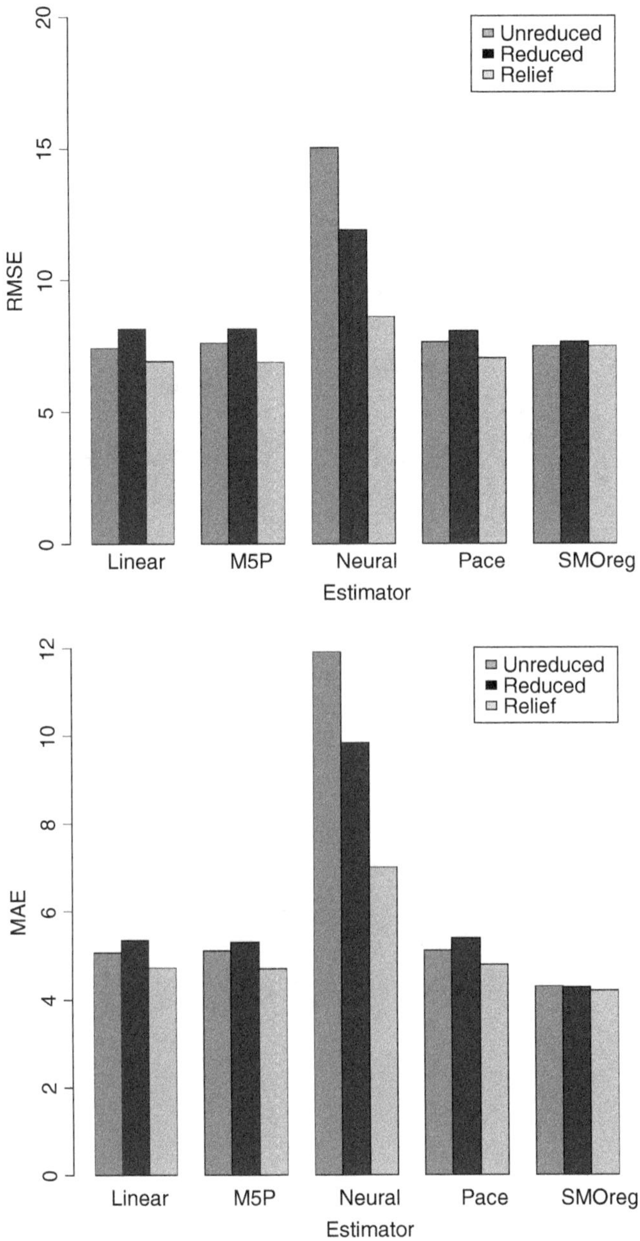

Figure 13.9 Unreduced, reduced, and RELIEF-reduced RMSEs and MAEs for species 3

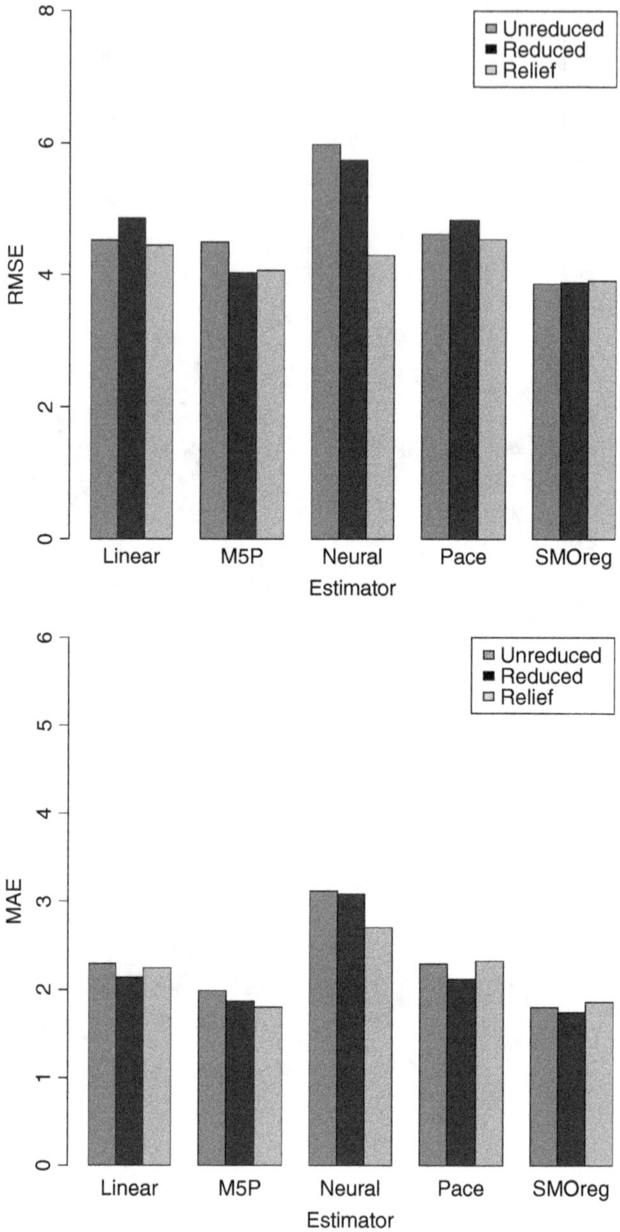

Figure 13.10 Unreduced, reduced, and RELIEF-reduced RMSEs and MAEs for species 4

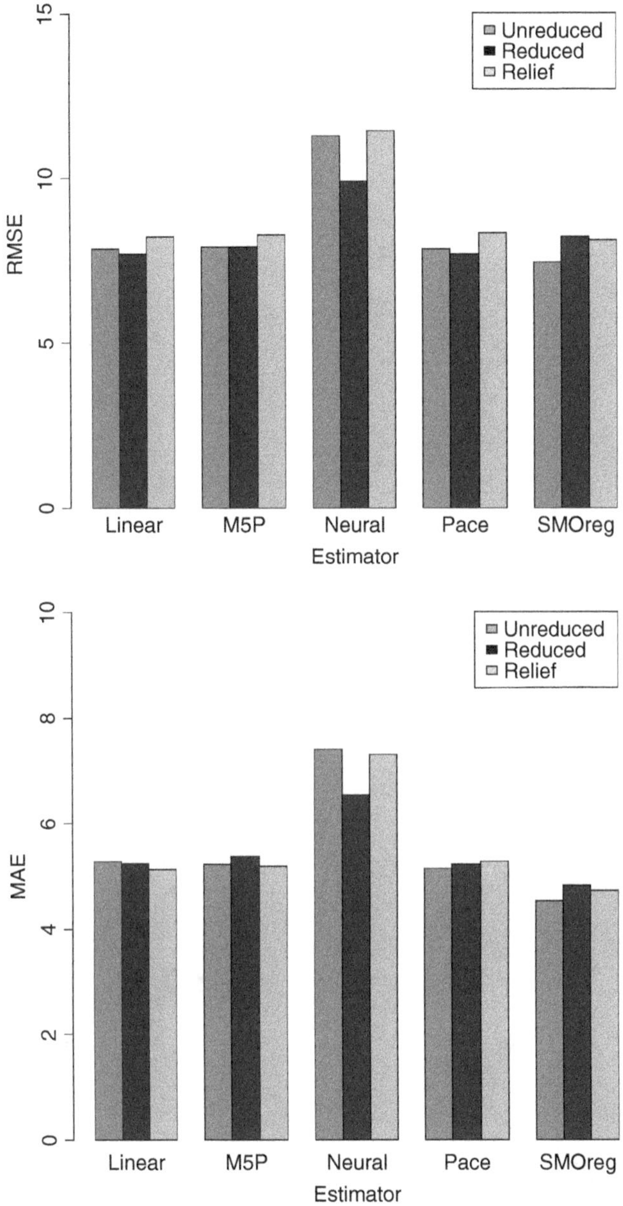

Figure 13.11 Unreduced, reduced and RELIEF-reduced RMSEs and MAEs for species 5

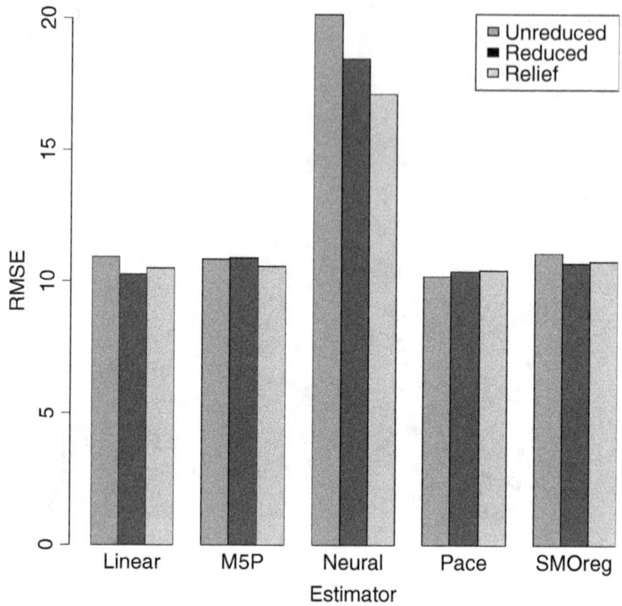

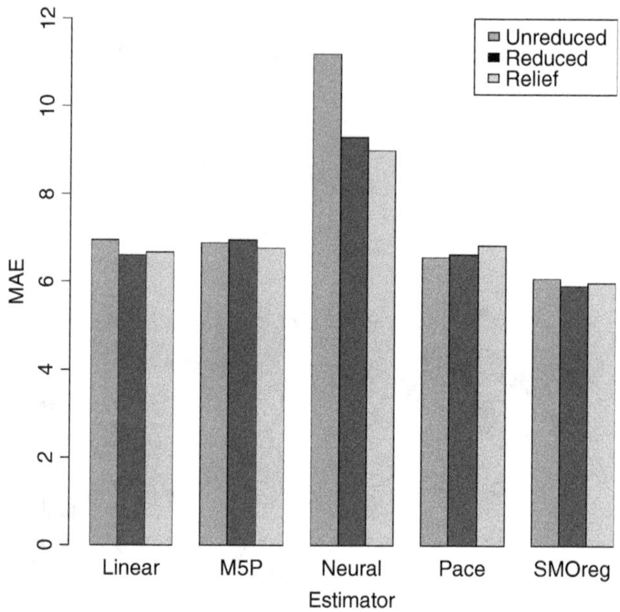

Figure 13.12 Unreduced, reduced, and RELIEF-reduced RMSEs and MAEs for species 6

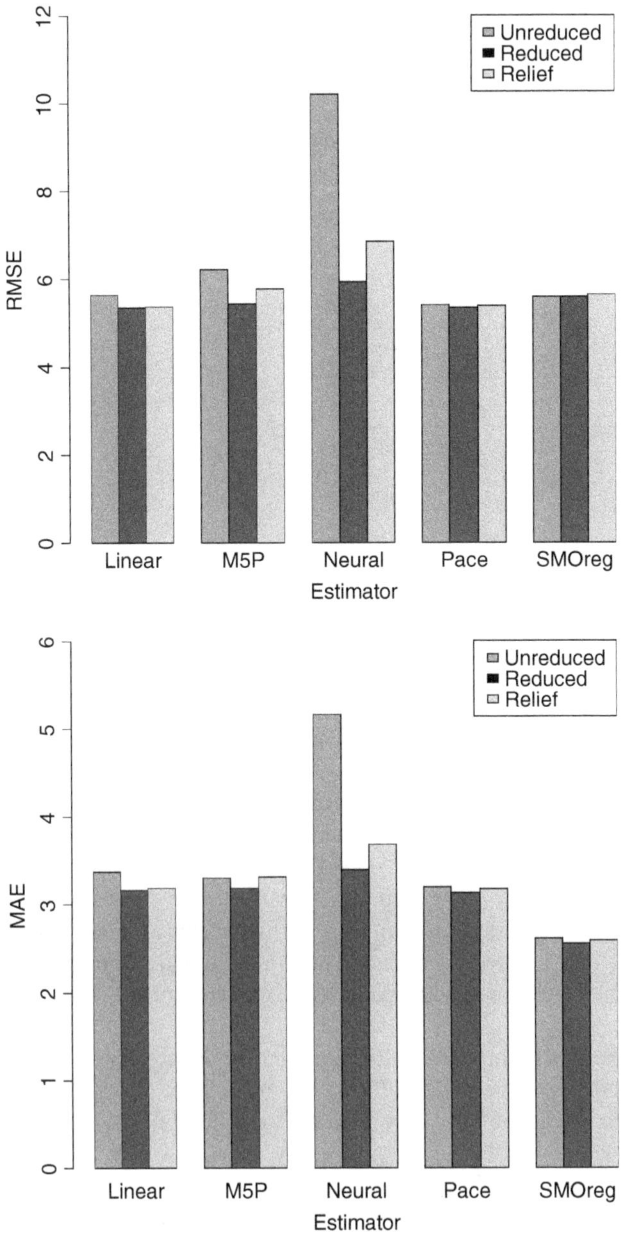

Figure 13.13 Unreduced, reduced, and RELIEF-reduced RMSEs and MAEs for species 7

TABLE 13.3 Each classifier's *P*-value for the FS methods

Classifier	RMSE		MAE	
	FRFS	RELIEF	FRFS	RELIEF
Linear	0.3792	0.6766	0.1620	0.6553
M5P	0.3588	0.5018	0.5416	0.4224
Neural	0.0061	0.0158	0.0065	0.0227
Pace	0.4735	0.4475	0.6175	0.2932
SMOreg	0.3109	0.2806	0.6260	0.3592

TABLE 13.4 Comparison of RMSEs for algae population estimators

Algae	Results from [49]		Best Predictor	
Species	Validation	Test	Unreduced	Reduced
1	13.05	18.14	16.22	15.88
2	13.30	7.14	10.27	10.22
3	5.57	5.33	7.43	7.65
4	3.98	6.27	3.86	3.88
5	6.33	4.93	7.48	7.70
6	16.36	9.54	10.23	10.29
7	5.37	3.55	5.42	5.34

to changes in their environment and, in turn, can influence the well-being of more complex life forms. Ecologies where algae are present are thus heavily dependent on chemical and physical balance in the environment. Growth in algae communities is associated with poor water clarity and various detrimental effects on other forms of river life, as well as humans. Thus measuring and reducing the impact that farming, manufacturing, and waste disposal have on river ecologies has attracted much attention recently, especially with respect to estimating and controlling river algae population fluctuations. Biologists are attempting to locate the chemical parameters that control the rapid growth of algae communities.

It would be desirable to have an intelligent computer-based system to locate these parameters and use this information to estimate population fluctuations. Such a system would serve in a number of ways. It would simplify the collection of measurements by isolating the absolutely necessary tests needed. The system would also decentralize the entire process, allowing individual testers to sample rivers and obtain results rapidly and in situ. This would, in turn, reduce monetary and time requirements.

This chapter has described an approximation-based fuzzy-rough estimator of algae populations. The approach integrates fuzzy-rough feature selection with potentially powerful estimator systems in a modular manner. The FRFS subsystem helps reduce the dimensionality of the domain with which the prediction subsystem has to cope. The FRFS algorithm has proved to be very useful in stripping out insignificant information, while retaining more important conditional attributes. Another desirable feature of this technique is the fact that unlike transformation-based attribute reduction approaches [163], it maintains the underlying semantics of the dataset, enabling human experts to glean the distilled knowledge. In all experimental studies there has been no attempt to optimize the fuzzifications or the classifiers employed. It can be expected that the results obtained with optimization would be even better than those already observed.

CHAPTER 14

APPLICATIONS V: FORENSIC GLASS ANALYSIS

The evaluation of glass evidence in forensic science is an important issue. Traditionally this has depended on the comparison of the physical and chemical attributes of an unknown fragment with a control fragment. A high degree of discrimination between glass fragments is now achievable due to advances in analytical capabilities. A random effects model using two levels of hierarchical nesting is applied to the calculation of a likelihood ratio (LR) as a solution to the problem of comparison between two sets of replicated continuous observations where it is unknown whether the sets of measurements shared a common origin. Replicate measurements from a population of such measurements allow the calculation of both within-group and between-group variances. Univariate normal kernel estimation procedures have been used for this, where the between-group distribution is considered to be nonnormal. However, the choice of variable for use in LR estimation is critical to the quality of LR produced. This chapter investigates the use of feature evaluation for the purpose of selecting the variable to model without the need for expert knowledge. Results are recorded for several selectors using normal, exponential, adaptive, and biweight kernel estimation techniques. Misclassification rates for the LR estimators are used to measure performance.

14.1 BACKGROUND

One of the less obvious, but frequent, sources of forensic evidence are traces of glass. This is regularly encountered at crime scenes, particularly those involving

Computational Intelligence and Feature Selection: Rough and Fuzzy Approaches, by Richard Jensen and Qiang Shen

motor vehicle accidents, car theft, and burglaries. Windows are a common point of entry into buildings for burglars and large quantities of broken glass are produced in traffic accidents. Such glass fragments may remain for a long time (depending on the type of crime) and do not degrade like biological evidence.

The forensic scientist's role in analyzing glass is to clearly and unambiguously determine the origin of the sample. Variation in the manufacture of glass allows considerable discrimination even with very small fragments. Consider the scenario where a crime has been committed involving the breakage of glass. It is likely that fragments will remain at the location of the offense (referred to as control fragments). Fragments may also have been transferred to the clothes and footwear of the offender. A suspect may be identified and, on subsequent examination, found to have glass fragments on their person (referred to as recovered fragments as their source is not known). In this case the purpose of fragment analysis is to evaluate the evidence for comparison of the proposition that the glass associated with the suspect is from the same source as the fragments from the crime scene with the proposition that the glass associated with the suspect is not from the same source as the fragments from the crime scene.

The first stage of the examination of this kind of evidence is the recovery of glass fragments from the suspect. This is most frequently achieved through shaking and/or brushing garments. The resulting debris is observed under an optical microscope and glass fragments separated manually. Most of these fragments have a linear dimension of less than 0.5 mm. Morphological features, such as thickness and color, do not provide enough information to deduce whether the fragments in question were derived from the same source as fragments found at the crime scene. Therefore it is necessary to determine their physicochemical properties. This is often achieved through the GRIM (glass refractive index measurements) method and instrumental methods of elemental assay (such as micro X-ray fluorescence (μ-XRF), laser ablation–inductively coupled plasma–mass spectrometry (LA-ICP-MS), and scanning electron microscope–energy dispersive using X ray (SEM-EDX).

The comparison between recovered and control glass fragments is then made on the basis of the analytical results. The increasing ability to collect and store data relevant for identification in a forensic context has led to a corresponding increase in methods for the numerical evaluation of evidence associated with particular evidence types. The comparison of two sets of glass fragments by numerical methods requires careful attention to the following considerations. First, the similarity of recovered glass fragment(s) to a control sample must be taken into account. Second, information must be considered about the rarity of the determined physicochemical characteristics (e.g., elemental concentrations) for control and recovered samples in the relevant population. Third, the level of association between different characteristics where more than one characteristic has been measured should be accounted for. Fourth, other possible sources of variation should be considered, including the variation of measurements of characteristics within the control items, within recovered items, and between control and recovered items.

Significance tests are often used for this purpose. However, these only take into account information concerning within-source variation and item similarity. This is a comparison of the samples based purely on their physicochemical properties. From a forensic viewpoint, knowledge concerning the sources of variability and rarity of the measured properties should also be considered to produce a more definitive evaluation. This motivates the use of two-level univariate models that can incorporate such information effectively [1].

As discussed in [2], problems are encountered when dealing with multivariate forensic data. For statistical analysis much more data are required than are generally accessible. Additionally, as the number of features increases, the number of means, variances, and covariances increase exponentially. The calculation of a full model is simply not practical. As a result of these issues a single informative variable is usually selected by an expert for use in the univariate model. The choice of such a variable is obviously a critical factor in the resulting quality of evidence evaluation. Unfortunately, it is not always known which single feature will provide the most information for this purpose. There are also situations where many competing variables co-exist; manual selection of which variable to use may result in subsequent analysis being too subjective. Through the use of feature evaluation methods this important decision can be made without expert knowledge. Indeed experts may not be available for making such decisions. The opposing side in a court case could argue that by selecting features, there is a loss of information. However, it is often the case that datasets contain many features that are purely redundant. The inclusion of such features will degrade the resulting quality of analysis. It is useful in this case to remove these or at least recommend these for potential removal to the forensic scientist.

14.2 ESTIMATION OF LIKELIHOOD RATIO

Commonly used significance tests, like the Student-t test for univariate data and Hotelling's T^2 [183,184], take into account only information about within-source variation and the similarity of the compared items. Thus the tests provide information on the similarity of items on the basis of their physicochemical properties. From the forensic point of view, this is inadequate. What is required is information concerning the value of the evidence of these measurements with relation to the proposition that the two samples of glass fragments did, or did not, come from the same source. This requires knowledge about the sources of variability and the rarity of the measured physicochemical properties in the relevant population. For instance, one would expect refractive index (RI) values from different locations on the same glass object to be very similar. However, equally similar RI values could well be observed from different glass items. Without a wider context it is not possible to ascribe meaning to the observed similarity. Therefore inferences about the source of glass fragments made purely on the basis of similarity of measurements are incomplete. This section outlines several techniques for the estimation of likelihood ratios that attempt to account for the variability and rarity issues through two-level models. Further details are given in [3].

14.2.1 Exponential Model

Consider a two-level random effects model for a random variable X such that $(X_{ij} \mid \mu, \sigma^2) \sim N(\mu, \sigma^2)$ and $(\mu \mid \alpha) \sim \exp(\alpha)$ with probability density function

$$f(\mu \mid \alpha) = \alpha \exp(-\alpha\mu)$$

Let $\{x_{ij}, i = 1, \ldots, m,\ j = 1, \ldots, k\}$ be a random sample from this model of k observations from each of m groups. Denote the m group means by $\overline{x}_1, \ldots, \overline{x}_m$, where $\overline{x}_i = \sum_{j=1}^{k} x_{ij}/k$. The overall mean is denoted $\overline{x}_.$ with $\overline{x}_. = \sum_{i=1}^{m}\sum_{j=1}^{k} x_{ij}/km$.

Data $\mathbf{y_1} = \{y_{1j}, j = 1, \ldots, n_c\}$ of n_c observations from one group from a crime scene (control data) and data $\mathbf{y_2} = \{y_{2j}, j = 1, \ldots, n_s\}$ of n_s observations from a group associated with a suspect (recovered data). The value V of the evidence of these data is to be determined.

The exponential distribution is investigated. This is because it is not easy to transform to a normal distribution and because a theoretical value for the likelihood ratio may be obtained against which various estimative procedures may be compared.

Some intermediate calculations and notation are required.

Let $\overline{y}_1 = \sum_{j=1}^{n_c} y_{1j}/n_c$ and $\overline{y}_2 = \sum_{j=1}^{n_s} y_{2j}/n_s$ denote the means of the crime and suspect data, respectively. Let $s_{y1}^2 = \sum_{j=1}^{n_c}(y_{1j} - \overline{y}_1)^2/(n_c - 1)$ and $s_{y2}^2 = \sum_{j=1}^{n_s}(y_{2j} - \overline{y}_2)^2/(n_s - 1)$ denote the variances of the crime and suspect data, respectively.

The within-group variance σ^2 of the underlying population is assumed known. Its value is taken to be $s_w^2 = \sum_{i=1}^{m}\sum_{j=1}^{k}(x_{ij} - \overline{x}_i)^2/(mk - m)$. The between-group variance of the underlying population is also assumed known. Its value is taken to be $s_b^2 = \sum_{i=1}^{m}(\overline{x}_i - \overline{x})^2/(m-1) - s_w^2/m$.

The expectation of μ, an exponentially distributed random variable, is $1/\alpha$. The parameter α is estimated by $(\bar{x})^{-1}$. The variance of μ is $1/\alpha^2$.

The value of the evidence $(\mathbf{y_1}, \mathbf{y_2})$ is given by

$$V = \frac{\int f(\mathbf{y_1}, \mathbf{y_2} \mid \mu, \sigma^2) f(\mu \mid \alpha) d\mu}{\int f(\mathbf{y_1} \mid \mu, \sigma^2) f(\mu \mid \alpha) d\mu \int f(\mathbf{y_2} \mid \mu, \sigma^2) f(\mu \mid \alpha) d\mu}$$

$$= \frac{\int f(\mathbf{y_1} \mid \mu, \sigma^2) f(\mathbf{y_2} \mid \mu, \sigma^2) f(\mu \mid \alpha) d\mu}{\int f(\mathbf{y_1} \mid \mu, \sigma^2) f(\mu \mid \alpha) d\mu \int f(\mathbf{y_2} \mid \mu, \sigma^2) f(\mu \mid \alpha) d\mu}$$

Some extra notation is helpful:

$$\sigma_{12}^2 = \sigma^2\left(\frac{1}{n_c} + \frac{1}{n_s}\right)$$

$$\sigma_3^2 = \frac{\sigma^2}{n_c + n_s} + \frac{1}{\alpha^2}$$

$$w = \frac{n_c \bar{y}_1 + n_s \bar{y}_2}{n_c + n_s}$$

The value V of the evidence is given as

$$V = \frac{1}{\alpha\,\sigma_{12}\,\sqrt{2\pi}} \exp\left[-\frac{1}{2\sigma_{12}^2} (\bar{y}_1 - \bar{y}_2)^2 + \frac{\alpha}{2}\left\{ 2(\bar{y}_1 + \bar{y}_2 - w) + \alpha\sigma_3^2 - \alpha\sigma^2 \left(\frac{1}{n_c} + \frac{1}{n_s} \right) \right\} \right]. \quad (14.1)$$

For estimations, the parameters α and σ^2 are replaced by their estimates from the population data $\{x_{ij}, i = 1, \ldots, m;\ j = 1, \ldots, k\}$, namely $(\bar{x})^{-1}$ and s_w^2, respectively. If the between-group distribution is assumed to be exponential, then an estimate of the value of evidence in a particular case with crime data $\mathbf{y}_1$ and suspect data $\mathbf{y}_2$ may be obtained with substitution of the appropriate numerical values for $\bar{y}_1$ and $\bar{y}_2$ in (14.1).

14.2.2 Biweight Kernel Estimation

The use of a kernel density estimate based on the normal distribution is difficult when there is an achievable lower bound to the range of the variable being modeled and the data are highly positively skewed so that much of the data are close to the lower bound. In the example to be discussed here, the lower bound is zero, and a kernel based on a normal distribution is very inaccurate close to this lower bound. A more appropriate approach for modeling a highly positively skewed distribution is the use of a biweight kernel [372] with a boundary kernel for use when the kernel comes close to the lower bound of the range of the random variable, in this case zero. The biweight kernel $K(z)$ is defined as

$$K(z) = \frac{15}{16}(1 - z^2)^2, \qquad |\, z \,| < 1 \quad (14.2)$$

This kernel is used to model the between-group distribution by way of the sample means $\{\bar{x}_1, \ldots, \bar{x}_m\}$. A general biweight kernel, with smoothing parameter h, and with a between-group variance of τ^2, is given by

$$\frac{1}{h\tau} K\left(\frac{\mu - \bar{x}}{h\tau} \right) = \frac{15}{16h\tau} \left\{ 1 - \left(\frac{\mu - \bar{x}}{h\tau} \right)^2 \right\}^2, \qquad \bar{x} - h\tau < \mu < \bar{x} + h\tau. \quad (14.3)$$

There are two candidates for the estimation of the between-group variance:

1. $s_b^2 = \sum_{i=1}^{m} (\bar{x}_i - \bar{x}_.)^2/(m-1) - s_w^2/k$
2. $1/(\bar{x})^2$

that is, the least-squares estimate and the method of moments estimate, respectively, of τ^2, the between-group variance.

The problem of a fixed lower bound at zero is tackled with a boundary kernel. When an observation $\bar{x}$ is close to zero, a different kernel, known as the boundary kernel [372], is used. Closeness is defined as $\bar{x} < h\tau$. For $\bar{x} > h\tau$, the biweight kernel (14.3) is used. For $\bar{x} < h\tau$, a boundary kernel

$$K_h(z) = \frac{\nu_2 - \nu_1 z}{\nu_0 \nu_2 - \nu_1^2} K(z) \tag{14.4}$$

is used, where $K(z)$ is as given in (14.2). For ease of notation, denote $h\tau$ by δ. The terms ν_0, ν_1, and ν_2 are constants, functions of δ. For the kernel (14.2) these are defined as

$$\nu_t = \int_{-1}^{\delta} z^t K(z) dz, \qquad t = 0, 1, 2$$

where the dependency in ν on δ is suppressed. These constants can be shown to be

$$\begin{aligned} \nu_2 &= \frac{1}{14} \left\{ 1 + \frac{1}{8} \delta^3 (35 - 42\delta^2 + 15\delta^4) \right\} \\ \nu_1 &= \frac{5}{32} \left\{ \delta^2 (3 - 3\delta^2 + \delta^4) - 1 \right\} \\ \nu_0 &= \frac{1}{2} + \frac{15}{16} \left(\delta - \frac{2}{3}\delta^3 + \frac{1}{5}\delta^5 \right) \end{aligned}$$

In practice, the factor $(\nu_2 - \nu_1 z)/(\nu_0 \nu_2 - \nu_1^2)$ is close to 1. An optimal value of the smoothing parameter h is given by

$$h_{opt} = \left(\frac{1}{7}\right)^{-2/5} \left(\frac{15}{21}\right)^{1/5} \left\{ \int f''(x)^2 dx \right\}^{-1/5} m^{-1/5}$$

[331]. Then it can be shown that, when $f(x) = \alpha \exp\{-\alpha x\}$,

$$h_{opt} = \left(\frac{70}{m}\right)^{1/5} \alpha^{-1}$$

which can be estimated by

$$h_{opt} = \left(\frac{70}{m}\right)^{1/5} \bar{x}.$$

14.2.3 Likelihood Ratio with Biweight and Boundary Kernels

First, consider the denominator and the factor that is associated with the crime sample $\{y_{1i}, i = 1, \ldots, n_c\}$. Denote this as D_c, which may be written as

$$D_c = \int f(y_{11}, \ldots, y_{1n_c} \mid \mu, \sigma^2) f(\mu \mid \alpha) d\mu$$

The factor associated with the suspect sample may be derived analogously, so denote this as D_s. The first term, D_c, in the denominator, with the biweight kernel (14.3) used for $f(\mu \mid \alpha)$, is given by

$$
\begin{aligned}
D_c &= \int f(\bar{y}_1 \mid \mu, \sigma^2) f(\mu \mid \alpha) d\mu \\
&= \frac{\sqrt{n_c}}{\sigma\sqrt{2\pi}} \int \exp\left\{-\frac{n_c}{2\sigma^2}(\bar{y}_1 - \mu)^2\right\} \left[\frac{15}{16\, m\, h\, \tau} \sum_{i=1}^{m} \left\{1 - \left(\frac{\mu - \bar{x}_i}{h\,\tau}\right)^2\right\}^2\right] d\mu \\
&= \frac{15\sqrt{n_c}}{16\, m\, \sigma\sqrt{2\pi}} \sum_{i=1}^{m} \int_{-1}^{1} (1 - z_i^2)^2 \exp\left\{-\frac{n_c}{2\sigma^2}(\bar{y}_1 - (\bar{x}_i + h\,\tau\, z_i))^2\right\} dz_i
\end{aligned}
$$

Similarly the second term, D_s, in the denominator, with the biweight kernel (14.3) used for $f(\mu \mid \alpha)$, is given by

$$
\begin{aligned}
D_s &= \int f(\bar{y}_2 \mid \mu, \sigma^2) f(\mu \mid \alpha) d\mu \\
&= \frac{15\sqrt{n_s}}{16\, m\, \sigma\sqrt{2\pi}} \sum_{i=1}^{m} \int_{-1}^{1} (1 - z_i^2)^2 \exp\left\{-\frac{n_s}{2\sigma^2}(\bar{y}_2 - (\bar{x}_i + h\,\tau\, z_i))^2\right\} dz_i
\end{aligned}
$$

The numerator, N_{cs}, is given by

$$
\begin{aligned}
N_{cs} &= f(\bar{y}_1 - \bar{y}_2) \int f(w \mid \mu) f(\mu \mid \alpha) d\mu \\
&= \frac{1}{\sigma_{12}\sqrt{2\pi}} \exp\left\{-\frac{1}{2\sigma_{12}^2}(\bar{y}_1 - \bar{y}_2)^2\right\} \frac{15}{16\, m\, \sigma_3\sqrt{2\pi}} \\
&\quad \times \sum_{i=1}^{m} \int_{-1}^{1} (1 - z_i^2)^2 \exp\left[-\frac{1}{2\sigma_3^2}\{w - (\bar{x}_i + h\,\tau\, z_i)\}^2\right] dz_i
\end{aligned}
$$

It can be shown that the likelihood ratio is given by the ratio of N_{cs} to the product of D_c and D_s. Numerical evaluation of the likelihood ratio may then be made with the substitution of σ by s_w, τ by s_b, and h by its optimal value $(70/m)^{1/5}\overline{x}$.

There is a boundary effect when an $(\bar{x}_i, i = 1, \ldots, m)$ is within $h\tau$ of zero. For these $\bar{x}_i$, the kernel expression

$$
\left\{1 - \left(\frac{\mu - \bar{x}_i}{h\tau}\right)^2\right\}^2 = \{1 - z_i^2\}^2
$$

has to be adjusted with the factor $(\nu_2 - \nu_1 z)/(\nu_0\nu_2 - \nu_1^2)$, where $z_i = (\mu - \bar{x}_i)/(h\tau)$ and ν_0, ν_1, ν_2 are as in (14.4), to give

$$
\frac{(\nu_2 - \nu_1 z_i)}{(\nu_0\nu_2 - \nu_1^2)} \{1 - z_i^2\}^2
$$

which can be written as

$$(a - bz_i)\left\{1 - z_i^2\right\}^2$$

where $a = \nu_2/(\nu_0\nu_2 - \nu_1^2)$ and $b = \nu_1/(\nu_0\nu_2 - \nu_1^2)$. Define an indicator function $\gamma(z_i)$ such that

$$\begin{aligned}\gamma(z_i) &= 1 \text{ if } x_i > h\tau \\ &= (a - bz_i) \text{ if } x_i < h\tau\end{aligned}$$

Then the likelihood ratio $N_{cs}/(D_c D_s)$ can be adapted to account for boundary effects to give a value for the evidence of

$$\frac{1}{\sigma_{12}\sqrt{2\pi}} \exp\left\{-\frac{1}{2\sigma_{12}^2}(\bar{y}_1 - \bar{y}_2)^2\right\} \frac{15}{16\, m\, \sigma_3\sqrt{2\pi}} \times \sum_{i=1}^{m} \int_{-1}^{1} \gamma(z_i)(1 - z_i^2)^2 \exp\left[-\frac{1}{2\sigma_3^2}\left\{w - (\bar{x}_i + h\,\tau\, z_i)\right\}^2\right] dz_i$$

divided by the product of

$$\frac{15\sqrt{n_c}}{16\, m\, \sigma\sqrt{2\pi}} \sum_{i=1}^{m} \int_{-1}^{1} \gamma(z_i)(1 - z_i^2)^2 \exp\left\{-\frac{n_c}{2\sigma^2}(\bar{y}_1 - (\bar{x}_i + h\,\tau\, z_i))^2\right\} dz_i$$

and

$$\frac{15\sqrt{n_s}}{16\, m\, \sigma\sqrt{2\pi}} \sum_{i=1}^{m} \int_{-1}^{1} \gamma(z_i)(1 - z_i^2)^2 \exp\left\{-\frac{n_s}{2\sigma^2}(\bar{y}_2 - (\bar{x}_i + h\,\tau\, z_i))^2\right\} dz_i$$

14.2.4 Adaptive Kernel

The value of the evidence, when the between-group distribution is taken to be nonnormal and is estimated by a normal kernel function as described in [1], equation (10.12), is adapted to allow for the correlation between the control and recovered data $\bar{y}_1$ and $\bar{y}_2$ if they are assumed, as in the numerator, to come from the same source. This expression is then extended to an adaptive kernel, where the smoothing parameter is dependent on x_i and is thus denoted h_i.

The numerator is

$$\frac{1}{m}(2\pi)^{-1}\left\{\left(\frac{n_c + n_s}{n_c\, n_s}\right)\sigma^2\right\}^{-1/2}\left\{\tau^2 + \frac{\sigma^2}{n_c + n_s}\right\}^{-1/2}(h_i^2\tau^2)^{-1/2} \times \left\{\left(\tau^2 + \frac{\sigma^2}{n_c + n_s}\right)^{-1} + (h_i^2\tau^2)^{-1}\right\}^{-1/2}$$

$$\times \exp\left\{-\frac{1}{2}(\overline{y}_1 - \overline{y}_2)^2 \left[\left(\frac{n_c + n_s}{n_c\, n_s}\right)\sigma^2\right]^{-1}\right\}$$

$$\times \sum_{i=1}^{m} \exp\left\{-\frac{1}{2}(w - \overline{x}_i)^2 \left(\tau^2 + \frac{\sigma^2}{n_c + n_s} + h_i^2\tau^2\right)^{-1}\right\}$$

The first term in the denominator is

$$\frac{1}{m}(2\pi)^{-1/2}\left\{\tau^2 + \frac{\sigma^2}{n_c}\right\}^{-1/2} (h_i 2\, \tau^2)^{-1/2}$$

$$\times \left\{\left(\tau^2 + \frac{\sigma^2}{n_c}\right)^{-1} + (h_i^2\tau^2)^{-1}\right\}^{-1/2}$$

$$\times \sum_{i=1}^{m} \exp\left\{-\frac{1}{2}(\overline{y}_1 - \overline{x}_i)^2 \left(\tau^2 + \frac{\sigma^2}{n_c} + h_i^2\tau^2\right)^{-1}\right\}$$

The second term in the denominator is

$$\frac{1}{m}(2\pi)^{-1/2}\left\{\tau^2 + \frac{\sigma^2}{n_s}\right\}^{-1/2} (h_i^2\, \tau^2)^{-1/2}$$

$$\times \left\{\left(\tau^2 + \frac{\sigma^2}{n_s}\right)^{-1} + (h_i^2\tau^2)^{-1}\right\}^{-1/2}$$

$$\times \sum_{i=1}^{m} \exp\left\{-\frac{1}{2}(\overline{y}_2 - \overline{x}_i)^2 \left(\tau^2 + \frac{\sigma^2}{n_s} + h_i^2\tau^2\right)^{-1}\right\}$$

The constant term in the ratio is then

$$\frac{m\left\{n_c\tau^2(h_i^2 + 1) + \sigma^2\right\}^{1/2}\left\{n_s\tau^2(h_i^2 + 1) + \sigma^2\right\}^{1/2}}{\sigma\left\{(n_c + n_s)\tau^2(h_i^2 + 1) + \sigma^2\right\}^{1/2}}$$

The remaining term, that involving $\overline{y}_1, \overline{y}_2$ and $\overline{x}_i$, is the ratio of

$$\exp\left\{-\frac{1}{2}(\overline{y}_1 - \overline{y}_2)^2 \left(\sigma^2\left(\frac{1}{n_c} + \frac{1}{n_s}\right)\right)^{-1}\right\}$$

$$\times \sum_{i=1}^{m} \exp\left\{-\frac{1}{2}(w - \overline{x}_i)^2 \left(\tau^2 + \frac{\sigma^2}{n_c + n_s} + h_i^2\tau^2\right)^{-1}\right\}$$

to

$$\sum_{i=1}^{m} \exp\left\{-\frac{1}{2}(\bar{y}_1 - \bar{x}_i)^2 \left(\tau^2 + \frac{\sigma^2}{n_c} + h_i^2\tau^2\right)^{-1}\right\}$$
$$\times \sum_{i=1}^{m} \exp\left\{-\frac{1}{2}(\bar{y}_2 - \bar{x}_i)^2 \left(\tau^2 + \frac{\sigma^2}{n_s} + h_i^2\tau^2\right)^{-1}\right\}$$

14.2.4.1 Adaptive Smoothing Parameter The adaptive smoothing parameter h_i is estimated using the procedure outlined in [331].

First, find a pilot estimate $\tilde{f}(x)$ that satisfies $\tilde{f}(x_i) > 0$ for all i. This is achieved by standard kernel density estimation with Gaussian kernels [331]. Then define the smoothing parameter h_i by

$$h_i = \left\{\frac{\tilde{f}(x_i)}{g}\right\}^{-\beta}$$

where g is the geometric mean of the $\tilde{f}(x_i)$:

$$\log g = m^{-1} \sum \log \tilde{f}(x_i)$$

and β is a sensitivity parameter, a number satisfying $0 \leq \beta \leq 1$.

14.3 APPLICATION

In order to evaluate the feature evaluation methods and LR estimators, a forensic dataset was obtained from the Forensic Research Institute in Krakow, Poland. An overview of the overall system can be found in Figure 14.1.

One large piece of glass from each of 200 glass objects was selected. Each of these 200 pieces was wrapped in a sheet of gray paper and further fragmented. The fragments from each piece were placed in a plastic Petri dish. Four glass fragments, of linear dimension less than 0.5 mm with surfaces as smooth and flat as possible, were selected for examination with the use of an SMXX Carl Zeiss (Jena, Germany) optical microscope (magnification 100×).

14.3.1 Fragment Elemental Analysis

The four selected glass fragments were placed on self-adhesive carbon tabs on an aluminum stub and then carbon coated using an SCD sputter (Bal-Tech, Switzerland). The prepared stub was mounted in the sample chamber of a scanning electron microscope. Analysis of the elemental content of each glass fragment was carried out using a scanning electron microscope (JSM-5800 Jeol, Japan),

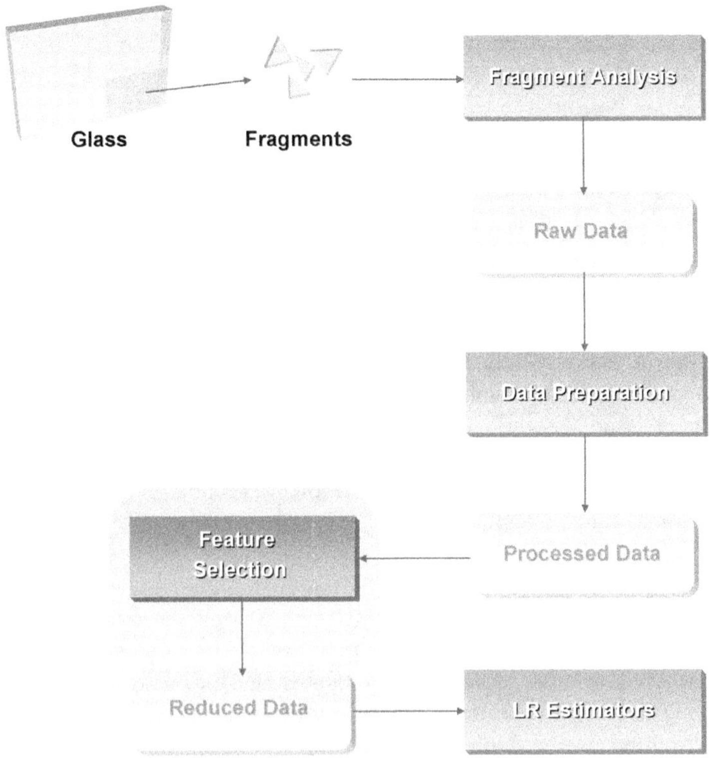

Figure 14.1 System overview

with an energy dispersive X-ray spectrometer (Link ISIS 300, Oxford Instruments Ltd., United Kingdom).

Three replicate measurements were taken from different areas on each of the four fragments, making 12 measurements from each glass object but only 4 independent measurements. Four means of the measurements were used for the analysis. The measurement conditions were accelerating voltages 20 kV, life time 50s, magnification 1000–2000×, and the calibration element was cobalt. The SEMQuant option (part of the software LINK ISIS, Oxford Instruments Ltd., United Kingdom) was used in the process of determining the percentage of particular elements in a fragment. The option applied a ZAF correction procedure that takes into account corrections for the effects of difference in the atomic number (Z), absorption (A), and X-ray fluorescence (F).

The selected analytical conditions allowed the determination of all elements except lithium (Li) and boron (B). However, only the concentrations of oxygen (O), sodium (Na), magnesium (Mg), aluminum (Al), silicon (Si), potassium (K), calcium (Ca), and iron (Fe) are considered further here as glass is essentially a silicon oxide with sodium and/or calcium added to create a commonly produced glass, and potassium, magnesium, aluminum, and iron added to stabilize its structure and modify its physicochemical properties. Histograms of the distributions of the data can be found in Figure 14.2.

14.3.2 Data Preparation

As two of the feature evaluation algorithms require fuzzy sets to be defined for each element in the dataset in order to maximize the use of information contained in the real-valued variables, further processing is required. Note that the attributes still take real values, and hence no discretization is performed. For this set of experiments, five fuzzy sets per feature were derived automatically based on the mean and standard deviation as seen in Figure 14.3. The value λ was set at 0.7. There is no theoretical argument as to why five sets should be chosen, although psychological research suggests that 5, 7, or 9 categories should be used, mimicking human cognition. To minimize computational effort, only five sets are defined for this application.

14.3.3 Feature Selection

Several feature evaluation methods outlined previously were applied to the processed glass data in order to select a single attribute for use in the univariate LR estimators. The dataset containing the full set of elements was processed by each method, resulting in a ranking of these features. The top ranked feature/element for each method was then selected, and the data reduced to this feature only.

14.3.4 Estimators

The performance of four procedures for estimating the likelihood ratio is compared. The procedures estimate the between-group distributions with a normal distribution, an exponential distribution, a normal adaptive kernel, and a biweight kernel with a boundary condition.

14.4 EXPERIMENTATION

In the experimentation two situations need to be considered: when the control and recovered data are from the same source and when they are from different sources. For same-source comparisons, the control and recovered data are taken from the same group by splitting the group into two equally sized, nonoverlapping halves (containing two measurements each). For different-source comparisons, the control and recovered data are entire groups selected from different sources.

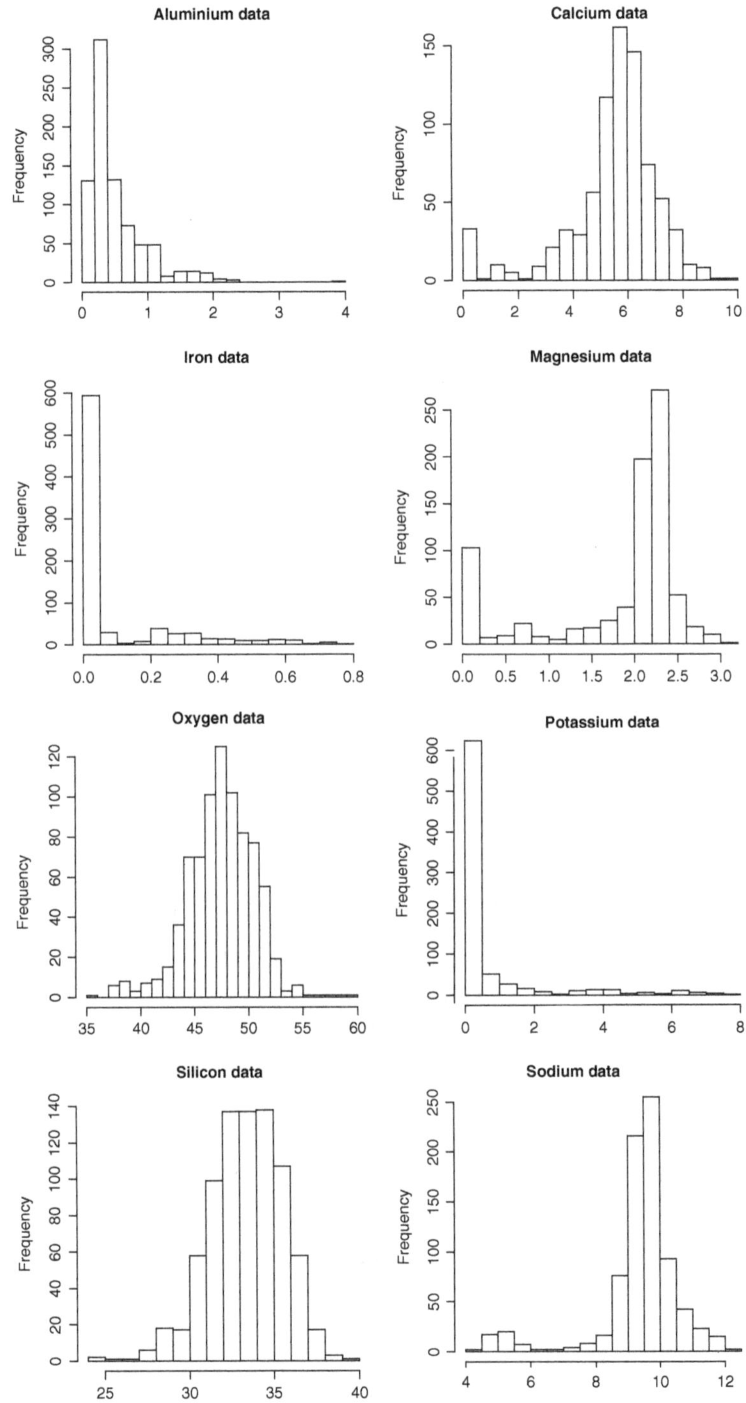

Figure 14.2 Data distributions for the eight elements

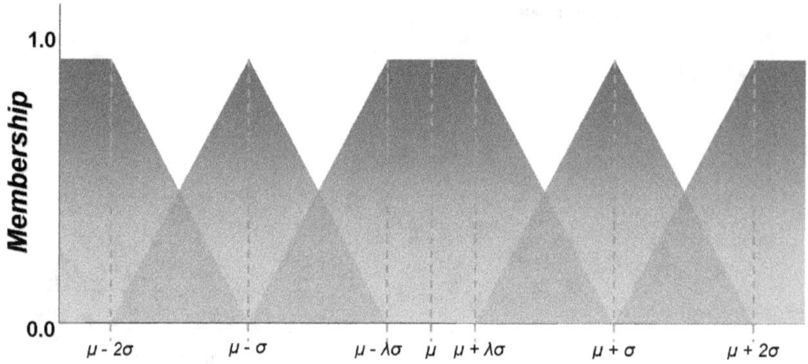

Figure 14.3 Fuzzy set construction

14.4.1 Feature Evaluation

Table 14.1 (summarized in Table 14.2) presents the ordering of features as determined by several leading measures of feature significance: fuzzy-rough feature evaluation (FRFS), fuzzy entropy (FuzEnt), χ^2, gain ratio (GR), information gain (IG), OneR, Relief-F, and symmetrical uncertainty (SU). It can be seen that FRFS, IG, and OneR select aluminum; FuzEnt and GR select sodium; χ^2 and SU select potassium; and Relief-F selects magnesium. Based on these selections and corresponding data reductions, the four estimators are applied.

14.4.2 Likelihood Ratio Estimation

There are 200 within-group comparisons of control and recovered data and $200 \times 199/2 = 19,900$ between-group comparisons For the 200 within-group comparisons, the likelihood ratio should be greater than 1 and for the 19,900 between-group comparisons, the likelihood ratio should be less than 1. Results are recorded for a normal kernel estimation (nn), an exponential kernel estimation (exp), an adaptive kernel estimation for $\beta = 0, 0.1, 0.2$, and 0.5 and a biweight kernel estimation (b).

TABLE 14.1 Ranking of Features

Element	FRFS	IG	OneR	FuzEnt	GR	Relief-F	χ^2	SU
O	0.003929	0.267	52.50	1.839151	0.211	0.0364	368.4882	0.1615
Na	0.025909	0.597	55.75	1.581143	0.391	0.0894	968.2267	0.3345
Mg	0.025426	0.718	52.13	1.658684	0.367	0.1234	1010.738	0.3589
Al	0.025996	0.843	57.63	1.647498	0.275	0.0661	1167.625	0.3301
Si	0.009267	0.129	41.38	1.801163	0.127	0.0309	150.6826	0.0846
K	0.003118	0.825	55.75	1.582525	0.338	0.0928	1578.554	0.3682
Ca	0.008341	0.511	53.25	1.600595	0.312	0.0881	686.8860	0.2774
Fe	0.022455	0.191	45.50	1.741598	0.145	0.0453	174.0253	0.1136

TABLE 14.2 Summary of Feature Ranking

Selection Method	Feature Ranking
FRFS	Al > Na > Mg > Fe > Si > Ca > O > K
IG	Al > K > Mg > Na > Ca > O > Si > Fe
OneR	Al > Na = K > Ca > O > Mg > Fe > Si
FuzEnt	Na > K > Ca > Al > Mg > Fe > Si > O
GR	Na > Mg > K > Ca > Al > O > Fe > Si
Relief-F	Mg > K > Na > Ca > Al > Fe > O > Si
χ^2	K > Al > Mg > Na > Ca > O > Fe > Si
SU	K > Mg > Na > Al > Ca > O > Fe > Si

Note: $A > B$ indicates that A is a more informative feature than B, and $A = B$ indicates identical ranking.

The results for the aluminum data (selected by FRFS, IG, and OneR) are shown in Table 14.3, magnesium data (selected by Relief-F) in Table 14.4, potassium data (selected by χ^2 and SU) in Table 14.5, and sodium data (selected by FuzEnt and GR) in Table 14.6. It can be seen that, overall, the results produced using the aluminum data are superior to those generated using the other considered elements. The within-group estimation of likelihood ratio is as good as or better than the others, and the between-group estimation is more accurate. The results also indicate that magnesium is a very informative feature for likelihood ratio estimation.

The level of false positives (important from a forensic viewpoint) is rather high. This could be the result of several factors. The SEM-EDX method delivers information about the main elemental content of glass; thus differences between glass objects are very small. Moreover soda-lime-silica glass (the most common) has a strict composition, and the production recipes, used in many factories, are very similar. Also the data are created mostly from glass used for car windows and building windows, which have extremely similar elemental concentrations. Sodium results are relatively bad as the concentration of Na is restricted by the definition of Na–Ca–Si glass. Potassium results are also unsatisfactory because it is only present in glass objects that have special optical properties, and so is absent or undetectable in many objects (this may explain the poor performance of the exponential model for this element). Aluminum and magnesium results are relatively superior as these elements are added to glass to create desired features of glass objects, and thus their content varies with glass properties.

For this dataset the feature ranking methods FRFS, IG, and OneR have selected the best feature for likelihood estimation. This is in keeping with other investigations that have shown the utility of FRFS, in particular, for this purpose [168]. Although Relief-F selected an alternative feature, the results achieved were almost as accurate. The results show that the normal kernel estimation procedure is inadequate at modeling the underlying data distributions in general. Excepting the case of sodium, it be seen that even an exponential distribution is a better model as the resulting likelihood ratios are more accurate.

TABLE 14.3 Summary of Likelihood Ratios for Aluminum (Al) Data (Chosen by FRFS, IG, and OneR)

Likelihood Ratio Range	nn	exp	β 0.0	β 0.1	β 0.2	β 0.5	b
			Within-group comparisons				
$0-1$	4	4	4	4	4	4	4
$1-10^1$	1	160	184	184	184	185	173
10^1-10^2	183	35	12	12	12	11	22
10^2-10^3	8	1	0	0	0	0	1
10^3-10^4	3	0	0	0	0	0	0
$>10^4$	1	0	0	0	0	0	0
Misclassification	2.0%	2.0%	2.0%	2.0%	2.0%	2.0%	2.0%
			Between-group comparisons				
$0-1$	10661	12548	12669	12662	12652	12518	13924
$1-10^1$	2958	7031	7172	7180	7190	7319	5866
10^1-10^2	6243	320	59	58	58	63	120
10^2-10^3	34	1	0	0	0	0	0
10^3-10^4	4	0	0	0	0	0	0
$>10^4$	0	0	0	0	0	0	0
Misclassification	46.4%	36.9%	36.3%	36.4%	36.4%	37.1%	30.0%

The point of the analysis is not to say whether some fragment is likely to have been derived from a glass object, but to make statements about to what extent the observations (in this case, of elemental concentrations) lend support to the proposition that the fragment came from the glass object in question, or to the proposition that the fragment came from some other glass object from the relevant population of objects. The likelihood ratio does not provide a direct attribution of source—it provides a measure of the strength of the evidence in support of one or other of the propositions in a case.

Small likelihood ratios would be regarded as very strong support that the particular fragment from which the observations were made had come from some source other than the broken glass object from the crime scene. However, this value, while providing very strong support, may not be persuasive enough, one way or the other, to affect the outcome of the case as a whole. The outcome is dependent on other evidence and court requirements.

14.5 GLASS CLASSIFICATION

The classification of glass fragments into their product-use type is another important task. By determining the type of glass recovered, an alibi may be corroborated

TABLE 14.4 Summary of Likelihood Ratios for Magnesium (Mg) Data (Chosen by Relief-F)

Likelihood Ratio Range	*nn*	exp	β 0.0	β 0.1	β 0.2	β 0.5	*b*
			Within-group comparisons				
0–1	6	6	6	6	6	6	6
$1–10^1$	0	29	157	157	157	157	151
$10^1–10^2$	165	165	37	37	37	37	43
$10^2–10^3$	29	0	0	0	0	0	0
$10^3–10^4$	0	0	0	0	0	0	0
$>10^4$	0	0	0	0	0	0	0
Misclassification	3.0%	3.0%	3.0%	3.0%	3.0%	3.0%	3.0%
			Between-group comparisons				
0–1	10955	11220	12020	12016	12003	11965	12408
$1–10^1$	1673	2032	6969	6868	6707	5937	7169
$10^1–10^2$	6983	6648	911	1016	1190	1998	323
$10^2–10^3$	289	0	0	0	0	0	0
$10^3–10^4$	0	0	0	0	0	0	0
$>10^4$	0	0	0	0	0	0	0
Misclassification	44.9%	43.6%	39.6%	39.6%	39.7%	39.9%	37.6%

or disproved, or evidence of product tampering may be detected. The glass data above contains information on the source of each glass fragment, and hence classifiers may be constructed from these training data to determine the source of unidentified fragments. The six glass sources are bulb glass, car window glass, headlamp glass, optic glass, glass containers, and building window glass. The task is to derive classifiers that correctly identify the source of glass based solely on chemical concentrations.

The results from the experimentation can be seen in Table 14.7. The classifiers used previously were employed for the purpose of predicting glass source. Again, these were evaluated using 10-fold cross-validation. Applying FRFS to the data resulted in all features being chosen for the classification task, and hence the results are equal to the unreduced approach. The tolerance method selected a smaller number of features (6) for the task. It can be seen that the FRFS method was correct in retaining the entire feature set as the reduction in dimensionality produced by the tolerance method resulted in a slight decrease in performance for all classifiers. However, it should be noted that there is a cost (both monetary and time) associated with measuring each chemical within fragments. The results show that (via tolerance rough sets) two of these can be eliminated with only a small drop in classification accuracy.

TABLE 14.5 Summary of Likelihood Ratios for Potassium (K) Data (Chosen by χ^2 and SU)

Likelihood Ratio Range	*nn*	exp	β 0.0	β 0.1	β 0.2	β 0.5	*b*
			Within-group comparisons				
0–1	4	43	4	4	4	4	4
$1-10^1$	174	138	181	181	181	181	170
10^1-10^2	13	3	15	15	15	15	26
10^2-10^3	4	9	0	0	0	0	0
10^3-10^4	2	4	0	0	0	0	0
$>10^4$	3	3	0	0	0	0	0
Misclassification	2.0%	21.5%	2.0%	2.0%	2.0%	2.0%	2.0%
			Between-group comparisons				
0–1	6463	7931	6989	6988	6986	6971	7861
$1-10^1$	2722	11881	12869	12870	12872	12887	11960
10^1-10^2	10696	44	40	41	41	41	78
10^2-10^3	8	33	2	1	1	1	1
10^3-10^4	8	9	0	0	0	0	0
$>10^4$	3	2	0	0	0	0	0
Misclassification	67.5%	60.1%	64.9%	64.9%	64.9%	65.0%	60.5%

14.6 SUMMARY

As a type of evidence, glass can be very useful contact trace material in a wide range of offenses including burglaries and robberies, murders, assaults, criminal damage, and thefts of and from motor vehicles. In all these situations there is the potential for glass fragment transfer, providing a link between suspect and crime. Hence the correct interpretation of glass evidence is critical in establishing such connections.

Previous work has been based on the use of a normal kernel estimation procedure for evidence evaluation. However, this may be inadequate when the data are positively skewed and an exponential distribution is thought to be a better model [3]. Three techniques that attempt to alleviate this problem were investigated and found to provide better likelihood ratio estimations.

In this chapter the role of feature evaluation as an aid to glass analysis was investigated. As the two-level models used were univariate, the task of the evaluation methods was to determine the most informative feature for use in the models. The results have shown that automated feature evaluation techniques can indeed aid the choice of variable for further modeling. This choice is a critical factor in the resulting quality of evidence evaluation. Situations are often encountered where many competing variables co-exist. The manual selection of which variable to use may result in subsequent analysis being too subjective.

TABLE 14.6 Summary of Likelihood Ratios for Sodium (Na) Data (Chosen by FuzEnt and GR)

Likelihood Ratio Range	*nn*	exp	β 0.0	β 0.1	β 0.2	β 0.5	*b*
			Within-group comparisons				
0–1	5	3	7	7	7	7	4
$1–10^1$	179	6	183	183	183	183	183
$10^1–10^2$	6	191	10	10	10	10	13
$10^2–10^3$	5	0	0	0	0	0	0
$10^3–10^4$	5	0	0	0	0	0	0
$>10^4$	0	0	0	0	0	0	0
Misclassification	2.5%	1.5%	3.5%	3.5%	3.5%	3.5%	2.0%
			Between-group comparisons				
0–1	8540	6543	9414	9414	9410	9345	7609
$1–10^1$	11239	2879	10452	10451	10455	10519	6614
$10^1–10^2$	83	10478	34	35	35	36	5677
$10^2–10^3$	18	0	0	0	0	0	0
$10^3–10^4$	20	0	0	0	0	0	0
$>10^4$	0	0	0	0	0	0	0
Misclassification	57.1%	67.1%	52.7%	52.7%	52.7%	53.0%	61.8%

TABLE 14.7 Results for Forensic Data

Method	Features	J48	JRip	PART	MODLEM
Unreduced	8	83.13	79.38	80.50	78.00
FRFS	8	83.13	79.38	80.50	78.00
Tolerance	6	82.00	78.25	78.00	74.13

Through the use of feature evaluation methods, this important decision can be made without expert assistance.

CHAPTER 15

SUPPLEMENTARY DEVELOPMENTS AND INVESTIGATIONS

This chapter presents other developments that have been initiated and future work in the areas of feature selection and rule induction, arising from some of the issues discussed in this book. For feature selection, RSAR-SAT is proposed based on the discernibility matrix approach to finding reducts and employing techniques from propositional satisfiability. Additionally the concept of fuzzy universal reducts is proposed as a way of enabling more accurate data reductions. A new decision tree induction method based on the fuzzy-rough metric is suggested, with some initial experimental investigations presented.

In addition to these developments, several areas of future work are discussed that utilize the properties of fuzzy-rough sets. A potential mechanism for fuzzy-rough rule induction and fuzzy-rough clustering are proposed. In crisp rough sets, rule induction and clustering a two successful areas of application, but these suffer the same drawbacks as crisp rough set feature selection. It is natural, then, to extend these methods via fuzzy-rough sets. There is also the possibility of using fuzzy-rough sets for the optimization of fuzzifications.

15.1 RSAR-SAT

The propositional satisfiability (SAT) problem [70] is one of the most studied NP-complete problems because of its significance in both theoretical research and practical applications. Given a Boolean formula (typically in

Computational Intelligence and Feature Selection: Rough and Fuzzy Approaches, by Richard Jensen and Qiang Shen

conjunctive normal form, CNF), the SAT problem requires an assignment of variables/features so that the formula evaluates to true, or a determination that no such assignment exists. In recent years search algorithms based on the well-known Davis–Logemann–Loveland algorithm (DPLL) [71] are emerging as some of the most efficient methods for complete SAT solvers. Such solvers can either find a solution or prove that no solution exists.

Stochastic techniques have also been developed in order to reach a solution quickly. These pick random locations in the space of possible assignments and perform limited local searches from them. However, as these techniques do not examine the entire search space, they are unable to prove unsatisfiability.

A CNF formula on n binary variables $x_1, \ldots, x_n$ is the conjunction of m clauses $C_1, \ldots, C_m$, each of which is the disjunction of one or more literals. A literal is the occurrence of a variable or its negation. A formula denotes a unique n-variable Boolean function $f(x_1, \ldots, x_n)$. Clearly, a function f can be represented by many equivalent CNF formulas. The satisfiability problem is concerned with finding an assignment to the arguments of $f(x_1, \ldots, x_n)$ that makes the function equal to 1, signaling that it is satisfiable, or proving that the function is equal to 0 and hence unsatisfiable [414]. By viewing the selection problem as a variant of SAT, with a bound on true assignments, techniques from this field can be applied to reduct search.

15.1.1 Finding Rough Set Reducts

The problem of finding the smallest feature subsets using rough set theory can be formulated as a SAT problem. Rough sets allows the generation from datasets of clauses of features in conjunctive normal form. If after assigning truth values to all features appearing in the clauses the formula is satisfied, then those features set to true constitute a valid subset for the data. The task is to find the smallest number of such features so that the CNF formula is satisfied. In other words, the problem here concerns finding a minimal assignment to the arguments of $f(x_1, \ldots, x_n)$ that makes the function equal to 1. There will be at least one solution to the problem (i.e., all x_i set to 1) for consistent datasets. Preliminary work has been carried out in this area [14], though this does not adopt a DPLL-style approach to finding solutions.

The DPLL algorithm for finding minimal subsets can be found in Figure 15.1 where a search is conducted in a depth-first manner. The key operation in this procedure is the unit propagation step, unitPropagate(F), in lines (6) and (7). Clauses in the formula that contain a single literal will only be satisfied if that literal is assigned the value 1 (for positive literals). These are called unit clauses. Unit propagation examines the current formula for unit clauses and automatically assigns the appropriate value to the literal they contain. The elimination of a literal can create new unit clauses, and thus unit propagation eliminates variables by repeated passes until there is no unit clause in the formula. The order of the unit clauses within the formula makes no difference to the results or the efficiency of the process.

```
DPLL(F)
Input: F, the formula containing the current set of clauses
Output: minimal feature subset
(1)  if F contains an empty clause
(2)     return  unsatisfiable
(3)  if F is empty
(4)     output current assignment
(5)     return  satisfiable
(6)  if F contains a unit clause {l}
(7)     F' ← unitPropagate(F)
(8)     return DPLL(F')
(9)  x ← selectLiteral(F)
(10) if DPLL(F∪{x}) is satisfiable
(11)    return  satisfiable
(12) else
(13)    return DPLL(F∪{−x})
```

Figure 15.1 Definition of the DPLL algorithm

Branching occurs at lines (9) to (12) via the function selectLiteral(F). Here the next literal is chosen heuristically from the current formula, assigned the value 1, and the search continues. If this branch eventually results in unsatisfiability, the procedure will assign the value 0 to this literal instead and continue the search. The importance of choosing good branching literals is well known—different branching heuristics produce drastically different sized search trees for the same basic algorithm, thus significantly affecting the efficiency of the solver. The heuristic currently used within RSAR-SAT is to select the variable that appears in the most clauses in the current set of clauses. Many other heuristics exist for this purpose [414] but are not considered here.

A degree of pruning can take place in the search by remembering the size of the currently considered subset and the smallest optimal subset encountered so far. If the number of variables currently assigned 1 equals the number of those in the presently optimal subset, and the satisfiability of F is still not known, then any further search down this branch will not result in a smaller optimal subset.

Although stochastic methods have been applied to SAT problems [311,137], these are not applicable here as they provide no guarantee of solution minimality. The DPLL-based algorithm will always find the minimal optimal subset. However, this will come at the expense of time taken to find it.

15.1.2 Preprocessing Clauses

The discernibility function can be simplified by replacing those variables that are simultaneously either present or absent in all clauses by single representative variables. For instance, in the formula below, variables a and f can be replaced by a single variable:

$$\{a \vee b \vee c \vee f\} \wedge \{b \vee d\} \wedge \{a \vee d \vee e \vee f\} \wedge \{d \vee c\}$$

The first and third clauses may be considered to be $\{\{a \vee f\} \vee b \vee c\}$ and $\{\{a \vee f\} \vee d \vee e\}$, respectively. Replacing $\{a \vee f\}$ with g results in

$$\{g \vee b \vee c\} \wedge \{b \vee d\} \wedge \{g \vee d \vee e\} \wedge \{d \vee c\}$$

If a reduct resulting from this discernibility function contains the new variable g, then this variable may be replaced by either a or f. Here $\{g, d\}$ is a reduct, and so $\{a, d\}$ and $\{f, d\}$ are reducts of the original set of clauses. Hence fewer attributes are considered in the reduct-determining process with no loss of information [350]. The complexity of this (optional) preprocessing step is $O(a * c + a^2)$, where a is the number of attributes and c is the number of clauses.

From the generation of the discernibility matrix, the core attributes are immediately determined. These may then be removed from the discernibility function as they will appear in every rough set reduct. Hence, if the union of the core attributes for a dataset results in a reduct, no search is required as this will be the minimal subset.

15.1.3 Evaluation

Initial experimentation has been carried out using the algorithm outlined previously [166]. The datasets have been obtained from [38]. Table 15.1 shows the average time taken for the preprocessing of each dataset. For RSAR, this involves constructing partitions for each attribute. For RSAR-SAT, the discernibility matrix is calculated and simplified. It can be seen from the table that

TABLE 15.1 Runtimes for RSAR and RSAR-SAT

Dataset	Number of Clauses	Number of Features	RSAR Setup (s)	SAT Setup (s)	RSAR (s)	SAT: Minimal (s)	SAT: Full (s)
M-of-N	6	13	0.164	2.333	0.171*	0.001	0.007
Exactly	6	13	0.146	2.196	0.304*	0.001	0.008
Exactly2	10	13	0.136	1.898	0.823*	0.001	0.008
Heart	12	13	0.085	0.380	0.207*	0.002	0.009
Vote	12	16	0.076	0.333	0.170*	0.004	0.009
Credit	200	20	0.148	3.873	1.988*	0.077	0.094
LED	167	24	0.125	68.20	0.097*	0.041	0.051
Letters	57	25	0.019	0.074	0.067*	0.024	0.116
Derm	1126	34	0.187	11.31	0.758*	0.094	0.456
Derm2	1184	34	0.133	6.796	0.897*	0.104	0.878
WQ	6534	38	0.168	87.85	9.590*	0.205	116.1
Lung	171	56	0.032	0.125	0.059	0.023	0.786
DNA	3861	58	0.139	30.40	1.644*	0.227	53.81

RSAR-SAT requires more preprocessing time. Included in this table are the number of clauses appearing in the resultant discernibility function for the RSAR-SAT method.

The average times of the execution of these algorithms are also presented in Table 15.1. The time taken for RSAR-SAT is split into two columns. The first indicates the average length of time taken to find the minimal subset, the second how long it takes to verify that this is indeed minimal. For RSAR an asterisk next to the time indicates that it found a nonminimal reduct.

The results show that RSAR-SAT is comparable to RSAR in the time taken to find reducts. However, RSAR regularly fails to find the smallest optimal subset, because of its being misled in the search process. For larger datasets the time taken for RSAR-SAT verification exceeds that of RSAR. Note that the verification stage involves simple chronological backtracking. There are ways in which this can be made more effective and less time-consuming.

DPLL resorts to chronological backtracking if the current assignment of variables results in the unsatisfiability of F. Much research has been carried out in developing solution techniques for SAT that draws on related work in solvers for constraint satisfaction problems (CSPs) [21,370]. Indeed the SAT problem can be translated to a CSP by retaining the set of Boolean variables and their $\{0, 1\}$ domains, and to translate the clauses into constraints. Each clause becomes a constraint over the variables in the constraint. Unit propagation can be seen to be a form of forward checking.

In CSPs more intelligent ways of backtracking have been proposed such as backjumping, conflict-directed backjumping, and dynamic backtracking. Many aspects of these have been adapted to the SAT problem solvers. In these solvers, whenever a conflict (deadend) is reached, a new clause is recorded to prevent the occurrence of the same conflict again during the subsequent search. Nonchronological backtracking backs up the search tree to one of the identified causes of failure, skipping over irrelevant variable assignments.

With the addition of intelligent backtracking, RSAR-SAT should be able to handle datasets containing large numbers of features. As seen in the preliminary results, the bottleneck in the process is the verification stage—the time taken to confirm that the subset is indeed minimal. This requires an exhaustive search of all subtrees containing fewer variables than the current best solution. Much of this search could be avoided through the use of more intelligent backtracking. This would result in a selection method that can cope with many thousands of features, while guaranteeing resultant subset minimality—something that is particularly sought after in feature selection.

15.2 FUZZY-ROUGH DECISION TREES

15.2.1 Explanation

A decision tree can be viewed as a partitioning of the instance space. Each partition, represented by a leaf, contains the objects that are similar in relevant respects and thus are expected to belong to the same class. The partitioning is carried out in a data-driven manner, with the final output representing the partitions as a tree. An important property of decision tree induction algorithms is that they attempt to minimize the size of the tree at the same time as they optimize a certain quality measure.

The general decision tree induction algorithm is as follows: The significance of features is computed using a suitable measure (in C4.5 this is the information gain metric [281]). Next the most discriminating feature according to this measure is selected and the dataset partitioned into sub-tables according to the values this feature may take. The chosen feature is represented as a node in the currently constructed tree. For each sub-table the procedure above is repeated, namely to determine the most discriminating feature and split the data into further sub-tables according to its values.

This is a similar process in the fuzzy case. However, a measure capable of handling fuzzy terms (instead of crisp values) must be used. Data are partitioned according to the selected feature's set of fuzzy terms. There must also be a way of calculating the number of examples that belong to a node. In the crisp case, this is clear; objects either contain a specific attribute value or they do not. In the fuzzy case, this distinction can no longer be made, as objects may belong to several fuzzy terms. A suitable stopping condition must also be chosen that will limit the number of nodes expanded.

Clearly, one important aspect of this procedure is the choice of feature significance measure. This measure influences the organization of the tree directly and will have profound effects on the resulting tree's accuracy. Given the utility of the fuzzy-rough dependency measure as an estimator of feature importance, it should be useful as a feature significance measure in decision tree construction. This section presents some initial investigations comparing the standard fuzzy ID3 approach [157,363] (F-ID3) using fuzzy entropy with an adapted fuzzy ID3 algorithm that employs the fuzzy-rough measure (FR-ID3).

15.2.2 Experimentation

To demonstrate the applicability of the proposed approach, both the fuzzy and fuzzy-rough decision tree induction methods were applied to a variety of benchmark datasets obtained from [38]. This section presents the results of experimentation carried out on these datasets.

In order for both decision tree inducers to operate, fuzzy sets must first be defined for real-valued attributes that appear in the data. For this, a simple fuzzification was carried out based on the statistical properties of the attributes themselves. It is expected that the classification performance would be greatly improved if the fuzzifications were optimized.

The datasets were then split into two halves of equal size, one for training and the other for testing, while maintaining the original class distributions. To show the general applicability of both approaches, the inducers were also applied to nonfuzzy data. Datasets WQ2class and WQ3class are the water treatment datasets with the original 13 classes collapsed into 2 or 3, respectively [165].

TABLE 15.2 Classification accuracies for fuzzy ID3 and fuzzy-rough ID3 (real-valued data)

Dataset	F-ID3		FR-ID3	
	Train	Test	Train	Test
Iris	0.973	0.947	0.973	0.947
Glass	0.514	0.495	0.523	0.476
Credit	0.890	0.849	0.742	0.641
WQ2class	0.824	0.787	0.824	0.782
WQ3class	0.816	0.696	0.801	0.722
Ionosphere	0.862	0.783	0.852	0.783
Olitos	0.678	0.590	0.695	0.656

TABLE 15.3 Number of rules produced (fuzzy data)

Dataset	F-ID3	FR-ID3
Iris	10	13
Glass	32	44
Credit	51	50
WQ2class	108	140
WQ3class	165	328
Ionosphere	22	28
Olitos	9	17

As can be seen from Table 15.2, both approaches perform similarly for the fuzzy data. The size of the resultant rulesets that produce these accuracies can be found in Table 15.3. In general, FR-ID3 produces slightly larger rulesets than the standard approach. There is a notable difference in performance for the Credit dataset, where FR-ID3 produces a reduced classification accuracy. From Table 15.3 it can be seen that this is the only case where FR-ID3 produces a smaller ruleset. The paired t-tests for training and testing results for F-ID3 and FR-ID3 produce the p-values 0.368 and 0.567, respectively.

The results of the application of fuzzy and fuzzy-rough ID3 to crisp data can be found in Table 15.4, with the resulting ruleset size in Table 15.5. The results show that FR-ID3 outperforms F-ID3 in general, as well as producing smaller

TABLE 15.4 Classification accuracies for fuzzy ID3 and fuzzy-rough ID3 (crisp data)

Dataset	F-ID3		FR-ID3	
	Train	Test	Train	Test
Derm	0.514	0.257	0.838	0.615
Derm2	0.932	0.818	0.972	0.906
DNA	0.575	0.348	0.475	0.386
Heart	0.826	0.772	0.826	0.793
WQ-disc	0.798	0.625	0.698	0.563

TABLE 15.5 Number of rules produced (crisp data)

Dataset	F-ID3	FR-ID3
Derm	53	46
Derm2	20	18
DNA	25	22
Heart	25	30
WQ-disc	28	28

rulesets. Here the paired t-tests for the training and testing results produce the p-values 0.695 and 0.283, respectively.

15.3 FUZZY-ROUGH RULE INDUCTION

One of the most successful areas of application of rough set theory is rule induction. The properties of rough sets allow the automatic derivation of rules from datasets based on the data contents alone. This is achieved primarily through the calculation of lower approximations (generating *certain* rules) and upper approximations (generating *possible* rules). However, current methods are only useful for datasets containing nominal values. The proposed fuzzy-rough method will use fuzzy-rough lower and upper approximations with extensions of other rough set concepts in order to induce fuzzy rules from continuous data.

In traditional rough set theory, rules are constructed by examining attribute-value pairs in the dataset and their relation to lower and upper approximations of concepts. Concepts are defined here as collections of objects with the same decision value in datasets. In order to locate the relevant pairs, rule inducers make use of the definitions of minimal complex and local covering for concepts [121]. These determine the smallest set of attribute-value pairs that may cover a concept.

A similar strategy can be adopted for fuzzy-rough rule induction through extensions of these definitions. In this case the building blocks of fuzzy rules are the pairs of attribute and fuzzy linguistic values. These values need to be determined beforehand either elicited from experts or derived automatically. The fuzzy lower approximations of concepts can be calculated based on given attribute subsets. Such approximations have been found to be useful in feature selection [162]. The approximations generated by subsets associate with an object a degree of membership to the approximation. These memberships indicate the ability of a subset (and corresponding attribute-value pairs) to approximate a concept. In crisp rough set rule induction the memberships are either 0 or 1, with those attribute-value pairs that produce a maximum membership used to form *certain* rules. In the fuzzy-rough case, an alternative strategy must be employed as memberships lie in the range [0, 1]. One possible way of accomplishing this is to use α-cuts to return crisp subsets. A more interesting approach would be to employ the information contained in the memberships to gauge the strength of the generated rules. Fuzzy upper approximations could be used to derive *possible* rules in a similar way.

```
FR-RULEINDUCE(ℂ)
Input: ℂ, the set of all conditional features;
Output: set of fuzzy rules
R ← ∅, A ← ∅
while concepts not covered
  foreach x ∈ (ℂ − R)
     A ← R ∪ x
     foreach concept X not covered
        calculate fuzzy lower approximation, A̲X
        remove objects covered by A̲X
        add generated attribute-value pairs to rulebase
  a ← best attribute according to γ′
  R ← R ∪ a
```

Figure 15.2 Fuzzy-rough rule induction algorithm

Clearly, one important factor in the process is the choice of search mechanism for the consideration of sets of attribute-value pairs. A simple approach would be a breadth-first search, though this would be too costly for large datasets. One possible technique can be found in Figure 15.2, which employs a greedy hill-climbing approach to induction.

The fuzzy-rough rule induction algorithm considers the addition of individual attributes to the current set of features (initially empty), generating rules from these. Fuzzy lower approximations of concepts are calculated for each attribute, removing objects from a concept if they are covered by the approximation. Once this is complete, the best attribute is selected via the fuzzy-rough measure of dependency, γ', and added to the set of features R. The process is then repeated for this new subset.

15.4 HYBRID RULE INDUCTION

As mentioned previously, rules can be generated through the use of minimal complexes. Let X be a concept, t an attribute-value pair (a, v), and T a set of attribute-value pairs. The block of t, denoted $[t]$, is the set of objects for which attribute a has value v. A concept X depends on a set of attribute-value pairs T, iff

$$\emptyset \neq \cap\{[t] | t \in T\} \subseteq X \tag{15.1}$$

T is a minimal complex of X iff X depends on T and no proper subset T' of T exists such that X depends on T' [121].

It is often the case that a minimal complex describes a concept only partially, and hence more than one minimal complex is required to cover a concept. A local covering $\mathbb{T}$ of a concept X is such a collection of minimal complexes, such that the union of all minimal complexes is exactly X and $\mathbb{T}$ is minimal (i.e., contains no spurious attribute-value pairs). The discovery of such local coverings forms the basis of most approaches to rough set rule induction [276].

A partitioning of the universe of discourse by a reduct will always produce equivalence classes that are subsets of the decision concepts and will cover each concept fully. Once a reduct has been found, rules may be extracted from the underlying equivalence classes. In the literature, reducts for the purpose of rule induction are termed global coverings.

The most widely used approach to rule induction is the LEM2 algorithm [121], which follows a heuristic strategy for creating an initial rule by choosing sequentially the "best" elementary conditions according to some heuristic criteria. Learning examples that match this rule are removed from consideration. The process is repeated iteratively while some learning examples remain uncovered. The resulting set of rules covers all learning examples. In [154] additional factors characterizing rules are taken into account: the strength of matched or partly matched rules (the total number of cases correctly classified by the rule during training), the number of nonmatched conditions, and the rule specificity (i.e., length of condition parts). All factors are combined and the strongest decision wins. If no rule is matched, the partly matched rules are considered and the most probable decision is chosen.

15.4.1 Hybrid Approach

It can be seen that the tasks of feature selection and rule induction (via global coverings) are very similar in rough set theory. Both are concerned with the discovery of reducts. For feature selection, once a reduct has been found, the original set of data features can be reduced to those appearing in the reduct only. For rule induction, the reduct's corresponding equivalence classes can be used to form rules that completely cover the decision concepts appearing in the data. A feature selection step often precedes rule induction in an attempt to speed up the induction process and make the resulting rules more comprehensible. If a dataset contains redundant or misleading features, any extracted knowledge may become opaque or even meaningless. Hence, in rough set theory, it seems natural to combine the two into a hybrid method that possesses the benefits of feature selection while inducing rulesets. For the purposes of combining the two approaches, rules are constructed from equivalence classes and a corresponding decision concept. An equivalence class corresponds to a set of attribute-value pairs, forming the consequent of the rule, the decision concept forms the consequent.

A rule is herein defined as a tuple $\langle E, F, D\rangle$, where E is an equivalence class, F is the set of features that generated the equivalence class, and D is the set of objects representing decision concept. This formulation is used as it provides a fast way of determining rule coverage (the cardinality of E) and rule specificity (the cardinality of F). To generate a human-readable form of the rules, the values of the features in F for any object appearing in E forms the antecedent, while the value of any object appearing in D for the decision feature(s) will form the consequent. A rule can be induced from any given equivalence class E if E is a subset of a decision concept D appearing in the dataset. From a rough set perspective, equivalence class E will appear in the lower approximation of D, and hence the features that generated E are predictive of this concept.

The strength of a rule, constructed from equivalence class E and decision concept D, can be defined as simply

$$\frac{|E|}{|D|} \tag{15.2}$$

For example, if $E = \{1\}$ and $D = \{1, 2, 3\}$ the strength is $\frac{1}{3}$. If $E = \{1, 2, 3\}$, the strength is 1, meaning all objects are covered for this decision concept.

15.4.2 Rule Search

The most basic approach to integrating rule induction and feature selection would be to generate rules from the global covering produced as a result of a typical rough set feature selection process. However, at this point, the generated rules would be too specific as each antecedent would include every feature appearing in the final reduct. A greedy hill-climbing approach based on the QuickReduct algorithm is proposed for the purpose (although any search method may be used) and can be seen in Figure 15.3. The progress through the search space is identical to that of QuickReduct, but at each stage of subset evaluation the resulting equivalence classes are considered for rule construction and inclusion in the

```
QUICKRULES(ℂ,𝔻)
Input: ℂ, the set of all conditional attributes; 𝔻, the set
of decision attributes
Output: R, the feature subset; a set of rules
    R ← {}
    while γ_R(𝔻) ≠ γ_ℂ(𝔻)
       T ← R
       foreach x ∈ (ℂ − R)
          if GAMMARULESx, R > γ_T(𝔻)
             T ← R ∪ {x}
       R ← T
    PRUNE()
    return R
```

Figure 15.3 QuickRules algorithm

```
GAMMARULES(a,R)
Input: a, the feature under consideration for inclusion in R;
R, the current reduct candidate
Output: γR(𝔻), the dependency
    Pos ← ∅
    foreach eq ∈ 𝕌/{R ∪ a} //each equivalence class
        foreach dec ∈ 𝕌/𝔻    //each decision concept
            if eq ⊆ dec  //lower approximation
                DETERMINEADD(eq, R ∪ {a}, dec)
                Pos ← Pos ∪ eq
    return |Pos| / |𝕌|         //dependency degree
```

Figure 15.4 GAMMARULES algorithm

current ruleset. The search continues until the maximum degree of dependency is reached for a subset. At this point, a global covering has been discovered and all concepts can be adequately described by the ruleset. (All equivalence classes will be a subset of a decision concept, and hence all concepts are covered.) Once the search has terminated, a simple pruning procedure is employed to remove superfluous features from rules. This is the same procedure as that employed by LEM2, except that it is employed at the end of the search rather than repeatedly during the search.

The definition of GAMMARULES can be found in Figure 15.4. Given an attribute a and the current reduct candidate R, this function constructs the positive region for $R \cup \{a\}$ and returns the resulting dependency degree. During this process, if an equivalence class is evaluated that is a subset of a decision concept (and hence included in the lower approximation of the concept), the function DETERMINEADD (Figure 15.5) is executed to determine whether a rule representing this is constructed. A rule will be constructed and added to the ruleset if the equivalence class has not been encountered before or if the equivalence class is a superset of an existing rule's equivalence class (i.e., has a larger rule strength). In the latter case the existing rule is removed as this covers fewer objects.

For this type of search (i.e., hill-climbing), equality of equivalence classes is not considered in the function. This is because the newer equivalence class will have been produced by the same or greater number of features, and so will have an equal or greater specificity.

Alternative heuristics can also be employed, for example, by replacing the GAMMARULES function with that of Figure 15.6 and minimizing entropy instead of maximizing dependency degree. Reducts (or super-reducts) are found when the entropy for a feature subset reaches zero, and hence all decision concepts are covered fully.

For any hill-climbing approach, the final subsets are not guaranteed to be minimal, but pruning during search would ensure this. It should be noted that in

```
DetermineAdd(eq, attrs, dec)
Input: eq, the equivalence class for consideration as a rule;
attrs, the features that generated the equivalence class;
dec, the corresponding decision concept
Output: updated ruleset
    add ← true; newRule ← ⟨eq, attrs, dec⟩
    foreach R ∈ Rules
        if eq ⊆ R.eq //covers fewer objects than existing rule
            add ← false; break
        else if R.eq ⊂ eq
            RemoveRule(R)
            add ← true
    if add == true
        AddRule(newRule)
    return
```

Figure 15.5 DetermineAdd algorithm

```
EntropyRules(a, R)
Input: a, the feature under consideration for inclusion in R;
R, the current reduct candidate
Output: entropy of subset R; updated ruleset
    E = 0
    foreach eq ∈ 𝕌/{R ∪ a} //each equivalence class
        p = |eq| / |𝕌|
        foreach dec ∈ 𝕌/𝔻      //each decision concept
            if eq ⊆ dec
                DetermineAdd(eq, R ∪ {a}, dec)
            E += −p × (|dec ∩ eq| / |eq|) × log2 (|dec ∩ eq| / |eq|)
    return E
```

Figure 15.6 EntropyRules algorithm

line with rough set ideology, no parameters are required for the operation of rule induction. Only the data are used, requiring no human input.

15.4.3 Walkthrough

To illustrate the operation of the proposed algorithms, the example dataset in Table 2.1, Section 2.3) is used and the rules induced via GammaRules.

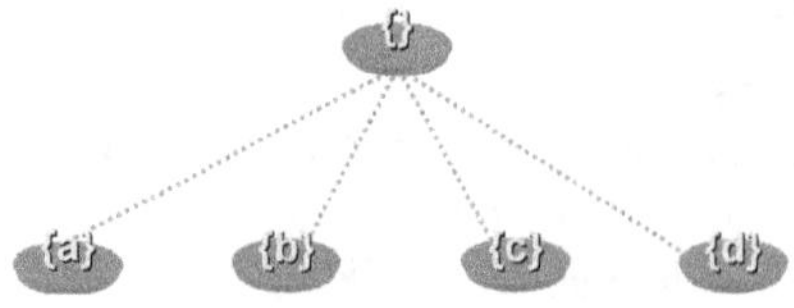

Figure 15.7 Search level one

The algorithm begins (as with QUICKREDUCT) by evaluating individual features, shown in Figure 15.7. First, feature a is considered and GAMMARULES(a, R) is evaluated. This will determine the equivalence classes of $\mathbb{U}/\{a\}$ and check if they are subsets of any decision concepts (here, $\{\{0\}, \{1, 3, 6\}, \{2, 4, 5, 7\}\}$). If this is the case, the algorithm determines whether to add a new rule representing this to the current set of rules. The equivalence classes are

$$\mathbb{U}/\{a\} = \{\{0, 3, 4\}, \{1, 7\}, \{2, 5, 6\}\}$$

It can be seen that no equivalence classes are subsets of decision concepts for this feature (the positive region is empty), and hence no rules are added. The resulting degree of dependency is zero. Next feature b is considered:

$$\mathbb{U}/\{b\} = \{\{0, 2, 4\}, \{1, 3, 6, 7\}, \{5\}\}$$

As $\{5\} \subseteq \{2, 4, 5, 7\}$, and the current ruleset is empty, a new rule is created and added: $\langle\{5\}, \{b\}, \{2, 4, 5, 7\}\rangle$. The equivalence class $\{5\}$ corresponds to feature b taking the value T, and decision concept $\{2, 4, 5, 7\}$ corresponds to feature e taking the value S. Hence the generated rule can be considered to be "**if** $b = T$, **then** $e = S$." The degree of dependency for feature b is $|\{5\}|/|\mathbb{U}| = \frac{1}{8}$. For feature c, the positive region is empty. For feature d,

$$\mathbb{U}/\{d\} = \{\{0, 3\}, \{1, 2, 5, 6\}, \{4, 7\}\}$$

As $\{4, 7\} \subseteq \{2, 4, 5, 7\}$, it is considered for inclusion in the ruleset. There is only one rule in the ruleset at this time, with equivalence class $\{5\}$; this is not a subset of $\{4, 7\}$, neither is $\{4, 7\}$ a subset of $\{5\}$, and so a new rule is created: $\langle\{4, 7\}, \{d\}, \{2, 4, 5, 7\}\rangle$. This rule can be interpreted as "**if** $d = R$, **then** $e = S$." The degree of dependency for this feature is $\frac{2}{8}$ and is the highest value for all considered features. The algorithm then selects this feature, adding it to the reduct candidate, and continues to the next level of search (as shown in Figure 15.8), as a subset producing the maximum dependency degree has not been found.

For subset $\{a, d\}$ the corresponding partition is $\mathbb{U}/\{a, d\} = \{\{0, 3\}, \{1\}, \{2, 5, 6\}, \{4\}, \{7\}\}$. $\{1\}$ is a subset of concept $\{1, 3, 6\}$, and a new rule representing this fact is added to the ruleset: $\langle\{1\}, \{a, d\}, \{1, 3, 6\}\rangle$. $\{4\}$ and $\{7\}$ are subsets

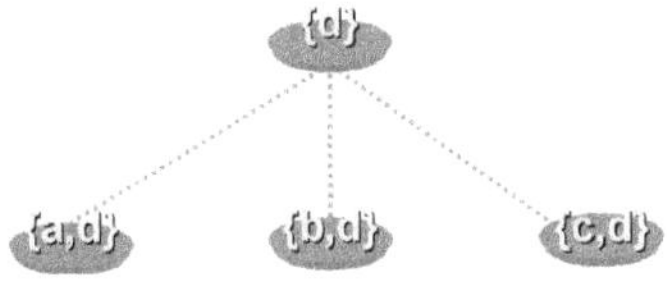

Figure 15.8 Search level two

of $\{2, 4, 5, 7\}$ but are not added to the ruleset as a more general rule exists $(\langle\{4, 7\}, \{d\}, \{2, 4, 5, 7\}\rangle)$.

For subset $\{b, d\}$, the partition is $\mathbb{U}/\{b, d\} = \{\{0\}, \{1, 6\}, \{2\}, \{3\}, \{4\}, \{5\}, \{7\}\}$. Equivalence class $\{0\} \subseteq \{0\}$ and is added to the ruleset: $\langle\{0\}, \{b, d\}, \{0\}\rangle$. $\{1, 6\}$ is a subset of decision concept $\{1, 3, 6\}$. From the existing rules it is seen that this equivalence class covers more objects than the existing rule $\langle\{1\}, \{a, d\}, \{1, 3, 6\}\rangle$ and this fact, therefore, is replaced with $\langle\{1, 6\}, \{b, d\}, \{1, 3, 6\}\rangle$. Rules are also created for equivalence classes $\{2\}$ and $\{3\}$—the remaining equivalence classes are already represented in the ruleset: $\langle\{2\}, \{b, d\}, \{2, 4, 5, 7\}\rangle$, $\langle\{3\}, \{b, d\}, \{1, 3, 6\}\rangle$.

For subset $\{c, d\}$, the partition is $\mathbb{U}/\{c, d\} = \{\{0\}, \{1, 6\}, \{2, 5\}, \{3\}, \{4\}, \{7\}\}$. A rule is generated for equivalence class $\{2, 5\}$, which then replaces the rules $\langle\{2\}, \{b, d\}, \{2, 4, 5, 7\}\rangle$ and $\langle\{5\}, \{d\}, \{2, 4, 5, 7\}\rangle$. Both $\{b, d\}$ and $\{c, d\}$ are reducts, so the algorithm terminates. The resulting ruleset is as follows:

$$\langle\{0\}, \{b, d\}, \{0\}\rangle$$
$$\langle\{3\}, \{b, d\}, \{1, 3, 6\}\rangle$$
$$\langle\{1, 6\}, \{b, d\}, \{1, 3, 6\}\rangle$$
$$\langle\{2, 5\}, \{c, d\}, \{2, 4, 5, 7\}\rangle$$
$$\langle\{4, 7\}, \{d\}, \{2, 4, 5, 7\}\rangle$$

which corresponds to the rules:

if $b = R$ **and** $d = T$, **then** $e = R$
if $b = S$ **and** $d = T$, **then** $e = T$
if $b = S$ **and** $d = S$, **then** $e = T$
if $c = R$ **and** $d = S$, **then** $e = S$
if $d = R$, **then** $e = S$

15.4.4 Experimentation

This section presents the initial experimental evaluation of the proposed methods for the task of pattern classification, over 24 benchmark datasets from [38] and

TABLE 15.6 Dataset details

Dataset	Conditional Features	Objects	Concepts
breast	9	699	2
car	6	1728	4
corral	6	64	2
derm	34	366	6
derm2	34	358	6
exactly	13	1000	2
exactly2	13	1000	2
heart	13	294	2
ionosphere	34	230	2
led	24	2000	9
lung	56	32	2
lymph	18	148	4
m-of-n	13	1000	2
monk3	6	432	2
mush	22	8124	2
prom	57	106	2
soybean-large	35	266	14
soybean-small	35	47	4
spect	22	267	2
tic-tac-toe	9	958	2
vote	16	300	2
water	38	521	13
zoo	16	101	7

[168] with several classifiers. The classifiers themselves are obtained from the WEKA toolkit [382] and ROSE software [276], and are evaluated using their default parameter settings.

The details of the benchmark datasets used can be found in Table 15.6. The number of conditional features ranges from 6 to 57 over the datasets, and the number of objects ranges from 32 to 8124. Additionally several datasets are multiple class.

In the experimentation, RS and Ent are the proposed dependency-based and entropy-based RIAs; RS-Prune and Ent-Prune are the same inducers but with a pruning postprocessing step. LEM and MLEM [276], are the leading rough set rule induction methods. Experimentation is also carried out with several non–rough set-based techniques, namely J48, JRip, PART, Prism, and naive Bayes [382] with default parameters selected. For each algorithm 10-fold cross-validation is repeated 10 times. The resulting classification accuracies and standard deviations can be seen in Table 15.7 for the rough set methods and Table 15.8 for the others. There was no data stratification for the rough set cross validation experimentation.

From Table 15.7. it can be seen that the performance of the proposed entropy-based method with pruning is superior to the other rough set-based rule inducers overall. RS-Prune also performs very well in general. It can be seen

TABLE 15.7 Classification accuracy: rough set methods

Dataset	RS	Ent	RS-Prune	Ent-Prune	LEM	MLEM
breast	95.04 ± 2.46	94.95 ± 2.25	95.95 ± 2.38	$\mathbf{96.37} \pm 2.43$	95.57 ± 2.51	94.57 ± 2.69
car	91.94 ± 1.57	87.39 ± 2.41	$\mathbf{96.84} \pm 1.28$	91.66 ± 2.00	94.56 ± 1.54	88.49 ± 3.45
corral	92.81 ± 10.2	93.76 ± 9.34	95.40 ± 8.74	97.12 ± 6.63	$\mathbf{100.0} \pm 0.00$	$\mathbf{100.0} \pm 0.00$
credit	68.12 ± 4.64	70.38 ± 4.29	81.66 ± 3.88	$\mathbf{84.07} \pm 4.04$	71.20 ± 4.79	71.20 ± 2.86
derm	84.38 ± 6.00	83.15 ± 6.31	88.94 ± 5.78	86.18 ± 5.63	$\mathbf{93.73} \pm 3.86$	84.96 ± 7.16
derm2	95.96 ± 3.39	94.87 ± 3.88	$\mathbf{96.29} \pm 3.39$	94.90 ± 3.66	92.74 ± 4.90	91.90 ± 6.21
exactly	78.00 ± 4.64	94.10 ± 7.32	87.34 ± 3.77	94.81 ± 5.62	$\mathbf{96.20} \pm 2.04$	95.70 ± 2.28
exactly2	70.08 ± 4.35	73.42 ± 4.48	83.44 ± 4.22	$\mathbf{84.70} \pm 3.75$	73.10 ± 3.86	75.00 ± 3.26
heart	75.44 ± 6.96	77.77 ± 6.49	84.48 ± 7.46	$\mathbf{85.02} \pm 6.28$	81.68 ± 7.27	79.57 ± 5.75
ionos	86.57 ± 7.44	85.70 ± 6.95	90.00 ± 6.44	$\mathbf{90.09} \pm 6.21$	86.09 ± 5.07	83.48 ± 9.49
led	93.60 ± 2.89	$\mathbf{100.0} \pm 0.000$	94.73 ± 2.56	$\mathbf{100.0} \pm 0.000$	$\mathbf{100.0} \pm 0.00$	$\mathbf{100.0} \pm 0.00$
lung	75.33 ± 21.19	81.25 ± 21.56	79.50 ± 19.37	$\mathbf{81.42} \pm 19.92$	55.00 ± 31.45	65.83 ± 30.38
lymph	75.97 ± 9.99	78.80 ± 10.45	86.59 ± 9.53	$\mathbf{86.73} \pm 8.74$	86.52 ± 9.40	82.43 ± 6.82
m-of-n	93.50 ± 2.41	99.89 ± 0.40	95.41 ± 2.07	99.89 ± 0.47	98.70 ± 1.55	$\mathbf{100.0} \pm 0.00$
monk3	$\mathbf{100.0} \pm 0.00$	$\mathbf{100.0} \pm 0.00$	$\mathbf{100.0} \pm 0.00$	$\mathbf{100.0} \pm 0.00$	$\mathbf{100.0} \pm 0.00$	$\mathbf{100.0} \pm 0.00$
mush	$\mathbf{100.0} \pm 0.00$	$\mathbf{100.0} \pm 0.00$	$\mathbf{100.0} \pm 0.00$	$\mathbf{100.0} \pm 0.00$	$\mathbf{100.0} \pm 0.00$	$\mathbf{100.0} \pm 0.00$
prom	78.53 ± 10.91	76.02 ± 12.83	83.66 ± 11.56	$\mathbf{83.72} \pm 11.01$	74.64 ± 12.61	82.64 ± 20.29
soybeanL	86.03 ± 5.87	85.94 ± 6.55	$\mathbf{89.87} \pm 6.06$	88.84 ± 6.10	84.87 ± 4.44	85.98 ± 5.51
soybeanS	96.45 ± 7.92	96.35 ± 8.16	96.20 ± 8.20	95.95 ± 8.46	$\mathbf{98.00} \pm 6.00$	$\mathbf{98.00} \pm 6.00$
spect	77.36 ± 7.60	77.66 ± 7.23	83.38 ± 8.09	$\mathbf{84.67} \pm 7.18$	74.56 ± 6.38	75.95 ± 7.74
tictactoe	82.55 ± 4.66	87.82 ± 3.08	90.18 ± 3.44	92.99 ± 2.89	$\mathbf{98.64} \pm 0.81$	$\mathbf{98.64} \pm 1.25$
vote	93.40 ± 4.24	94.20 ± 4.20	95.37 ± 3.82	$\mathbf{96.80} \pm 3.41$	92.67 ± 3.59	93.00 ± 4.82
water	59.02 ± 6.02	64.75 ± 6.27	74.63 ± 6.16	$\mathbf{78.79} \pm 5.83$	61.44 ± 6.51	12.26 ± 15.25
zoo	97.20 ± 5.33	95.70 ± 6.85	$\mathbf{97.90} \pm 4.33$	97.00 ± 5.60	92.09 ± 8.70	95.09 ± 4.92

TABLE 15.8 Classification accuracy: Other classifiers

Dataset	J48	JRip	PART	Prism	Naive Bayes
breast	94.34 ± 3.13	93.99 ± 2.73	93.81 ± 3.09	91.62 ± 3.15	**96.58** ± 2.19
car	92.34 ± 2.01	87.55 ± 2.98	**94.88** ± 1.75	90.13 ± 1.94	85.73 ± 2.67
corral	92.40 ± 10.67	96.79 ± 8.60	**99.00** ± 7.11	97.69 ± 7.45	84.04 ± 14.12
credit	71.78 ± 3.85	71.06 ± 3.96	71.09 ± 4.27	58.18 ± 4.62	**74.99** ± 3.78
derm	94.27 ± 3.14	89.57 ± 4.96	**95.65** ± 3.12	73.13 ± 6.75	70.95 ± 7.09
derm2	94.45 ± 3.23	88.44 ± 4.90	**95.51** ± 3.04	87.18 ± 5.64	88.07 ± 5.50
exactly	83.12 ± 9.08	69.59 ± 5.36	**92.42** ± 4.21	76.91 ± 5.05	68.65 ± 4.12
exactly2	75.05 ± 1.86	73.79 ± 2.19	72.81 ± 3.93	61.56 ± 4.89	**75.80** ± 4.33
heart	80.09 ± 7.04	80.56 ± 7.26	79.88 ± 7.50	65.43 ± 8.03	**83.01** ± 7.42
ionos	83.12 ± 9.08	69.59 ± 5.36	81.91 ± 7.12	80.83 ± 6.83	**91.30** ± 4.81
led	**100.00** ± 0.00	**100.00** ± 0.00	**100.00** ± 0.00	**100.00** ± 0.00	**100.0** ± 0.000
lung	**79.25** ± 21.50	77.92 ± 19.48	76.92 ± 22.75	63.67 ± 26.60	78.42 ± 21.12
lymph	77.84 ± 9.33	78.79 ± 8.95	**82.07** ± 10.19	74.47 ± 9.74	80.03 ± 11.09
m-of-n	**100.00** ± 0.00	97.97 ± 1.79	99.95 ± 0.22	99.93 ± 0.26	95.26 ± 2.44
monk3	**100.00** ± 0.00	98.77 ± 2.36	**100.00** ± 0.00	**100.00** ± 0.00	97.22 ± 2.13
mushr	**100.00** ± 0.00	**100.00** ± 0.02	**100.00** ± 0.00	**100.00** ± 0.02	99.67 ± 0.19
prom	79.04 ± 12.68	78.33 ± 10.60	84.13 ± 11.60	66.24 ± 12.02	**88.13** ± 9.18
soybeanL	88.64 ± 5.23	84.77 ± 5.91	**90.16** ± 4.85	75.53 ± 7.93	79.72 ± 7.68
soybeanS	97.65 ± 7.12	97.65 ± 7.12	**100.00** ± 0.00	97.65 ± 7.12	90.15 ± 15.69
spect	81.35 ± 6.16	**82.42** ± 5.73	79.86 ± 6.38	75.22 ± 5.48	80.63 ± 7.41
tictactoe	85.73 ± 3.23	**97.61** ± 1.53	93.33 ± 3.29	96.22 ± 1.87	69.69 ± 4.21
vote	94.27 ± 3.79	**94.53** ± 3.72	93.23 ± 4.75	91.17 ± 4.70	89.27 ± 5.16
water	68.29 ± 6.31	69.32 ± 6.00	66.52 ± 6.33	54.12 ± 7.64	**75.30** ± 5.73
zoo	92.61 ± 7.33	88.53 ± 7.99	93.41 ± 7.28	**93.75** ± 5.95	86.74 ± 10.95

that these methods tend to overfit the training data without pruning. It should be noted that both LEM and MLEM incorporate a degree of pruning at each stage of induction. Ent-Prune achieves the highest classification accuracy for 14 of the 24 datasets; for LEM, 8 datasets; MLEM, 7 datasets; and RS-Prune, 6 datasets.

When compared with several leading classifiers (Table 15.8), the proposed techniques perform very well, often performing better than or close to the optimal for these methods. Note that the results for the non–rough set-based classifiers were produced using data that was stratified, maintaining the distribution of class labels.

15.5 FUZZY UNIVERSAL REDUCTS

A single given dataset may contain more than one optimal feature subset. For the simple example given in Section 5.1, there are two minimal subsets. For larger problems, there may be many more. Typically only one of these subsets is chosen to perform the necessary dataset reduction. The problem here is how to discern between reducts in order to choose the best one to reduce the data.

A solution to this problem may be *fuzzy universal reducts*. Instead of committing to a single reduct, a fuzzy universal reduct captures the frequency information of features within all reducts for a dataset. Those features that appear often in subsets are determined to be more significant than those that appear less frequently. It is thought that by capturing this information, a better subset of features can be selected for reducing data. The techniques suggested here can be applied to the fuzzy-rough as well as the crisp rough set problem.

The set of all reducts, T, for a dataset can be calculated in the following way:

$$T = \{X : X \subseteq \mathbb{C}, \gamma_X(\mathbb{D}) = \gamma_{\mathbb{C}}(\mathbb{D})\} \tag{15.3}$$

A fuzzy universal reduct R for a dataset is defined ($\forall x \in \mathbb{C}$) as

$$\mu_R(x) = \frac{|\{x|x \in S, \forall S \in T\}|}{|T|} \tag{15.4}$$

An attribute a belongs fully to the reduct ($\mu_R(a) = 1$) if it appears in all reducts (a member of the *core*). By applying α-cuts, subsets of the attributes may be selected for dataset reduction. However, to calculate all reducts for nontrivial datasets is not feasible. Therefore some way is needed to approximate these memberships. It is thought that the use of pheromone in ant colony optimization based feature selection might act as such an approximater.

The basic idea is that during the course of their search, ants traverse the graph of nodes (features) depositing artificial pheromone. The amount of pheromone on an edge is an indicator of the utility of that edge in previous searches for optimal subsets. A useful node will have high amounts of pheromone on edges connected to it. By combining the values of the pheromone levels on edges connected to a node, a measure of that node's significance can be obtained.

Another means of estimating memberships to a fuzzy universal reduct could be obtained through a genetic algorithm based feature selection process. As the algorithm searches for optimal reducts, it encounters and evaluates many subsets along the way. By recording the information concerning the frequency of occurrence of features (and how "good" the subset is), an estimate of feature importance can be determined.

15.6 FUZZY-ROUGH CLUSTERING

Fuzzy c-means [31] and rough c-means [275] have been used with some success for the purpose of clustering. By allowing objects to belong to clusters to different extents, a degree of flexibility is afforded. However, the success of these methods is highly dependent on the choice of initial prototypes; a greater amount of robustness is required. The properties of fuzzy-rough sets can provide this stability by making use of two fuzzy approximations for each concept (lower and upper).

In order to provide such flexible clustering with fuzzy-rough sets, two alternative approaches could be considered: fuzzy-rough c-means and general fuzzy-rough clustering.

15.6.1 Fuzzy-Rough c-Means

Fuzzy-rough sets are a natural construct for modeling clusters. The fuzzy lower approximation contains objects that definitely belong to a concept; the upper approximation contains those that possibly belong. Each concept is then defined by two fuzzy sets, which are comprised of unions of equivalence classes (by the definitions of upper and lower approximations). These provide more information than ordinary fuzzy clusters by the notion of a fuzzy boundary between approximations where objects belong to concepts with a fuzzy uncertainty.

From the initial data several prototype fuzzy-rough sets need to be defined, in a manner similar to fuzzy c-means. One possible approach to this is to generate fuzzy-rough sets for each object and perform a degree of cluster merging. Merging can be easily performed via the union of fuzzy-rough sets. From these initial clusters the fuzzy-rough sets can be iteratively moved/adapted according to object membership (the degree of belongingness to the concept) and concept separation (distance between fuzzy-rough sets). For this purpose a distance measure to determine fuzzy-rough cluster closeness will need to be defined, which could be achieved through the definition of a center-of-gravity for a fuzzy-rough set. Although such definitions exist for fuzzy sets, there currently is no counterpart for fuzzy-rough sets. Suitably close clusters may also be merged as a result of cluster proximity. Refinement and adaptation will continue until no further adjustments of concepts result, or an acceptable level of data coverage is acquired.

Note that the output of this process is a collection of fuzzy-rough sets that represent concepts, not fuzzy clusters as in current methods. Fuzzy-rough concepts contain additional information as objects can now belong to fuzzy upper and lower approximations of concepts, rather than individual fuzzy clusters.

15.6.2 General Fuzzy-Rough Clustering

In the same way that feature selection can be viewed as a search through the space of possible subsets, the clustering problem may be viewed as a search through the space of possible clusters. It may therefore be useful to consider the clustering process as consisting of three procedures (mirroring feature selection): generation, evaluation, and validation. The generation procedure implements a search method that generates clusters for evaluation. This could be achieved through the use of a variety of algorithms: hill-climbing, simulated annealing, genetic algorithms, and so forth. The suitability of the potential clusters produced by the generation procedure is calculated via the evaluation function. The stopping criteria determine when to stop the search for clusters as a suitably accurate clustering has been reached. If the criteria are not met, the process repeats, with appropriate modification of the evaluated clusters. For use, the resulting set of clusters may be validated.

In rough set applications the decision values are known and are used to evaluate how well a given feature subset approximates each concept. However, this can be reversed for the purposes of clustering, so that concepts can be approximated in relation to conditional attributes. Hence the fuzzy-rough dependency degree can be used as an evaluation function to provide a measure of the accuracy of a given clustering. An additional criterion to be included in evaluation is that of the number of clusters. The process should attempt to minimize the number of clusters while maximizing the fuzzy-rough measure. This should avoid the need for a user-defined number of clusters to be specified initially.

15.7 FUZZIFICATION OPTIMIZATION

Most fuzzy methods in data mining require fuzzifications to be defined beforehand by expert opinion, or they can be derived automatically from the statistical properties of the data. Hence the extracted fuzzy knowledge may not be globally optimal due to these origins. By tuning membership functions after knowledge extraction (e.g., in the form of rules or clusters), performance can be improved significantly [224]. Another use for the fuzzy-rough dependency degree is in this area of optimizing fuzzifications. Given sets of fuzzy partitions, the metric can be used to gauge the quality of these partitions through fuzzy-rough lower approximations. A higher value of the measure indicates a better "fit" to the data. The fitness optimization can be performed for subsets of attributes, although only the partition optimization of individual attributes is required.

As there are many real-valued parameters that must be optimized for a fuzzy partitioning, a suitable optimization mechanism should be chosen. Particle swarm optimization (PSO) [175], a population-based stochastic optimization technique, may be particularly useful. In PSO the system is initialized with a population of random solutions, called particles. Optima are searched for by updating generations, with particles moving through the parameter space toward the current local and global optimum particles. At each time step the velocities of all particles are changed depending on the current optima. Although there are similarities with GAs, PSO systems tend to require fewer (possibly erroneous) design choices, such as the choice of evolutionary operators.

For this application the particles will represent potential membership function definitions defined by sets of parameters. The initial population of particles could be generated by random parameter deviations from the original membership functions. Extra constraints will need to be enforced in order to restrict search to meaningful fuzzifications. For example, only convex fuzzy sets should be considered. Particles are rated according to the fuzzy-rough dependency degree to provide a measure of fitness. From this, the local and global optima can be determined and used to adjust particle velocities.

15.8 SUMMARY

This chapter has presented several potential directions for crisp and fuzzy-rough developments. One particular area of interest is that of reformulating the search for a reduct as a propositional satisfiability problem. Solution techniques from this domain can then be applied to locate the smallest reducts. In addition to this and other areas of further research in feature selection, the possibility of inducing decision trees using the fuzzy-rough metric was discussed. The initial results show that the proposed method performs comparably to fuzzy ID3 for fuzzy datasets, and better than it does for crisp data. Further experimentation is to be carried out on a fuller range of datasets in the future.

Other developments have been proposed that use fuzzy-rough set-based methodologies. Fuzzy-rough rule induction could be useful in extracting fuzzy rules from data in a manner similar to that of crisp rough set rule induction. By the extension of crisp concepts to the fuzzy-rough case, traditional rule induction algorithms can be extended for use with real-valued and noisy data. With these extensions new algorithms can be constructed for the purpose. Fuzzy-rough clustering can also be developed in a similar way, using fuzzy-rough methods. A related area is that of fuzzification optimization, where fuzzy-rough sets could prove beneficial in fine-tuning fuzzy set membership functions.

APPENDIX A

METRIC COMPARISON RESULTS: CLASSIFICATION DATASETS

This section contains the full results for the tables in Section 8.7.2. The datasets were created by generating around 30 random feature values for 400 objects. Two or three features (referred to as x, y, or z) are chosen to contribute to the final Boolean classification by means of an inequality. The tables show the rating given to the features from the corresponding metric. For the data presented in the first table, the first feature, x, is used to determine the classification. The values of features y and z are derived from x: $y = \sqrt{x}$, $z = x^2$. Table cells are shaded to highlight the top ranked features determined by the feture metrics. Darker shading indicates a higher ranking.

Computational Intelligence and Feature Selection: Rough and Fuzzy Approaches, by Richard Jensen and Qiang Shen

TABLE A.1 Feature evaluation for $x > 0.5$, $y = \sqrt{x}$, $z = x^2$

Feature	FR	Re	IG	GR	1R	χ^2
x	0.5257	0.31758	0.997	1	99.5	200
y	0.5296	0.24586	0.997	1	99.5	200
z	0.5809	0.32121	0.997	1	99.5	200
3	0.0	−0.00276	0	0	55.5	0
4	0.0	0.00148	0	0	47.5	0
5	0.0	−0.00268	0	0	44.5	0
6	0.0	−0.00221	0	0	58.5	0
7	0.0	−0.01002	0	0	52.5	0
8	0.0	−0.00649	0	0	57.5	0
9	0.0	0.00889	0	0	49.0	0
10	0.0	−0.00222	0	0	53.0	0
11	0.0	0.00182	0	0	59.5	0
12	0.0	0.00144	0	0	42.5	0
13	0.0	0.00475	0	0	50.0	0
14	0.0	0.01006	0	0	65.5	0
15	0.0	0.00613	0	0	55.5	0
16	0.0	0.00488	0	0	47.0	0
17	0.0	0.00563	0	0	56.5	0
18	0.0	0.01427	0	0	50.0	0
19	0.0	−0.00467	0	0	53.5	0
20	0.0	−0.01785	0	0	54.5	0
21	0.0	0.00327	0	0	50.0	0
22	0.0	0.00350	0	0	48.5	0
23	0.0	0.01339	0	0	51.5	0
24	0.0	0.00464	0	0	49.5	0
25	0.0	0.01334	0	0	59.0	0
26	0.0	0.01715	0	0	48.5	0
27	0.0	−0.01742	0	0	49.0	0
28	0.0	0.00685	0	0	60.5	0
29	0.0	−0.00206	0	0	53.5	0
30	0.0	0.00164	0	0	51.5	0
31	0.0	0.00171	0	0	49.0	0
32	0.0	−0.00325	0	0	51.0	0

TABLE A.2 Feature evaluation for $(x+y)^2 > 0.25$

Feature	FR	Re	IG	GR	1R	χ^2
x	0.2330	0.1862	0.2328	0.1579	86.75	128.47
y	0.2597	0.1537	0.1687	0.169	87.75	71.97
2	0.0	0.0132	0	0	84.5	0
3	0.0	0.0307	0	0	85.25	0
4	0.0	0.0320	0	0	86.0	0
5	0.0	0.0112	0	0	85.5	0
6	0.0	0.0127	0	0	86.0	0
7	0.0	0.0248	0	0	86.0	0
8	0.0	0.0219	0	0	86.0	0
9	0.0	0.0364	0	0	85.5	0
10	0.012	0.0345	0.0344	0.0576	86.5	23.638
11	0.0	0.0180	0	0	85.5	0
12	0.0	0.0246	0	0	85.75	0
13	0.0	0.0312	0	0	86.0	0
14	0.0	0.0182	0	0	86.0	0
15	0.0	0.0216	0	0	86.0	0
16	0.0	0.0245	0	0	86.0	0
17	0.0	0.0188	0	0	85.25	0
18	0.0	0.0145	0	0	85.5	0
19	0.0	0.0292	0	0	86.0	0
20	0.0	0.0132	0	0	86.0	0
21	0.0	0.0148	0	0	84.5	0
22	0.0	0.0116	0	0	86.0	0
23	0.0	0.0248	0	0	86.0	0
24	0.0	0.0190	0	0	86.0	0
25	0.0	0.0290	0	0	85.25	0
26	0.0	0.0222	0	0	86.0	0
27	0.0	0.0292	0	0	85.75	0
28	0.0	0.0307	0	0	84.75	0
29	0.0	0.0255	0	0	86.0	0
30	0.0	0.0163	0	0	86.0	0
31	0.0	0.0221	0	0	84.5	0

TABLE A.3 Feature evaluation for $(x+y)^2 > 0.5$

Feature	FR	Re	IG	GR	1R	χ^2
x	0.209	0.140067	0.241	0.156	79.0	119.56
y	0.2456	0.151114	0.248	0.165	78.25	122.34
2	0.0	0.008450	0	0	76.0	0
3	0.0	0.009063	0	0	73.75	0
4	0.0	0.005004	0	0	70.25	0
5	0.0	0.013202	0	0	74.75	0
6	0.0	0.011766	0	0	72.25	0
7	0.0	0.029141	0	0	73.5	0
8	0.0	0.007746	0	0	74.25	0
9	0.0	0.007245	0	0	73.5	0
10	0.0	0.018969	0	0	76.25	0
11	0.0	0.008741	0	0	75.5	0
12	0.0	0.012712	0	0	72.5	0
13	0.0	0.009962	0	0	72.25	0
14	0.0	−0.000115	0	0	75.0	0
15	0.0	0.003541	0	0	73.5	0
16	0.0	0.012629	0	0	75.0	0
17	0.0	0.019661	0	0	73.75	0
18	0.0	0.013886	0	0	76.0	0
19	0.0	0.011437	0	0	73.25	0
20	0.0	0.008366	0	0	74.25	0
21	0.0	0.017771	0	0	72.25	0
22	0.0	0.003630	0	0	74.5	0
23	0.0	0.013811	0	0	75.5	0
24	0.0	0.017560	0	0	74.5	0
25	0.0	0.003648	0	0	73.5	0
26	0.0	0.013574	0	0	72.75	0
27	0.0	0.009583	0	0	73.75	0
28	0.0	−0.000367	0	0	75.25	0
29	0.0	−0.000397	0	0	75.25	0
30	0.0	0.011544	0	0	76.25	0
31	0.0	0.007605	0	0	74.75	0

TABLE A.4 Feature evaluation for $(x+y)^3 < 0.125$

Feature	FR	Re	IG	GR	1R	χ^2
x	0.2445	0.1486	0.134	0.134	87.75	57.46
y	0.2441	0.1659	0.159	0.164	87.25	73.39
2	0.0	0.0229	0	0	88.5	0
3	0.0	0.0232	0	0	89.0	0
4	0.0	0.0322	0	0	88.25	0
5	0.0	0.0301	0	0	89.0	0
6	0.0	0.0252	0	0	89.0	0
7	0.0	0.0203	0	0	89.0	0
8	0.0	0.0341	0	0	89.0	0
9	0.0	0.0289	0	0	89.0	0
10	0.0	0.0339	0	0	88.5	0
11	0.0	0.0313	0	0	89.0	0
12	0.0	0.0287	0	0	89.0	0
13	0.0	0.0545	0	0	89.0	0
14	0.0	0.0458	0	0	89.0	0
15	0.0	0.0378	0	0	89.0	0
16	0.0	0.0289	0	0	89.0	0
17	0.0	0.0332	0	0	89.0	0
18	0.0	0.0306	0	0	89.0	0
19	0.0	0.0397	0	0	88.25	0
20	0.0	0.0247	0	0	89.0	0
21	0.0	0.0163	0	0	89.0	0
22	0.0	0.0330	0	0	89.0	0
23	0.0	0.0276	0	0	89.0	0
24	0.0	0.0189	0	0	88.75	0
25	0.0	0.0279	0	0	88.75	0
26	0.0	0.0252	0	0	88.75	0
27	0.0	0.0157	0	0	89.0	0
28	0.0	0.0304	0	0	89.0	0
29	0.0	0.0285	0	0	89.0	0
30	0.0	0.0315	0	0	88.75	0
31	0.0	0.0290	0	0	89.0	0

TABLE A.5 Feature evaluation for $x * y * z > 0.125$

Feature	FR	Re	IG	GR	1R	χ^2
x	0.1057	0.0750547	0.169	0.123	64.25	73.65
y	0.0591	0.1079423	0.202	0.226	66.75	88.04
z	0.1062	0.0955878	0.202	0.160	67.5	84.28
3	0.0	0.0031390	0	0	56.75	0
4	0.0	−0.0156922	0	0	60.75	0
5	0.0	0.0088234	0	0	58.5	0
6	0.0	−0.0076636	0	0	53.25	0
7	0.0	0.0050098	0	0	57.5	0
8	0.0	0.0006841	0	0	55.75	0
9	0.0	−0.0015287	0	0	54	0
10	0.0	0.0031223	0	0	53	0
11	0.0	0.0021915	0	0	57.75	0
12	0.0	0.0027260	0	0	61.75	0
13	0.0	0.0108794	0	0	57.75	0
14	0.0	0.0008456	0	0	59.25	0
15	0.0	−0.0002930	0	0	60	0
16	0.0	−0.0018220	0	0	57.5	0
17	0.0	0.0019899	0	0	61.75	0
18	0.0	0.0090028	0	0	57.5	0
19	0.0	0.0043929	0	0	60.25	0
20	0.0	0.0006062	0	0	53.75	0
21	0.0	−0.0075626	0	0	53.75	0
22	0.0	0.0185202	0	0	57.0	0
23	0.0	−0.0056034	0	0	59.25	0
24	0.0	0.0116144	0	0	57.75	0
25	0.0	0.0001139	0	0	55.75	0
26	0.0	−0.0010561	0	0	56.25	0
27	0.0	0.0002921	0	0	54.5	0
28	0.0	0.0062014	0	0	51.75	0
29	0.0	−0.0092218	0	0	59.25	0
30	0.0	0.0000525	0	0	61.75	0
31	0.0	−0.0011460	0	0	57.0	0
32	0.0	−0.0059597	0	0	57.0	0

TABLE A.6 Feature evaluation for $x * y * z^2 > 0.125$

Feature	FR	Re	IG	GR	1R	χ^2
x	0.1511	0.09800	0.1451	0.0947	76.5	65.43
y	0.1101	0.05571	0.0909	0.1080	78.0	35.36
z	0.2445	0.14736	0.2266	0.2271	79.75	93.81
3	0.0	0.00725	0	0	77.5	0
4	0.0	0.00652	0	0	78.5	0
5	0.0	0.01793	0	0	77.75	0
6	0.0	0.00716	0	0	78.0	0
7	0.0	0.02053	0	0	76.5	0
8	0.0	0.00339	0	0	78.25	0
9	0.0	0.01114	0	0	77.0	0
10	0.0	0.00409	0	0	77.75	0
11	0.0	0.01595	0	0	77.75	0
12	0.0	0.01640	0	0	77.75	0
13	0.0	0.01224	0	0	78.5	0
14	0.0	0.00170	0	0	75.5	0
15	0.0	0.00735	0	0	78.75	0
16	0.0	0.00575	0	0	78.0	0
17	0.0	0.01831	0	0	78.25	0
18	0.0	0.00508	0	0	76.0	0
19	0.0	0.01943	0	0	79.0	0
20	0.0	0.00929	0	0	78.5	0
21	0.0	0.00493	0	0	77.75	0
22	0.0	0.00579	0	0	75.75	0
23	0.0	0.01252	0	0	76.25	0
24	0.0	0.01957	0	0	79.0	0
25	0.0	0.01700	0	0	78.25	0
26	0.0	0.01175	0	0	76.5	0
27	0.0	0.01055	0	0	76.5	0
28	0.0	0.01405	0	0	78.0	0
29	0.0	0.02123	0	0	77.75	0
30	0.0	0.00884	0	0	77.5	0
31	0.0	0.01270	0	0	77.75	0
32	0.0	0.00806	0	0	77.5	0

APPENDIX B

METRIC COMPARISON RESULTS: REGRESSION DATASETS

This section contains the full results for three regression datasets. The datasets were created by generating around 30 random feature values for 400 objects. The decision value is created by means of a function involving two or more variables denoted x, y, and, optionally, z if there is a third variable in the function. The tables show the rating given to the features from the corresponding metric.

Computational Intelligence and Feature Selection: Rough and Fuzzy Approaches, by Richard Jensen and Qiang Shen

TABLE B.1 Feature evaluation for $f(x, y) = x * y$

Feature	FR	Re
x	0.156714	0.075208
y	0.183391	0.113038
2	0	−0.009917
3	0	−0.005152
4	0	0.002461
5	0	−0.000395
6	0	0.002699
7	0	0.003785
8	0	0.001662
9	0	−0.004740
10	0	0.001569
11	0	0.001495
12	0	−0.011035
13	0	−0.002583
14	0	−0.001381
15	0	−0.008143
16	0	−0.006520
17	0	−0.006878
18	0	−0.007850
19	0	−0.010232
20	0	−0.012889
21	0	−0.002477
22	0	−0.006587
23	0	−0.013506
24	0	−0.002338
25	0	−0.005290
26	0	−0.010280
27	0	−0.008658
28	0	−0.013362
29	0	0.000171
30	0	−0.006342
31	0	−0.012110
32	0	−0.006963

TABLE B.2 Feature evaluation for $f(x,y,z) = x * y * z$

Feature	FR	Re
x	0.227428	0.076988
y	0.165914	0.036674
z	0.203991	0.046234
3	0	−0.017056
4	0	0.004989
5	0.027450	−0.006416
6	0	−0.006045
7	0	0.003952
8	0	−0.001892
9	0	−0.010275
10	0	0.002356
11	0	−0.002328
12	0	−0.008287
13	0	−0.002969
14	0	−0.000050
15	0	0.002246
16	0	−0.006986
17	0	−0.001557
18	0	−0.007045
19	0	−0.002284
20	0	−0.006091
21	0	−0.015512
22	0.014446	−0.002249
23	0	−0.003250
24	0	0.000463
25	0	−0.010833
26	0	−0.002515
27	0	0.003845
28	0	−0.006644
29	0	−0.004005
30	0	−0.004674
31	0	−0.002399
32	0	0.001246

TABLE B.3 Feature evaluation for $f(x, y, z) = x * y * z^2$

Feature	FR	Re
x	0.128874	0.042855
y	0.123908	0.046120
z	0.170640	0.087663
3	0	−0.003307
4	0	0.000906
5	0	−0.005428
6	0	−0.009834
7	0	−0.014792
8	0	−0.006332
9	0	−0.008469
10	0	−0.000288
11	0	−0.005352
12	0	0.002245
13	0	−0.009150
14	0	−0.011524
15	0	−0.010116
16	0	−0.003124
17	0.005709	0.009794
18	0	−0.002233
19	0.005229	−0.011230
20	0	−0.005987
21	0	0.003280
22	0.011279	−0.010511
23	0	−0.004181
24	0.003855	−0.005324
25	0	0.000940
26	0	0.008235
27	0.019171	−0.013884
28	0	−0.006893
29	0	−0.011087
30	0	−0.011827
31	0	−0.007989
32	0	−0.018503

REFERENCES

1. C. G. G. Aitken and F. Taroni. *Statistics and the Evaluation of Evidence for Forensic Scientists*, 2nd ed. New York: Wiley. 2004.
2. C. G. G. Aitken, G. Zadora, and D. Lucy. A two-level model for evidence evaluation. *J. Forensic Sci.* 52: 412–419. 2007. Forthcoming.
3. C. G. G. Aitken, Q. Shen, R. Jensen, and B. Hayes. The evaluation of evidence for exponentially distributed data. *Comput. Stat. Data Anal.* 12(12): 5682–5693. 2007.
4. H. Almuallim and T. G. Dietterich. Learning with many irrelevant features. In *9th National Conference on Artificial Intelligence.* Cambridge: MIT Press, pp. 547–552. 1991.
5. J. J. Alpigini, J. F. Peters, J. Skowronek, and N. Zhong, eds. *Rough Sets and Current Trends in Computing. Proceedings. 3rd International Conference*, Malvern, PA, October 14–16, 2002. Lecture Notes in Computer Science 2475. Berlin: Springer. 2002.
6. K. K. Ang and C. Quek. Stock trading using RSPOP: a novel rough set-based neuro-fuzzy approach. *IEEE Trans. Neural Net.* 17(5): 1301–1315. 2006.
7. C. Apté, F. Damerau, and S. M. Weiss. Automated learning of decision rules for text categorization. *ACM Trans. Inf. Sys.* 12(3): 233–251. 1994.
8. S. Asharaf and M. N. Murty. An adaptive rough fuzzy single pass algorithm for clustering large data sets. *Pattern Recog.* 36(12): 3015–3018. 2004.
9. S. Asharaf, S. K. Shevade, and N. M. Murty. Rough support vector clustering. *Pattern Recog.* 38(10): 1779–1783. 2005.
10. J. Atkinson-Abutridy, C. Mellish, and S. Aitken. Combining information extraction with genetic algorithms for text mining. IEEE Intelligent Systems 19(3): 22–30. 2004.

Computational Intelligence and Feature Selection: Rough and Fuzzy Approaches, by Richard Jensen and Qiang Shen

11. G. Attardi, A. Gullí, and F. Sebastiani. Automatic Web Page Categorization by Link and Context Analysis. In *Proceedings of 1st European Symposium on Telematics, Hypermedia and Artificial Intelligence*, pp. 105–119. 1999.
12. W. H. Au and K. C. C. Chan. An effective algorithm for discovering fuzzy rules in relational databases. In *Proceedings of the 7th IEEE International Conference on Fuzzy Systems*. Piscataway, NJ: IEEE Press, pp. 1314–1319. 1998.
13. T. Baeck. *Evolutionary Algorithms in Theory and Practice.* Oxford: Oxford University Press. 1996.
14. A. A. Bakar, M. N. Sulaiman, M. Othman, and M. H. Selamat. Propositional satisfiability algorithm to find minimal reducts for data mining. *Int. J. Comput. Math.* 79(4): 379–389. 2002.
15. M. Balasubramanian, E. L. Schwartz, J. B. Tenenbaum, V. de Silva, and J. C. Langford. The isomap algorithm and topological stability. *Science* 295(5552): 7. 2002.
16. J. F. Baldwin, J. Lawry, and T. P. Martin. A mass assignment based ID3 algorithm for decision tree induction. *Int. J. Intell. Sys.* 12(7): 523–552. 1997.
17. J. K. Baltzersen. An attempt to predict stock market data: a rough sets approach. Diploma thesis. Knowledge Systems Group, Department of Computer Systems and Telematics, Norwegian Institute of Technology, University of Trondheim. 1996.
18. P. Baranyi, T. D. Gedeon, and L. T. Kóczy. A general interpolation technique in fuzzy rule bases with arbitrary membership functions. In *Proceedings of IEEE International Conference on Systems, Man, and Cybernetics*. Piscataway, NJ: IEEE Press, pp. 510–515. 1996
19. P. Baranyi, D. Tikk, Y. Yam, and L. T. Kóczy. A new method for avoiding abnormal conclusion for α-cut based rule interpolation. In *Proceedings of FUZZ-IEEE'99*, Seoul, Korea. Piscataway, NJ: IEEE Press, pp. 383–388. 1999.
20. L. Bardossy and L. Duckstein. *Fuzzy Rule-Based Modeling with Application to Geophysical, Biological and Engineering Systems.* Roca Baton: CRC Press. 1995.
21. R. Bayardo Jr. and R. Schrag. Using CSP look-back techniques to solve exceptionally hard SAT instances. In *Proceedings of International Conference on Principles and Practice of Constraint Programming*, Lecture Notes in Computer Science, 1118, Springer-Verlag, pp. 46–60. 1996.
22. J. Bazan. A comparison of dynamic and non-dynamic rough set methods for extracting laws from decision tables. In *Rough Sets in Knowledge Discovery.* Heidelberg: Physica, pp. 321–365. 1998.
23. J. Bazan, A. Skowron, and P. Synak. Dynamic reducts as a tool for extracting laws from decision tables. In Z. W. Ras, M. Zemankova, eds., *Proceedings of 8th Symposium on Methodologies for Intelligent Systems*. Lecture Notes in Artificial Intelligence 869. Berlin: Springer, pp. 346–355. 1994.
24. J. Bazan, A. Skowron, and P. Synak. Market data analysis: A rough set approach. ICSResearch Reports 6/94. Warsaw University of Technology. 1994.
25. T. Beaubouef, F. E. Petry, and G. Arora. Information measures for rough and fuzzy sets and application to uncertainty in relational databases. In [258] 1999.
26. R. Bellman. *Adaptive Control Processes: A Guided Tour.* Princeton: Princeton University Press. 1961.
27. M. J. Beynon. An investigation of β-reduct selection within the variable precision rough sets model. In *Proceedings of 2nd International Conference on Rough Sets and*

Current Trends in Computing (RSCTC 2000), Lecture Notes in Computer Science, 2005, Springer-Verlag, pp 114–122. 2000.

28. M. J. Beynon. Reducts within the variable precision rough sets model: A further investigation. *Eur. J. Oper. Res.* 134(3): 592–605. 2001.
29. M. J. Beynon. Stability of continuous value discretisation: An application within rough set theory. *Int. J. Approx. Reason.* 35: 29–53. 2004.
30. J. C. Bezdek, Fuzzy mathematics in pattern classification. PhD thesis. Center for Applied Mathematics, Cornell University. 1973.
31. J. C. Bezdek. *Pattern Recognition with Fuzzy Objective Function Algorithms.* New York: Plenum Press. 1981.
32. R. B. Bhatt and M. Gopal. FRID: Fuzzy-rough interactive dichotomizers. *IEEE International Conference on Fuzzy Systems (FUZZ-IEEE'04).* Piscataway, NJ: IEEE Press, pp. 1337–1342. 2004.
33. R. B. Bhatt and M. Gopal. On fuzzy-rough sets approach to feature selection. *Pattern Recog. Lett.* 26(7): 965–975. 2005.
34. R. B. Bhatt and M. Gopal. On the compact computational domain of fuzzy-rough sets. *Pattern Recog. Lett.* 26(11): 1632–1640. 2005.
35. H. Bian and L. Mazlack. Fuzzy-rough nearest-neighbor classification approach. *Proceeding of the 22nd International Conference of North American Fuzzy Information Processing Society*. Piscataway, NJ: IEEE Press, pp. 500–505. 2003.
36. C. M. Bishop. *Neural Networks for Pattern Recognition.* Oxford: Oxford University Press. 1995.
37. A. T. Bjorvand and J. Komorowski. Practical applications of genetic algorithms for efficient reduct computation. In *Proceedings of 15th IMACS World Congress on Scientific Computation, Modelling and Applied Mathematics*, Vol. 4, IMACS, Wissenschaft & Technik Verlag, pp. 601–606. 1997.
38. C. L. Blake and C. J. Merz. UCI Repository of machine learning databases. Irvine: University of California. 1998. Available at http://www.ics.uci.edu/~mlearn/.1
39. D. Boixader, J. Jacas, and J. Recasens. Upper and lower approximations of fuzzy sets. *Int. J. Gen. Sys.* 29(4): 555–568. 2000.
40. E. Bonabeau, M. Dorigo, and G. Theraulez. *Swarm Intelligence: From Natural to Artificial Systems.* New York: Oxford University Press. 1999.
41. B. Bouchon-Meunier, C. Marsala, and M. Rifqi. Interpolative reasoning based on graduality. In *Proceedings of FUZZ-IEEE'2000*, San Antonio, TX. Piscataway, NJ: IEEE Press, pp. 483–487. 2000.
42. U. M. Braga-Neto and E. R. Dougherty. Is cross-validation valid for small-sample microarray classification? *Bioinformatics* 20(3): 374–380. 2004.
43. C. Bregler and S. M. Omoundro. Nonlinear image interpolation using manifold learning. In G. Tesauro, D. S. Touretzky, and T. K. Leen, eds., *Advances in Neural Information Processing Systems* 7, The MIT Press, pp. 973–980. 1995.
44. L. Breiman, J. H. Friedman, R. A. Olshen, and C. J. Stone. *Classification and Regression Trees.* Monterey, CA: Wadsworth. 1984.
45. M. A. Carreira-Perpiñán. Continuous latent variable models for dimensionality reduction and sequential data reconstruction. PhD thesis. University of Sheffield, UK. 2001.

46. W. Cedeño and D. K. Agrafiotis. Using particle swarms for the development of QSAR models based on k-nearest neighbor and kernel regression. *J. Comput. Aid. Mol. Des.* 17: 255–263. 2003.
47. J. Cendrowska. PRISM: An algorithm for inducing modular rules. *Int. J. Man-Machine Studies* 27 (4): 349–370. 1987.
48. S. Chakrabarti, B. Dom, P. Raghaven, S. Rajagopalan, D. Gibson, and J. Kleinberg. Automatic resource compilation by analyzing hyperlink structure and associated text. *Proceedings of the 7th International World Wide Web Conference*, Elsevier Science B.V., Amsterdam, pp. 65–74. 1998.
49. R. Chan. Protecting rivers and streams by monitoring chemical concentrations and algae communities. ERUDIT: 3rd International Competition of Data Analysis by Intelligent Techniques (runner up). 1999.
50. K. Chan and A. Wong. APACS: A System for Automatic Analysis and Classification of Conceptual Patterns. *Comput. Intell.* 6(3): 119–131. 1990.
51. S. Chen, S. L. Lee, and C. Lee. A new method for generating fuzzy rules from numerical data for handling classification problems. *Appl. Art. Intell.* 15(7): 645–664. 2001.
52. W. C. Chen, N. B. Chang, and J. C. Chen. Rough set-based hybrid fuzzy-neural controller design for industrial wastewater treatment. *Water Res.* 37(1): 95–107. 2003.
53. D. Chen, W. X. Zhang, D. Yeung, and E. C. C. Tsang, Rough approximations on a complete completely distributive lattice with applications to generalized rough sets. *Infor. Sci.* 176(13): 1829–1848. 2006.
54. A. Chouchoulas and Q. Shen. Rough set-aided keyword reduction for text categorisation. *Appl. Art. Intell.* 15 (9): 843–873. 2001.
55. S. Chimphlee, N. Salim, M. S. B. Ngadiman, W. Chimphlee, and S. Srinoy. Independent component analysis and rough fuzzy based approach to Web usage mining. In *Proceedings of the Artificial Intelligence and Applications*. ALTA Press, Anaheim, pp. 422–427. 2006.
56. A. Chouchoulas, J. Halliwell, and Q. Shen. On the implementation of rough set attribute reduction. *Proceedings of 2002 UK Workshop on Computational Intelligence*, pp. 18–23. 2002.
57. P. Clark and R. Boswell. *Machine Learning—Proceedings of the Fifth European Conference* (EWSL-91). Berlin: Springer, pp. 151–163. 1991.
58. P. Clark and T. Niblett. The CN2 induction algorithm. *Mach. Learning J.* 3(4): 261–283. 1989.
59. W. W. Cohen. Fast effective rule induction. In *Machine Learning: Proceedings of the 12th International Conference*, Morgan Kaufmann, pp. 115–123. 1995.
60. O. Cordón, M. J. del Jesus, and F. Herrera. Evolutionary approaches to the learning of fuzzy rule-based classification systems. In L. C. Jain, ed., *Evolution of Enginnering and Information Systems and Their Applications.* Roca Baton: CRC Press, pp. 107–160. 1999.
61. O. Cordón, F. Gomide, F. Herrera, F. Hoffmann, and L. Magdalena. Ten years of genetic fuzzy systems: Current framework and new trends. *Fuzzy Sets and Systems* 141: 5–31. 2001.

62. O. Cordón, F. Herrera, F. Hoffmann, and L. Magdalena. *Genetic Fuzzy Systems: Evolutionary Tuning and Learning of Fuzzy Knowledge Bases*. Singapore: World Scientific. 2001.
63. O. Cordón, M. J. del Jesus, F. Herrera, L. Magdalena, and P. Villar. A multiobjective genetic learning process for joint feature selection and granularity and contexts learning in fuzzy rule-based classification systems. In J. Casillas, O. Cordón, F. Herrera, and L. Magdalena, eds., *Interpretability Issues in Fuzzy Modeling*. Berlin: Springer, pp. 79–99. 2003.
64. C. Cornelis, M. De Cock, and A. Radzikowska. Vaguely quantified rough sets. *Proceedings of 11th International Conference on Rough Sets, Fuzzy Sets, Data Mining and Granular Computing*. Lecture Notes in Artificial Intelligence 4482. Berlin: Springer, pp. 87–94. 2007.
65. C. Cornelis and R. Jensen. A noise-tolerant approach to fuzzy-rough feature selection. *17th IEEE International Conference on Fuzzy Systems* (FUZZ-IEEE'08), submitted.
66. E. Cox. *The Fuzzy Systems Handbook: A Practitioner's Guide to Building, Using and Maintaining Fuzzy Systems*. San Diego: Academic Press. 1999.
67. M. Dash and H. Liu. Feature selection for classification. *Intell. Data Anal.* 1(3): 131–156. 1997.
68. M. Dash, K. Choi, P. Scheuermann, and H. Liu. Feature selection for clustering—A filter solution. In *Proceedings of IEEE International Conference on Data Mining (ICDM)*, Piscataway, NJ: IEEE Press, pp. 115–122. 2002.
69. M. Dash and H. Liu. Consistency-based search in feature selection. *Art. Intell.* 151(1–2): 155–176. 2003.
70. M. Davis and H. Putnam. A computing procedure for quantification theory. *J. ACM* 7(3): 201–215. 1960.
71. M. Davis, G. Logemann, and D. Loveland. A machine program for theorem proving. *Comm. ACM* 5: 394–397. 1962.
72. M. De Cock, C. Cornelis, and E.E. Kerre. Fuzzy rough sets: beyond the obvious. *Proceedings of IEEE International Conference on Fuzzy Systems*, Vol. 1. Piscataway, NJ: IEEE Press, pp. 103–108, 2004.
73. M. De Cock and C. Cornelis. Fuzzy rough set based Web query expansion. *Proceedings of Rough Sets and Soft Computing in Intelligent Agent and Web Technology*. International Workshop at WIIAT2005 (2005 IEEE/WIC/ACM International Joint Conference on Web Intelligence and Intelligent Agent Technology), Piscataway, NJ: IEEE Press, pp. 9–16, 2005.
74. M. De Cock, C. Cornelis, and E. E. Kerre. Fuzzy rough sets: The forgotten step. *IEEE Trans. Fuzzy Sys.* 15(1): 121–130. 2007.
75. S. Deerwester, S. T. Dumais, T. K. Landauer, G. W. Furnas, and R. A. Harshman. Indexing by latent semantic analysis. *J. Soc. Info. Sci.* 41(6): 391–407. 1990.
76. P. Dempster. A generalization of Bayesian inference. *J. Roy. Stat. Soc.* B 30: 205–247. 1968.
77. P. Devijver and J. Kittler. *Pattern Recognition: A Statistical Approach*. Englewood Cliffs, NJ: Prentice Hall. 1982.

78. P. Diaconis and D. Freedman. Asymptotics of graphical projection pursuit. *An. Stat.* 12: 793–815. 1984.

79. A. I. Dimitras, R. Slowinski, R. Susmaga, and C. Zopounidis. Business failure prediction using rough sets. *Eur. J. Oper. Res.* 114: 263–280. 1999.

80. A. I. Dimitras, S. H. Zanakis, and C. Zopounidis. A survey of business failure with an emphasis on prediction methods and industrial applications. *Eur. J. Oper. Res.* 90: 487–513. 1996.

81. J. Dong, N. Zhong, and S. Ohsuga. Using rough sets with heuristics for feature selection. In *New Directions in Rough Sets, Data Mining, and Granular-Soft Computing.* 7th International Workshop (RSFDGrC 99), pp. 178–187. 1999.

82. M. Dorigo, V. Maniezzo, and A. Colorni. The ant system: Optimization by a colony of cooperating agents. *IEEE Trans. Sys. Man Cyber.* B 26(1): 29–41. 1996.

83. G. Drwal and A. Mrózek. System RClass—software implementation of the rough classifier. In *Proceedings of 7th International Symposium on Intelligent Information Systems*, pp. 392–395. 1998.

84. G. Drwal. Rough and fuzzy-rough classification methods implemented in RClass system. In *Proceedings of 2nd International Conference on Rough Sets and Current Trends in Computing*. Lecture Notes in Computer Science, 2005, Springer-Verlag, pp 152–159. 2000.

85. D. Dubois and H. Prade. Rough fuzzy sets and fuzzy rough sets. *Int. J. Gen. Sys.* 17: 191–209. 1990.

86. D. Dubois and H. Prade. Putting rough sets and fuzzy sets together. In [340], pp. 203–232. 1992.

87. S. T. Dumais. Combining evidence for effective information filtering. *AAAI Spring Symposium on Machine Learning in Information access Technical Papers*. 1996.

88. J. C. Dunn. A fuzzy relative of the ISODATA process and its use in detecting compact well-separated clusters. *J. Cyber.* 3(3): 32–57. 1973.

89. I. Düntsch and G. Gediga. Rough set data analysis. In A. Kent and J. G. Williams, eds., *Ency. Comput. Sci. Technol*. 43 (28): 281–301. 2000.

90. I. Düntsch and G. Gediga. *Rough Set Data Analysis: A road to Non-invasive Knowledge Discovery.* Bangor: Methodos. 2000.

91. S. Dzeroski, B. Cestnik, and I Petrovski. Using the m-estimate in rule induction. *Journal of Computing and Information Technology*, 1(1): 37–46. 1993.

92. A. L. Edwards, *An Introduction to Linear Regression and Correlation.* San Francisco: Freeman. 1976.

93. T. Endou and Q. Zhao. Generation of comprehensible decision trees through evolution of training data. In *Proceedings of the 2002 IEEE World Congress on Computational Intelligence*. Piscataway, NJ: IEEE Press, pp. 1221–1225. 2002.

94. ERUDIT (European Network for Fuzzy Logic and Uncertainty Modeling in Information Technology). *Protecting Rivers and Streams by Monitoring Chemical Concentrations and Algae Communities.* 3rd International Competition. 1999.

95. B. S. Everitt. *An Introduction to Latent Variable Models.* Monographs on Statistics and Applied Probability. London: Chapman Hall, 1984.

96. K. Farion, W. Michalowski, R. Slowinski, S. Wilk, and S. Rubin. S. Tsumoto et al., eds., *Rough Set Methodology in Clinical Practice: Controlled Hospital Trial of the MET System.* Springer-Verlag Heiddberg RSCTC, LNAI 3066, pp. 805–814. 2004.

97. U. Fayyad, G. Piatetsky-Shapiro, and P. Smyth. From data mining to knowledge discovery in databases. *Art. Intell.*, 17(3): 37–54. 1996.
98. L. Feng, J.-J. Qiu, and Y. J. Cao. Performance of the novel rough fuzzy-neural network on short-term load forecasting. In *Proceedings of the Power Systems Conference and Exposition*, IEEE Press pp. 543–547. 2004.
99. R. Fisher. The use of multiple measurementsin taxonomic problems. *An. Eugen.* 7: 179–188. 1936.
100. B. Flury and H. Riedwyl. *Multivariate Statistics: A Practical Approach.* Englewood Cliffs, NJ: Prentice Hall. 1988.
101. D. B. Fogel. The advantages of evolutionary computation. In *Proceedings of the Biocomputing and Emergent Computation Conference*, World Scientific, pp. 1–11. 1997.
102. G. Forman. An Extensive Empirical Study of Feature Selection Metrics for Text Classification. *J. Mach. Learning Res.* 3: 1289–1305. 2003.
103. J. H. Friedman and J. W. Tukey. A projection pursuit algorithm for exploratory data analysis. *IEEE Trans. Comput.* C 23(9): 881–890. 1974.
104. J. H. Friedman and W. Stuetzle. Projection pursuit regression. *J. Am. Stat. Assoc.* 76: 817–823. 1981.
105. J. H. Friedman. Multivariate Adaptive Regression Splines. *An. Stat.* 19(1): 1–141. 1991.
106. N. Fuhr. Probabilistic Models in Information Retrieval. *Comput. J.* 35(3): 243–55. 1992.
107. N. Fuhr, N. Gövert, M. Lalmas, and F. Sebastiani. Categorisation tool: Final prototype. Deliverable 4.3, Project LE4-8303 "EUROSEARCH," Commission of the European Communities. 1999.
108. M. Galea and Q. Shen. Fuzzy rules from ant-inspired computation. In *Proceedings of the IEEE International Conference on Fuzzy Systems*, Vol.3, Piscataway, NJ: IEEE Press, pp. 1691–1696. 2004.
109. M. Galea and Q. Shen. Iterative vs simultaneous fuzzy rule induction. In *Proceedings of the IEEE International Conference on Fuzzy Systems*, Vol. 4. Piscataway, NJ: IEEE Press, pp. 767–772. 2005.
110. M. Galea, Q. Shen and J. Levine. Evolutionary approaches to fuzzy modelling for classification. *Knowledge Eng. Rev.* 19(1): 27–59. 2004.
111. D. Gering. Linear and nonlinear data dimensionality reduction. Technical report. Massachusetts Institute of Technology. 2002.
112. F. Ginter, T. Pahikkala, S. Pyysalo, J. Boberg, J. Järvinen and T. Salakoski. In S. Tsumoto et al., eds., *Extracting Protein–Protein Interaction Sentences by Applying Rough Set Data Analysis.* Springer Verlag, LNAI 3066, pp. 780–785. 2004.
113. J. Gleick. *Chaos: Making New Science.* Penguin Books. 1988.
114. R. Golan. *Stock market analysis utilizing rough set theory. PhD thesis*. Department of Computer Science, University of Regina, Canada. 1995.
115. R. Golan and D. Edwards. Temporal rules discovery using datalogic/R+ with stock market data. In *Proceedings of International Workshop on Rough Sets and Knowledge Discovery*, Springer Verlag, pp. 74–81. 1993.
116. A. Gonzalez, R. Perez, and J. L. Verdegay. Learning the structure of a fuzzy rule: A genetic approach. *Fuzzy Sys. Art. Intell.* 3(1): 57–70. 1994.

117. S. Greco, M. Inuiguchi, and R. Slowinski. Fuzzy rough sets and multiple-premise gradual decision rules. *Int. J. Approx. Reason.* 41: 179–211. 2005.

118. S. Greco, M. Inuiguchi, and R. Slowinski. A new proposal for fuzzy rough approximations and gradual decision rule representation. *Transactions on Rough Sets II.* Lecture Notes in Computer Science, Vol. 3135. Berlin Springer, pp. 319–342. 2004.

119. J.W. Grzymala-Busse. LERS—A system for learning from examples based on rough sets. In R. Slowinski, ed., *Intelligent Decision Support.* Dordrecht: Kluwer Academic, pp. 3–18. 1992.

120. J. W. Grzymala-Busse. A comparison of three strategies to rule induction from data with numerical attributes. *Proceedings of the International Workshop on Rough Sets in Knowledge Discovery*. European Joint Conferences on Theory and Practice of Software. Elsevier, pp. 132–140. 2003.

121. J. W. Grzymala-Busse. Three strategies to rule induction from data with numerical attributes. *Transactions on Rough Sets II*, Berlin: Springer, pp. 54–62. 2004.

122. L. Han, J.F. Peters, S. Ramanna, and R. Zhai. Classifying faults in high voltage power systems: A rough-fuzzy neural computational approach. *Proceedings of 7th International Workshop on Rough Sets, Data Mining, and Granular-Soft Computing*, Springer-Verlag, pp. 47–54. 1999.

123. J. Han, X. Hu, and T. Y. Lin. Feature subset selection based on relative dependency between attributes. *Rough Sets and Current Trends in Computing: 4th International Conference*, Uppsala, Sweden, June 1–5, pp. 176–185. 2004.

124. H. Handels, T. Roß, J. Kreusch, H. H. Wolff, and S. Pöpple. Feature selection for optimized skin tumor recognition using genetic algorithms. *Art. Intell. Med.* 16(3): 283–297. 1999.

125. J. A. Hartigan and M. A. Wong. A k-means clustering algorithm. *Appl. Stat.* 28(1): 100–108. 1979.

126. I. Hayashi, T. Maeda, A. Bastian, and L.C. Jain. Generation of fuzzy decision trees by fuzzy ID3 with adjusting mechanism of AND/OR operators. In *Proceedings of 7th IEEE International Conference on Fuzzy Systems*. Piscataway, NJ: IEEE Press, pp. 681–685. 1998.

127. F. Herrera. Genetic fuzzy systems: Status, critical considerations and future directions. *Int. J. Comput. Intell. Res.* 1(1): 59–67. 2005.

128. M. P. Hippe. In J. J. Alpigini et al., eds., *Towards the Classification of Musical Works: A Rough Set Approach.* LNAI 2475, Springer Verlag, pp. 546–553. 2002.

129. S. Hirano and S. Tsumoto. Rough clustering and its application to medicine. *J. Info. Sci.* 124: 125–137, 2000.

130. K. Hirota, ed. *Industrial Applications of Fuzzy Technology.* Tokyo: Springer. 1993.

131. T. B. Ho, S. Kawasaki, and N. B. Nguyen. Documents clustering using tolerance rough set model and its application to information retrieval. *Studies In Fuzziness and Soft Computing, Intelligent Exploration of the Web*, Physica-Verlag Heidelberg, pp. 181–196. 2003.

132. U. Höhle. Quotients with respect to similarity relations. *Fuzzy Sets Sys.* 27(1): 31–44. 1988.

133. J. Holland. *Adaptation In Natural and Artificial Systems.* Ann Arbor: University of Michigan Press, 1975.

134. J. Holland and J. S. Reitman. Cognitive systems based on adaptive algorithms. In D. A. Waterman F. Hayes-Roth. *Pattern-Directed Inference Systems.* New York: Academic Press. p. 49. 1978.

135. R. C. Holte. Very simple classification rules perform well on most commonly used datasets. *Machine Learn.* 11(1): 63–90. 1993.

136. T. P. Hong, Y. L. Liou, and S. L. Wang. Learning with hierarchical quantitative attributes by fuzzy rough sets. In *Proceedings of Joint Conference on Information Sciences, Advances in Intelligent Systems Research*, Atlantic Press. 2006.

137. H. H. Hoos and T. Stützle. Towards a characterisation of the behaviour of stochastic local search algorithms for SAT. *Artificial Intelligence* 112: 213–232. 1999.

138. F. Hoppner, R. Kruse, F. Klawonn, and T. Runkler. *Fuzzy Cluster Analysis: Methods for Classification, Data Analysis and Image Recognition.* New York: Wiley. 2000.

139. W. H. Hsiao, S. M. Chen, and C. H. Lee. A new interpolative reasoning method in sparse rule-based systems. *Fuzzy Sets Sys.* 93: 17–22. 1998.

140. Q. Hu, D. Yu, and Z. Xie. Information-preserving hybrid data reduction based on fuzzy-rough techniques. *Pattern Recog. Lett.* 27(5): 414–423, 2006.

141. Q. Hu, D. Yu, Z. Xie, and J. Liu. Fuzzy probabilistic approximation spaces and their information measures. *IEEE Trans. Fuzzy Sys.* 14(2): 191–201. 2006.

142. Y.-X. Huang, C.-G. Zhou, S.-X. Zou, Y. Wang, and Y.-C. Liang. A rough-set-based fuzzy-neural-network system for taste signal identification. *Proceedings of International Symposium on Neural Networks*, IEEE Press pp. 337–343. 2004.

143. H. H. Huang, Y. H. Kuo, and H. C. Yang. Fuzzy-rough set aided sentence extraction summarization. *Proceedings of 1st International Conference on Innovative Computing, Information and Control* Vol. 1, IEEE Press pp. 450–453. 2006.

144. Z. Huang and Q. Shen. Fuzzy interpolative reasoning via scale and move transformation. *IEEE Trans. Fuzzy Sys.* 14(2): 340–359. 2006.

145. Z. Huang and Q. Shen. Fuzzy interpolative and extrapolative reasoning: A practical approach. *IEEE Trans. Fuzzy Sys.*, 16(1): 13–28. 2008. Forthcoming.

146. E. Hunt, J. Martin, and P. Stone. *Experiments in Induction.* New York: Academic Press. 1966.

147. T. Hvidsten, A. Læ greid, and J. Komorowski. Learning rule-based models of biological process from gene expression time profiles using gene ontology. *Bioinformatics* 19: 1116–1123. 2003.

148. Internet News. "1.5 Million Pages Added to Web Each Day, Says Research Company," September 1st, 1998. Available at http://www.internetnews.com/bus-news/article.php/37891.

149. H. Ishibuchi, K. Nozaki, and H. Tanaka. Distributed representation of fuzzy rules and its application to pattern classification. *Fuzzy Sets Sys.* 52(1): 21–32. 1992.

150. H. Ishibuchi, K. Nozaki, N. Yamamoto, and H. Tanaka. Selecting fuzzy if-then rules for classification problems using genetic algorithms. *IEEE Trans. Fuzzy Sys.* 3(3): 260–270. 1995.

151. H. Ishibuchi, T. Nakashima, and T. Murata. Performance evaluation of fuzzy classifier systems for multi-dimensional pattern classification problems. *IEEE Trans. Sys. Man Cyber.* B 29(5): 601–618. 1999.

152. H. Ishibuchi and T. Yamamoto. Fuzzy rule selection by multi-objective genetic local search algorithms and rule evaluation measures in data mining. *Fuzzy Sets Sys.* 141(1): 59–88. 2004.

153. H. Ishibuchi and T. Yamamoto. Multi-objective evolutionary design of fuzzy rule-based systems. *IEEE International Conference on Systems, Man and Cybernetics*, Vol. 3. Piscataway, NJ: IEEE Press, pp. 2362–2367. 2004.

154. P. Jan, J. W. Grzymala-Busse, and S. H. Zdzislaw. Melanoma prediction using data mining system LERS. *Proceedings of 25th Annual International Computer Software and Applications Conference*, Chicago, IL. IEEE pp. 615–620, 2001.

155. J.-S. R. Jang, Y. C. Lee, and C.-T. Sun. Functional equivalence between radial basis function networks and fuzzy inference systems. *IEEE Trans. Neural Net.* 4(1): 156–159. 1993.

156. J.-S. R. Jang, C.-T. Sun, and E. Mizutani. *Neuro-Fuzzy and Soft Computing, Matlab Curriculum.* Englewood Cliffs, NJ: Prentice Hall. 1997.

157. C. Z. Janikow. Fuzzy decision trees: Issues and methods. *IEEE Trans. Sys. Man Cyber.* B 28(1): 1–14. 1998.

158. J. Jelonek and J. Stefanowski. Feature subset selection for classification of histological images. *Art. Intell. Med.* 9(3): 227–239. 1997.

159. J. Jelonek, E. Lukasik, A. Naganowski, and R. Slowinski. Inducing jury's preferences in terms of acoustic features of violin sounds. In L. Rutkowski et al., eds., *Lecture Notes in Artificial Intelligence 3070*, Springer-Verlag, pp. 492–497. 2004.

160. R. Jensen. Performing Feature Selection with ACO. In A. Abraham, C. Grosan and V. Ramos, eds., *Swarm Intelligence and Data Mining.* Berlin: Springer, pp. 45–73. 2006.

161. R. Jensen and Q. Shen. Using Fuzzy Dependency-Guided Attribute Grouping in Feature Selection. In *Rough Sets, Fuzzy Sets, Data Mining, and Granular Computing, 9th International Conference.* Berlin: Springer, pp. 250–255. 2003.

162. R. Jensen and Q. Shen. Fuzzy-rough attribute reduction with application to Web categorization. *Fuzzy Sets Sys.* 141(3): 469–485. 2004.

163. R. Jensen and Q. Shen. Semantics-Preserving Dimensionality Reduction: Rough and Fuzzy-Rough Based Approaches. *IEEE Trans. Knowledge Data Eng.* 16(12): 1457–1471. 2004.

164. R. Jensen and Q. Shen. Fuzzy-rough data reduction with ant colony optimization. *Fuzzy Sets Sys.* 149(1): 5–20. 2005.

165. R. Jensen and Q. Shen. Fuzzy-rough feature significance for fuzzy decision trees. In *Proceedings of 2005 UK Workshop on Computational Intelligence*. pp. 89–96. 2005.

166. R. Jensen, Q. Shen, and Andrew Tuson. Finding Rough Set Reducts with SAT. In *Proceedings of 10th International Conference on Rough Sets, Fuzzy Sets, Data Mining and Granular Computing*. Springer-Verlag, LNAI 3641, pp. 194–203, 2005.

167. R. Jensen and Q. Shen. Rough set-based feature selection: A review. In A. E. Hassanien, Z. Suraj, D. Slezak, and P. Lingras, eds., *Rough Computing: Theories, Technologies and Applications.* IGI Global Press pp. 70–107. 2007.

168. R. Jensen and Q. Shen. Fuzzy-Rough Sets Assisted Attribute Selection. *IEEE Trans. Fuzzy Sys.* 15(1): 73–89. 2007.

169. R. Jensen and Q. Shen. New approaches to fuzzy-rough feature selection. *IEEE Trans. Fuzzy Sys.*, forthcoming.

170. Y. Jin. Fuzzy modeling of high-dimensional systems: Complexity reduction and interpretability improvement. *IEEE Trans. Fuzzy Sys.* 8(2): 212–221. 2000.

171. G. H. John, R. Kohavi, and K. Pfleger. Irrelevant features and the subset selection problem. *Proceedings of 11th International Conference on Machine Learning*. Morgan Kaufmann, pp. 121–129. 1994.

172. L. T. Jolliffe. *Principal Component Analysis.* Berlin: Springer. 1986.

173. L. Kallel, B. Naudts, and A. Rogers, eds. *Theoretical Aspects of Evolutionary Computing.* Berlin: Springer. 2001.

174. L. Kaufman and P. J. Rousseeuw. *Finding Groups in Data.* New York: Wiley. 1990.

175. J. Kennedy and R. C. Eberhart. Particle swarm optimization. *Proceedings of IEEE International Conference on Neural Networks*. Piscataway, NJ: IEEE Press, pp. 1942–1948. 1995.

176. K. Kira and L. A. Rendell. The feature selection problem: Traditional methods and a new algorithm. In *Proceedings of Ninth National Conference on Artificial Intelligence*. AAAI Press, pp. 129–134. 1992.

177. S. Kirkpatrick, C. Gelatt and M. Vecchi. Optimization by simulated annealing. *Science* 220(4598): 671–680. 1983.

178. W. W. Koczkodaj, M. Orlowski, and V. W. Marek. Myths about rough set theory. *Comm. ACM* 41(11): 102–103. 1998.

179. L. T. Kóczy and K. Hirota. Approximate reasoning by linear rule interpolation and general approximation. *Int. J. Approx. Reason.* 9: 197–225. 1993.

180. D. Koller and M. Sahami. Toward optimal feature selection. In *Proceedings of International Conference on Machine Learning*, Morgan Kaufmann, Springer-Verting pp. 284–292. 1996.

181. J. Komorowski, Z. Pawlak, L. Polkowski, and A. Skowron. Rough sets: A tutorial. In [258], pp. 3–98. 1999.

182. I. Kononenko. Estimating attributes: Analysis and extensions of RELIEF. *Proceedings of European Conference on Machine Learning*, Springer-Verlag pp. 171–182. 1994.

183. R. D. Koons and J. Buscaglia. The forensic significance of glass composition and refractive index measurements. *J. Forensic Sci.* 44(3): 496–503. 1999.

184. R. D. Koons and J. Buscaglia. Interpretation of glass composition measurements: the effects of match criteria on discrimination capability. *J. Forensic Sci.* 47(3): 505–512. 2002.

185. B. Kosko. Fuzzy entropy and conditioning. *Info Sci.* 40(2): 165–174. 1986.

186. B. Kostek, P. Szczuko, and P. Zwan. *Processing of musical data employing rough sets and artificial neural networks*. In S. Tsumoto et al., eds., LNAI Springer-Verlag, pp. 539–548. 2004.

187. M. A. Kramer. Nonlinear principal component analysis using autoassociative neural networks. *AIChE J.* 37(2): 233–243. 1991.

188. J. B. Kruskal. Multidimensional scaling by optimizing goodness of fit to a non-metric hypothesis. *Psychometrika* 29(1): 1–27. 1964.

189. M. Kryszkiewicz. Maintenance of reducts in the variable precision rough sets model. ICS Research Report 31/94. Warsaw University of Technology. 1994.

190. M. Kudo and J. Skalansky. Comparison of algorithms that select features for pattern classifiers. *Pattern Recog.* 33(1): 25–41. 2000.

191. L.I. Kuncheva. Fuzzy rough sets: application to feature selection. *Fuzzy Sets Sys.* 51(2): 147–153. 1992.

192. N. Kwak and C. H. Choi. Input feature selection for classification problems. *IEEE Trans. Neural Net.* 13(1): 143–159. 2002.

193. I. M. F. Langlands. Accounting issues surrounding the split capital investment trust crisis of 2001 and 2002. MA thesis. University of Edinburgh. 2004.

194. P. Langley. Selection of relevant features in machine learning. In *Proceedings of AAAI Fall Symposium on Relevance*. Washington: AAAI, pp. 1–5. 1994.

195. K. Larson and M. Czerwinski. Web page design: Implications of memory, structure and scent for information retrieval. In *Proceedings of the ACM SIGCHI Conference on Human Factors in Computing Systems*, ACM Press/Addison-Wesley Publishing, pp. 25–32. 1998.

196. N. Lavrac. Computational logic and machine learning: A roadmap for inductive logic programming. Technical report. J. Stefan Institute, Ljubljana, Slovenia. 1998.

197. M. H. Law, M. A. T. Figueiredo, and A. K. Jain. Feature selection in mixture-based clustering. *Adv. Neural Info. Processing Sys.* 15: 609–616. 2002.

198. E. L. Lawler, J. K. Lenstra, A. H. G. Rinnooy Kan, and D. B. Shmoys. *The Travelling Salesman Problem.* New York: Wiley. 1985.

199. L. Lazareck1 and S. Ramanna. Classification of swallowing sound signals: A rough set approach. In S. Tsumoto et al., eds., RSCTC, LNAI Springer Verlag, pp. 679–684. 2004.

200. R. Leardi, R. Boggia and M. Terrile. Genetic Algorithms as a Strategy for Feature Selection. *J. Chemomet.* 6(5): 267–28. 1992.

201. R. Leardi and A. L. Gonzalez. Genetic algorithms applied to feature selection in PLS regression: How and when to use them. *Chemomet. Intell. Lab. Sys.* 41(2): 195–207. 1998.

202. W. S. Li, Q. Vu, D. Agrawal, Y. Hara, and H. Takano. PowerBookmarks: A system for personalizable Web information organization, sharing, and management. *Computer Networks: Int. J. Comput. Telecomm. Network.* 31(11–16): 1375–1389. 1999.

203. P. Lingras. Rough neural networks. In *Proceedings of International Conference on Information Processing and Management of Uncertainty in Knowledge-Based Systems*. Springer-Verlag, pp. 1445–1450. 1996.

204. P. Lingras. Unsupervised learning using rough Kohonen neural network classifiers. In *Proceedings of Symposium on Modelling, Analysis and Simulation*, CESA'96 IMACS Multiconference. Elsevier Science Amsterdam. pp. 753–757. 1996.

205. P. Lingras. Fuzzy-rough and rough-fuzzy serial combinations in neurocomputing. *Neurocomputing*, 36: 29–44, 2001.

206. P. Lingras and C. Davies. Applications of rough genetic algorithms. *Comput. Intell.* 17(3): 435–445. 2001.

207. P. Lingras, R. Yan, and A. Jain. Clustering of Web users: K-Means vs. fuzzy C-means. In *Proceedings of First Indian International Conference on Artificial Intelligence*, IICAI pp. 1062–1073, 2003.

208. P. Lingras, R. Yan, and C. West. Comparison of Conventional and Rough K-Means Clustering. In *Rough Sets, Fuzzy Sets, Data Mining, and Granular Computing, 9th International Conference*, Berlin: Springer, pp. 130–137. 2003.

209. P. Lingras, M. Hogo, and M. Snorek. Interval set clustering of Web users using modified Kohonen self-organizing maps based on the properties of rough sets. *Web Intell. Agent Sys.* 2(3): 217–230. 2004.

210. P. Lingras and C. West. Interval set clustering of Web users with rough K-means. *J. Intell. Info. Sys.* 23(1): 5–16. 2004.

211. H. Liu and R. Setiono. Chi2: Feature selection and discretization of numeric attributes. In *Proceedings of 7th IEEE International Conference on Tools with Artificial Intelligence*. Piscataway, NJ: IEEE Press, pp. 336–391. 1995.

212. H. Liu and R. Setiono. A probabilistic approach to feature selection—A filter solution. In *Proceedings of 9th International Conference on Industrial and Engineering Applications of AI and ES*. Springer Verlag, pp. 284–292. 1996.

213. H. Liu and R. Setiono. Feature selection and classification—A probabilistic wrapper approach. In *Proceedings of 9th International Conference on Industrial and Engineering Applications of AI and ES*. Springer Verlag, pp. 419–424. 1996.

214. H. Liu, H. Motoda, eds. *Feature Extraction, Construction and Selection: A Data Mining Perspective.* Dordrecht: Kluwer Academic. 1998.

215. H. Liu and L. Yu. Toward integrating feature selection algorithms for classification and clustering. *IEEE Trans. Knowledge Data Eng.* 17(3): 1–12. 2005.

216. A. Lozowski, T. J. Cholewo, and J. M. Zurada. Crisp rule extraction from perceptron network classifiers. In *Proceedings of IEEE International Conference on Neural Networks*. Piscataway, NJ: IEEE Press, pp. 94–99. 1996.

217. Y. S. Maarek and I. Z. Ben Shaul. Automatically organizing bookmarks per contents. *Comput. Net. ISDN Sys*. 28 (7–11): 1321–1333. 1996.

218. N. Mac Parthalain, R. Jensen, and Q. Shen. Fuzzy entropy-assisted fuzzy-rough feature selection. In *Proceedings of 2006 IEEE International Conference on Fuzzy Systems*. Piscataway, NJ: IEEE Press, pp. 423–430, 2006.

219. J. B. MacQueen. Some methods for classification and analysis of multivariate observations. In *Proceedings of 5th Berkeley Symposium on Mathematical Statistics and Probability*, Vol. 1, Berkeley: University of California Press, pp. 281–297. 1967.

220. B. Mandelbrot. *The Fractal Geometry of Nature.* San Francisco: Freeman. 1982.

221. V. Maniezzo and A. Colorni. The ant system applied to the quadratic assignment problem. *Knowledge Data Eng.* 11(5): 769–778. 1999.

222. K. Mannar and D. Ceglarek. Continuous failure diagnosis for assembly systems using rough set approach. *An. CIRP* 53: 39–42. 2004.

223. K. V. Mardia, J. T. Kent, and J. M. Bibby. *Multivariate Analysis.* New York: Academic Press. 1979.

224. J. G. Marin-Blǽzquez and Q. Shen. From approximative to descriptive fuzzy classifiers. *IEEE Trans. Fuzzy Sys.* 10(4): 484–497. 2002.

225. J. G. Marin-Blǽzquez and Q. Shen. Regaining comprehensibility of approximative fuzzy models via the use of linguistic hedges. In Casillas et al., eds., *Interpretability Issues in Fuzzy Modelling.* Studies in Fuzziness and Soft Computing, Vol. 128, Springer, Berlin, pp. 25–53. 2003.

226. C.-W. Mao, S.-H. Liu, and J.-S. Lin. Classification of multispectral images through a rough-fuzzy neural network. *Optical Eng.* 43: 103–112. 2004.

227. S. J. Messick and R. P. Abelson. The additive constant problem in multidimensional scaling. *Psychometrika* 21: 1–17. 1956.

228. J. S. Mi and W. X. Zhang. An axiomatic characterization of a fuzzy generalization of rough sets. *Info. Sci*. 160 (1–4): 235–249. 2004.

229. R. S. Michalski, I. Mozetic, J. Hong, and N. Lavrac. The Multi-Purpose Incremental Learning System AQ15 and Its Testing Application to Three Medical Domains. (Proc of AAAI-86) AAAI Press, pp. 1041–1047. 1986

230. H. Midelfart, H. J. Komorowski, K. Nørsett, F. Yadetie, A. K. Sandvik, and A. Lægreid. Learning rough set classifiers from gene expressions and clinical data. *Fundamenta Informaticae* 53(2): 155–183. 2002.

231. A. J. Miller. *Subset Selection in Regression.* London: Chapman and Hall. 1990.

232. T. Mitchell. *Machine Learning.* New York: McGraw-Hill. 1997.

233. P. Mitra and S. Mitra. Staging of Cervical Cancer with Soft Computing. *IEEE Trans. Biomed. Eng.* 47(7): 934–940. 2000.

234. D. Mladenic. Text-learning and related intelligent agents: A survey. *IEEE Intell. Sys.* 14(4): 44–54. 1999.

235. D. Mladenic and M. Grobelnik. Feature selection for unbalanced class distribution and Naive Bayes. In *Proceedings of 16th International Conference on Machine Learning*. Morgan Kaufmann, pp. 258–267. 1999.

236. M. Modrzejewski. Feature selection using rough sets theory. In *Proceedings of 11th International Conference on Machine Learning*. Morgan Kaufmann, pp. 213–226. 1993.

237. N. N. Morsi and M. M. Yakout. Axiomatics for fuzzy rough sets. *Fuzzy Sets Sys.* 100(1–3): 327–342. 1998.

238. A. Moukas and P. Maes. Amalthaea: An evolving multi-agent information filtering and discovery system for the WWW. *J. Autonomous Agents Multi-Agent Sys.* 1(1): 59–88. 1998.

239. I. Moulinier. A framework for comparing text categorization approaches. In *Proceedings of AAAI Spring Symposium on Machine Learning in Information Access*. Washington: AAAI, pp. 61–68. 1996.

240. S. Muggleton. Inverse entailment and Progol. *New Gen. Comput.* 13(3–4): 245–286. 1995.

241. S. Muggleton and L. De Raedt. Inductive logic programming. *J. Logic Program.* 19/20: 629–679. 1994.

242. S. Muggleton and C. Feng. Efficient induction of logic programs. In *Proceedings of 1st Conference on Algorithmic Learning Theory*. Springer/Ohmsha pp. 368–381. 1990.

243. T. Nakashima, H. Ishibuchi, and T. Murata. Evolutionary algorithms for constructing linguistic rule-based systems for high-dimensional pattern classification problems. In *Proceedings of 1998 IEEE International Conference on Evolutionary Computation*. Piscataway, NJ: IEEE Press, pp. 752–757. 1998.

244. S. Nanda and S. Majumdar. Fuzzy rough sets. *Fuzzy Sets Sys.* 45: 157–160, 1992.

245. D. Nauck, F. Klawonn, R. Kruse, and F. Klawonn. *Foundations of Neuro-Fuzzy Systems.* New York: Wiley. 1997.

246. D. Nauck and R. Kruse. A neuro-fuzzy method to learn fuzzy classification rules from data. *Fuzzy Sets Sys.* 89(3): 277–288. 1997.

247. R. T. Ng and J. Han. Efficient and effective clustering methods for spatial data mining. In *Proceedings of International Conference Very Large Data Bases*. San Francisco: Morgan Kaufmann, pp. 144–155. 1994.

248. H. T. Ng, W. B. Goh, and K. L. Low. Feature selection, perceptron learning, and a usability case study for text categorisation. In *Proceedings of SIGIR-97, 20th ACM International Conference on Research and Development in Information Retrieval*. ACM, pp. 67–73. 1997.

249. H. S. Nguyen and A. Skowron. Boolean reasoning for feature extraction problems. In *Proceedings of 10th International Symposium on Methodologies for Intelligent Systems*. Springer-Verlag pp. 117–126. 1997.

250. S. H. Nguyen and H. S. Nguyen. Some efficient algorithms for rough set methods. In *Proceedings of the Conference of Information Processing and Management of Uncertainty in Knowledge-Based Systems*. Springer-Verlag pp. 1451–1456. 1996.

251. S. H. Nguyen, T. T. Nguyen, and H. S. Nguyen. *Rough set approach to sunspot classification problem*. In D Ślęzak et al., eds., LNAI Springer-Verlag, pp. 263–272. 2005.

252. M. Ningler, G. Stockmanns, G. Schneider, O. Dressler, and E. F. Kochs. Rough set-based classification of EEG-signals to detect intraoperative awareness: Comparison of fuzzy and crisp discretization of real value attributes. In S. Tsumoto et al., eds., LNAI 3066, pp. 825–834. 2004.

253. A. Ohrn. Discernibility and rough sets in medicine: Tools and applications. Department of Computer and Information Science. Norwegian University of Science and Technology, Trondheim,Norway. Report 133/1999. 1999.

254. L. S. Oliveira, N. Benahmed, R. Sabourin, F. Bortolozzi, and C.Y. Suen. Feature subset selection using genetic algorithms for handwritten digit recognition. In *Proceedings of 14th Brazilian Symposium on Computer Graphics and Image Processing*. IEEE Computer Society, pp. 362–369. 2001.

255. P. Paclik, R. P. W. Duin, G. M. P. van Kempen, and R. Kohlus. On feature selection with measurement cost and grouped features. In Proceedings of 4th International Workshop on Statistical Techniques in Pattern Recognition. Berlin: Springer, pp. 461–469. 2002.

256. S. K. Pal. *Pattern Recognition Algorithms for Data Mining*. London: Chapman and Hall, 2004.

257. S. K. Pal and S. Mitra. *Neuro-Fuzzy Pattern Recognition: Methods in Soft Computing*. New York: Wiley. 1999.

258. S. K. Pal and A. Skowron, eds. *Rough-Fuzzy Hybridization: A New Trend in Decision Making*. Berlin: Springer. 1999.

259. S. K. Pal, S. Mitra, and P. Mitra. Rough-fuzzy MLP: Modular evolution, rule generation, and evaluation. *IEEE Trans. Knowledge Data Eng.* 15(1): 14–25. 2003.

260. Z. Pawlak. Rough sets. *Int. J. Comput. Info. Sci.* 11(5): 341–356. 1982.

261. Z. Pawlak. *Rough Sets: Theoretical Aspects of Reasoning About Data.* Dordrecht: Kluwer Academic. 1991.

262. Z. Pawlak and A. Skowron. Rough membership functions. In R. Yager, M. Fedrizzi, and J. Kacprzyk, eds., *Advances in the Dempster-Shafer Theory of Evidence.* New York: Wiley, pp. 251–271. 1994.

263. Z. Pawlak. Some Issues on rough sets. *LNCS Trans. Rough Sets* (1): 1–53. 2003.

264. K. Pearson. On the criterion that a given system of deviations from the probable in the case of a correlated system of variables is such that it can reasonably supposed to have arisen from random sampling. *Philoso. Mag.* Series 5: 157–175. 1900.

265. W. Pedrycz. *Fuzzy Modelling: Paradigms and Practice.* Norwell, MA: Kluwer Academic Press. 1996.

266. W. Pedrycz. Shadowed sets: Bridging fuzzy and rough sets. In [258], pp. 179–199. 1999.

267. W. Pedrycz and F. Gomide. *An Introduction to Fuzzy Sets: Analysis and Design.* Cambridge: MIT Press. 1998.

268. W. Pedrycz and G. Vukovich. Feature analysis through information granulation. *Pattern Recog.* 35(4): 825–834. 2002.

269. J. F. Peters, A. Skowron, Z. Suraj, W. Rzasa, and M. Borkowski. Clustering: A rough set approach to constructing information granules. In *Proceedings of 6th International Conference on Soft Computing and Distributed Processing*. Amsterdam: IOS Press, pp. 57–61, 2002.

270. J. F. Peters, Z. Suraj, S. Shan, S. Ramanna, W. Pedrycz, and N. J. Pizzi. Classification of meteorological volumetric radar data using rough set methods. *Pattern Recog. Lett.* 24 (6): 911–920. 2003.

271. A. Petrosino and M. Ceccarelli. Unsupervised texture discrimination based on rough fuzzy sets and parallel hierarchical clustering. In *Proceedings of IEEE International Conference on Pattern Recognition*. Piscataway, NJ: IEEE Press, pp. 1100–1103, 2000.

272. J. Platt. Fast Training of Support Vector Machines using Sequential Minimal Optimization. In B. Schlkopf, C. Burges, and A. Smola, eds., *Advances in Kernel Methods: Support Vector Learning.* Cambridge: MIT Press, pp. 185–208. 1998.

273. L. Polkowski, T. Y. Lin, and S. Tsumoto, eds. *Rough Set Methods and Applications: New Developments in Knowledge Discovery in Information Systems*, Vol. 56. Studies in Fuzziness and Soft Computing. Heidelberg: Physica. 2000.

274. L. Polkowski. *Rough Sets: Mathematical Foundations. Advances in Soft Computing.* Heidelberg: Physica. 2002.

275. H. A. do Prado, P. M. Engel, and H. C. Filho. *Rough Clustering: An Alternative to Find Meaningful Clusters by Using the Reducts from a Dataset Source.* Berlin: Springer, pp. 234–238. 2002.

276. B. Predki and Sz. Wilk. Rough set based data exploration using ROSE system. In Z. W. Ras and A. Skowron, eds., *Foundations of Intelligent Systems.* Berlin: Springer, pp. 172–180. 1999.

277. W. Z. Qiao, M. Mizumoto, and S. Y. Yan. An improvement to Kóczy and Hirota's interpolative reasoning in sparse fuzzy rule bases. *Int. J. Approx. Reason.* 15: 185–201. 1996.

278. K. Qina and Z. Pei. On the topological properties of fuzzy rough sets. *Fuzzy Sets Sys.* 151(3): 601–613. 2005.

279. J. R. Quinlan. Induction of decision trees. *Machine Learn.* 1: 81–106. 1986.

280. J. R. Quinlan. Learning logical definitions from relations. *Machine Learn.* 5(3): 239–266. 1990.

281. J. R. Quinlan. *C4.5: Programs for Machine Learning*. San Mateo, CA: Morgan Kaufmann Publishers. 1993.

282. A. M. Radzikowska and E. E. Kerre. A comparative study of fuzzy rough sets. *Fuzzy Sets Sys.* 126(2): 137–155. 2002.

283. A. M. Radzikowska and E. E. Kerre. Fuzzy rough sets based on residuated lattices. In *Transactions on Rough Sets II*. Berlin: Springer, pp. 278–296. 2004.

284. R. Rallo, J. Ferré-Giné, and F. Giralt. Best feature selection and data completion for the design of soft neural sensors. In *Proceedings of AIChE 2003, 2nd Topical Conference on Sensors*, San Francisco. 2003.

285. B. Raman and T.R. Ioerger. Instance-based filter for feature selection. *J. Machine Learn. Res.* 1: 1–23. 2002.

286. K. Rasmani and Q. Shen. Modifying weighted fuzzy subsethood-based rule models with fuzzy quantifiers. In *Proceedings of 13th International Conference on Fuzzy Systems*, NJ: IEEE Press, pp. 1687–1694. 2004.

287. K. Rasmani and Q. Shen. Data-driven fuzzy rule generation and its application for student academic performance evaluation. *Appl. Intell.* 25(3): 305–319. 2006.

288. T. Rauma. Knowledge acquisition with fuzzy modeling. In *Proceedings of 5th IEEE International Conference on Fuzzy Systems*, Vol. 3. Piscataway, NJ: IEEE Press, pp. 1631–1636. 1996.

289. M. W. Richardson. Multidimensional psychophysics. *Psycholog. Bull.* 35: 659–660. 1938.

290. W. Romao, A. A. Freitas, and R. C. S. Pacheco. A genetic algorithm for discovering interesting fuzzy prediction rules: Applications to science and technology data. In *Proceedings of the Genetic and Evolutionary Computation Conference*. Morgan Kaufmann, pp. 343–350. 2002.

291. The ROSETTA homepage. Available at http://rosetta.lcb.uu.se/general/.

292. RSES: Rough Set Exploration System. Available at http://logic.mimuw.edu.pl/~rses/.

293. S. T. Roweis and L. K. Saul. Nonlinear dimensionality reduction by locally linear embedding. *Science* 290(5500): 2323–2326. 2000.

294. M. Ruggiero. Turning the key. *Futures* 23(14): 38–40. 1994.

295. M. E. Ruiz and P. Srinivasan. Hierarchical neural networks for text categorization. In *Proceedings of SIGIR-99, 22nd ACM International Conference on Research and Development in Information Retrieval*. NY: ACM, pp. 281–282. 1999.

296. D. Rumelhant, E. Hinton, and R. Williams. Learning internal representations by error propagating. In E. Rumelhant and J. McCkekkand, eds., *Parallel Distributed Processing*. Cambridge: MIT Press. 1986.

297. T. Runkler. Selection of appropriate defuzzification methods using application specific properties. *IEEE Trans. Fuzzy Sys.* 5(1): 72–79. 1997.

298. S. Russell and P. Norvig. *Artificial Intelligence: A Modern Approach*. Englewood Cliffs, NJ: Prentice Hall. 1995.

299. Y. Saeys, S. Degroeve, D. Aeyels, P. Rouze, and Y. Van De Peer. Feature selection for splice site prediction: A new method using EDA-based feature ranking. *BMC Bioinform.* 5(64): 2004.

300. G. Salton, A. Wong, and C.S. Yang. A vector space model for automatic indexing. *Comm. ACM* 18(11): 613–620. 1975.

301. G. Salton, E. A. Fox, and H. Wu. Extended Boolean information retrieval. *Comm. ACM* 26(12): 1022–1036. 1983.

302. G. Salton, *Introduction to Modern Information Retrieval.* New York: McGraw-Hill. 1983.

303. G. Salton, and C. Buckley. Term weighting approaches in automatic text retrieval. Technical report TR87-881. Department of Computer Science, Cornell University. 1987.

304. M. Sarkar. Fuzzy-rough nearest neighbors algorithm. In *Proceedings of the IEEE International Conference on Systems, Man and Cybernetics*. Piscataway, NJ: IEEE Press, pp. 3556–3561. 2000.

305. M. Sarkar. Ruggedness measures of medical time series using fuzzy-rough sets and fractals. *Pattern Recog. Lett.* 27(5): 447–454. 2006.

306. J. C. Schlimmer. Efficiently inducing determinations—A complete and systematic search algorithm that uses optimal pruning. *International Conference on Machine Learning.* Morgan Kaufmann, pp. 284–290. 1993.

307. B. Schölkopf. *Support Vector Learning.* Munich: Oldenbourg Verlag. 1997.

308. M. Schroeder. *Fractals, Chaos, Power Laws: Minutes from an Infinite Paradise.* New York: Freeman. 1991.

309. F. Sebastiani. Machine learning in automated text categorisation. *ACM Comput. Sur.* 34(1): 1–47. 2002.

310. M. Sebban and R. Nock. A hybrid filter/wrapper approach of feature selection using information theory. *Pattern Recog.* 35(4): 835–846. 2002.

311. B. Selman and H. A. Kautz. Domain-independant extensions to GSAT: Solving large structured variables. In *Proceedings of 13th International Joint Conference on Artificial Intelligence*. Springer, pp. 290–295. 1993.

312. R. Setiono and H. Liu. Neural network feature selector. *IEEE Trans. Neural Net.* 8(3): 645–662. 1997.

313. M. Setnes and H. Roubos. GA-fuzzy modeling and classification: Complexity and performance. *IEEE Trans. Fuzzy Sys.* 8(5): 509–522. 2000.

314. H. Sever. The status of research on rough sets for knowledge discovery in databases. In *Proceedings of 2nd International Conference on Nonlinear Problems in Aviation and Aerospace*, Cambridge: European Conference Publications, Vol. 2. pp. 673–680. 1998.

315. G. Shafer. *A Mathematical Theory of Evidence.* Princeton: Princeton University Press. 1976.

316. D. Shan, N. Ishii, Y. Hujun, N. Allinson, R. Freeman, J. Keane, and S. Hubbard. Feature weights determining of pattern classification by using a rough genetic algorithm with fuzzy similarity measure. In *Proceedings of the Intelligent Data Engineering and Automated Learning*. Springer pp. 544–550, 2002.

317. C. Shang and Q. Shen. Rough feature selection for neural network based image classification. *Int. J. Image Graph.* 2(4): 541–556. 2002.

318. C. Shang and Q. Shen. Aiding classification of gene expression data with feature selection: A comparative study. *Comput. Intell. Res.* 1(1): 68–76. 2006.

319. Q. Shen. Rough feature selection for intelligent classifiers. *LNCS Trans. Rough Sets* 7: 244–255. 2007.

320. Q. Shen and A. Chouchoulas. FuREAP: A fuzzy-rough estimator of algae population. *Art. Intell. Eng.* 15(1): 13–24. 2001.

321. Q. Shen and A. Chouchoulas. A fuzzy-rough approach for generating classification rules. *Pattern Recog.* 35(11): 341–354. 2002.

322. Q. Shen and R. Jensen. Selecting informative features with fuzzy-rough sets and its application for complex systems monitoring. *Pattern Recog.* 37(7): 1351–1363. 2004.

323. Q. Shen and R. Jensen. Rough Sets, their Extensions and Applications. *Int. J. Auto. Comput.* 4(3): 217–228. 2007.

324. Q. Shen and R. Jensen. Approximation-based feature selection and application for algae population estimation. in *Appl. Intell.*, forthcoming 28(2): 167–181.

325. L. Shen, F. E. H. Tay, L. Qu, and Y. Shen. Fault diagnosis using rough sets theory. *Comput. Indus.* 43: 61–72. 2000.

326. R. N. Shepard. The analysis of proximities: Multidimensional scaling with an unknown distance function, I, II. *Psychometrika* 27: 125–140, 219–246. 1962.

327. Y. Shi and M. Mizumoto. Some considerations on Kóczy's interpolative reasoning method. In *Proceedings of FUZZ-IEEE'95*, Yokohama, Japan. Piscataway, NJ: IEEE Press, pp. 2117–2122. 1995.

328. H. Shütze, D. A. Hull and J. O. Pederson. A comparison of classifiers and document representations for the routing problem. In *Proceedings of SIGIR-95, 18th ACM International Conference on Research and Development in Information Retrieval*. ACM Press pp. 229–237. 1995.

329. W. Siedlecki and J. Sklansky. On automatic feature selection. *Int. J. Pattern Recog. Art. Intell.* 2(2): 197–220. 1988.

330. W. Siedlecki and J. Sklansky. A note on genetic algorithms for large-scale feature selection. *Pattern Recog. Lett.* 10(5): 335–347. 1989.

331. B. W. Silverman. *Density Estimation for Statistics and Data Analysis*. London: Chapman and Hall. 1986.

332. S. Singh, M. Singh, and M. Markou. Feature Selection for Face Recognition based on Data Partitioning. In *Proceedings of 15th International Conference on Pattern Recognition*. IEEE Computer Society, pp. 680–683. 2002.

333. A. Skowron and C. Rauszer. The discernibility matrices and functions in information systems. In [340], pp. 331–362. 1992.

334. A. Skowron and J. W. Grzymala-Busse. From rough set theory to evidence theory. In R. Yager, M. Fedrizzi, and J. Kasprzyk, eds., *Advances in the Dempster-Shafer Theory of Evidence*. New York: Wiley 1994.

335. A. Skowron and J. Stepaniuk. Tolerance approximation spaces. *Fundamenta Informaticae* 27(2): 245–253. 1996.

336. A. Skowron, Z. Pawlak, J. Komorowski, and L. Polkowski. A rough set perspective on data and knowledge. In *Handbook of Data Mining and Knowledge Discovery*. Oxford: Oxford University Press, pp. 134–149. 2002.

337. A. Skowron, S. K. Pal. Rough sets, pattern recognition and data mining. *Pattern Recog. Lett.* 24(6): 829–933. 2003.

338. D. Ślęzak. Approximate reducts in decision tables. In *Proceedings of 6th International Conference on Information Processing and Management of Uncertainty in Knowledge-Based Systems*. Springer Verlag, pp. 1159–1164. 1996.

339. D. Ślęzak. Normalized decision functions and measures for inconsistent decision tables analysis. *Fundamenta Informaticae* 44(3): 291–319. 2000.

340. R. Slowinski, ed. *Intelligent Decision Support.* Dordrecht: Kluwer Academic. 1992.

341. R. Slowinski and C. Zopounidis. Rough set sorting of firms according to bankruptcy risk. In M. Paruccini, ed., *Applying Multiple Criteria Aid for Decision to Environmental Management.* Dordrecht: Kluwer Academic, pp. 339–357. 1994.

342. R. Slowinski and C. Zopounidis. Application of the rough set approach to evaluation of bankruptcy risk. *Int. J. Intell. Sys. Account. Fin. Manag.* 4(1): 27–41. 1995.

343. R. Slowinski and D. Vanderpooten. Similarity relation as a basis for rough approximations. *Adv. Machine Intell. Soft Comput.* 4: 17–33. 1997.

344. R. Slowinski, C. Zopounidis, A. I. Dimitras, and R. Susmaga. Rough set predictor of business failure. In R. A. Ribeiro, H. J. Zimmermann, R. R. Yager, and J. Kacprzyk, eds. *Soft Computing in Financial Engineering.* Wurzburg: Physica, pp. 402–424. 1999.

345. P. Smets and P. Magrez. Implication in fuzzy logic. *Int. J. Approx. Reason.* 1(4): 327–347. 1987.

346. A. J. Smola and B. Schölkopf. *A Tutorial on Support Vector Regression.* NeuroCOLT2 Technical Report Series. 1998.

347. M. G. Smith and L. Bull. Feature construction and selection using genetic programming and a genetic algorithm. In *Proceedings of 6th European Conference on Genetic Programming*. Springer, pp. 229–237. 2003.

348. S. F. Smith. A learning system based on genetic adaptive algorithms. PhD thesis. Computer Science Department, University of Pittsburgh. 1980.

349. P. Srinivasan, M. E. Ruiz, D. H. Kraft, and J. Chen. Vocabulary mining for information retrieval: Rough sets and fuzzy sets. *Info. Process. Manag.* 37(1): 15–38. 1998.

350. J. A. Starzyk, D. E. Nelson, and K. Sturtz. Reduct Generation in Information Systems. *Bull. Int. Rough Set Soc.* 3(1–2): 19–22. 1999.

351. J. Stefanowski. On rough set based approaches to induction of decision rules. In A. Skowron, L. Polkowski, eds., *Rough Sets in Knowledge Discovery*, Vol. 1. Heidelberg: Physica, pp. 500–529. 1998.

352. J. Stefanowski and A. Tsoukiàs. Valued tolerance and decision rules. Proc. of *Rough Sets Curr. Trends Comput.* Springer Verlag 212–219. 2000.

353. M. Stone. Cross-validatory choice and assessment of statistical predictions. *J. Roy. Stat. Soc.* B 36: 111–147. 1974.

354. R. W. Swiniarski. Rough set expert system for online prediction of volleyball game progress for US olympic team. In *Proceedings of 3rd Biennial European Joint Conference on Engineering Systems Design Analysis*. pp. 15–20. 1996.

355. R. W. Swiniarski and A. Skowron. Rough set methods in feature selection and recognition. *Pattern Recog. Lett.* 24(6): 833–849. 2003.

356. A. Szladow and D. Mills. Tapping financial databases. *Bus. Credit* 95(7): 8. 1993.

357. F. E. H. Tay and L. Shen. Economic and financial prediction using rough sets model. *Eur. J. Oper. Res.* 141: 641–659. 2002.

358. J. B. Tenenbaum, V. de Silva, and J. C. Langford. A global geometric framework for nonlinear dimensionality reduction. *Science* 290(5500): 2319–2323. 2000.

359. H. Thiele. Fuzzy rough sets versus rough fuzzy sets—An interpretation and a comparative study using concepts of modal logics. Technical report no. CI-30/98, University of Dortmund. 1998.

360. W. S. Torgerson. "Multidimensional Scaling." *Psychometrika* 17:401–419. 1952.

361. G. C. Y. Tsang, D. Chen, E. C. C. Tsang, J. W. T. Lee, and D. S. Yeung. On attributes reduction with fuzzy rough sets. In *Proceedings of the 2005 IEEE International Conference on Systems, Man and Cybernetics*, Vol. 3. Piscataway, NJ: IEEE Press, pp. 2775–2780, 2005.

362. C. Traina Jr, A. Traina, L. Wu, and C. Faloutsos. Fast feature selection using the fractal dimension. In *Proceedings of 15th Brazilian Symposium on Databases*. CEFET, pp. 158–171. 2000.

363. M. Umano, H. Okamota, I. Hatono, H. Tamura, F. Kawachi, S. Umedzu, and J. Kinoshita. Generation of fuzzy decision trees by fuzzy ID3 algorithm and its application to diagnosis by gas in oil. In *Proceedings of the 1994 Japan–USA Symposium on Flexible Automation*. MI: Ann Arbor, pp. 1445–1448. 1994.

364. J. J. Valdés and A. J. Barton. Relevant attribute discovery in high dimensional data based on rough sets and unsupervised classification: Application to leukemia gene expressions. In D Ślęzak et al., eds., Proc. of RSFDGrC 2005 Springer pp. 362–371. 2005.

365. C. J. van Rijsbergen. *Information Retrieval.* London: Butterworths. 1979. Available at http://www.dcs.gla.ac.uk/Keith/Preface.html.

366. G. Vass, L. Kalmar and L. T. Kóczy. Extension of the fuzzy rule interpolation method. In *Proceedings of the International Conference on Fuzzy Sets Theory and Its Applications*. pp. 1–6. 1992.

367. M. A. Vila, J. C. Cubero, J. M. Medina, and O. Pons. Using OWA operators in flexible query processing. In R. R. Yager & J. Kacprzyk *The Ordered Weighted Averaging Operators: Theory, Methodology and Applications.* London: Kluwer Academic, pp. 258–274. 1997

368. K. E. Voges, N. K. L. Pope, and M. R. Brown. Cluster Analysis of Marketing Data: A Comparison of K-Means, Rough Set, and Rough Genetic Approaches. In C.S. Newton, eds., *Heuristics and Optimization for Knowledge Discovery*, edited by H.A. Abbas, R.A. Sarker, PA,Idea Group Publishing, pp. 208–216, 2002.

369. M. Wallace, Y. Avrithis, and S. Kollias. Computationally efficient sup-t transitive closure for sparse fuzzy binary relations. *Fuzzy Sets Sys.* 157(3): 341–372, 2006.

370. T. Walsh. SAT v. CSP. In *Proceedings of the International Conference on Principles and Practice of Constraint Programming*. Springer, pp. 441–456. 2000.

371. D. Walter and C. K. Mohan. ClaDia: A fuzzy classifier system for disease diagnosis. In Proceedings of the Congress on Evolutionary Computation, pp. 1429–1435. 2000.

372. M. Wand and M. Jones. *Kernel Smoothing.* London: Chapman and Hall. 1995.

373. L. X. Wang and J. M. Mendel. Generating fuzzy rules by learning from examples. *IEEE Trans. Sys. Man Cyber.* 22(6): 1414–1427. 1992.

374. J. Wang and J. Wang. Reduction Algorithms based on discernibility matrix: The ordered attributes method. *J. Comput. Sci. Technol.* 16(6): 489–504. 2001.

375. Y. Wang and I. H. Witten. Inducing model trees for continuous classes. In M. van Someren and G. Widmer, eds., *Proceedings of Poster Papers: Ninth European Conference on Machine Learning*. London: Springer Verlag, pp. 128–137. 1997.

376. Y. Wang. A new approach to fitting linear models in high dimensional spaces. PhD thesis. Department of Computer Science, University of Waikato. 2000.

377. X. Wang, J. Yang, X. Teng, and N. Peng. *Fuzzy-rough set based nearest neighbor clustering classification algorithm. Proc. of FSKD 2005* Berlin: Springer, pp. 370–373, 2005.

378. X. Z. Wang, Y. Ha, and D. Chen. On the reduction of fuzzy rough sets. In *Proceedings of 2005 International Conference on Machine Learning and Cybernetics*, Vol. 5. IEEE Press, pp. 3174–3178, 2005.

379. Z. Wang, X. Shao, G. Zhang, and H. Zhu. Integration of Variable Precision Rough Set and Fuzzy Clustering: An Application to Knowledge Acquisition for Manufacturing Process Planning. In *Proceedings of 10th International Conference*. Springer, RSFDGrC 2005, pp. 585–593. 2005.

380. D. Whitley. An overview of evolutionary algorithms: practical issues and common pitfalls. *Info. Software Technol.* 43(14): 817–831. 2001.

381. I. H. Witten and E. Frank. Generating accurate rule sets without global optimization. In *Machine Learning: Proceedings of 15th International Conference*. San Francisco: Morgan Kaufmann, 1998.

382. I. H. Witten and E. Frank. *Data Mining: Practical Machine Learning Tools with Java Implementations.* San Francisco: Morgan Kaufmann, 2000.

383. A. Wojna. Analogy-based reasoning in classifier construction. *Trans. Rough Sets*, 4: 277–374, 2005.

384. J. Wróblewski. Finding minimal reducts using genetic algorithms. In *Proceedings of 2nd Annual Joint Conference on Information Sciences*. pp. 186–189. 1995.

385. W. Z. Wu, J. S. Mi, and W. X. Zhang. Generalized fuzzy rough sets. *Info. Sci.* 151: 263–282, 2003.

386. W. Z. Wu and W. X. Zhang. Constructive and axiomatic approaches of fuzzy approximation operators. *Info. Sci.* 159(3–4): 233–254, 2004.

387. W. Z. Wu. A study on relationship between fuzzy rough approximation operators and fuzzy topological spaces. In L. Wang and Y. Jin, eds., FSKD 2005, Berlin: Springer, pp. 167–174, 2005.

388. W. Z. Wu, Y. Leung, and J. S. Mi. On characterizations of (I,T)-fuzzy rough approximation operators. *Fuzzy Sets Sys.* 154(1): 76–102, 2005.

389. M. Wygralak. Rough sets and fuzzy sets—Some remarks on interrelations. *Fuzzy Sets Sys.* 29(2): 241–243. 1989.

390. E. P. Xing. *Feature Selection in Microarray Analysis: A Practical Approach to Microarray Data Analysis.* Dordrecht: Kluwer Academics. 2003.

391. M. Xiong, W. Li, J. Zhao, L. Jin, and E. Boerwinkle. Feature (gene) selection in gene expression-based tumor classification. *Mol. Genet. Metabol.* 73(3): 239–247. 2001.

392. N. Xiong and L. Litz. Reduction of fuzzy control rules by means of premise learning - method and case study. *Fuzzy Sets Sys.* 132(2): 217–231. 2002.

393. Yahoo.www.yahoo.com.

394. S. Yan, M. Mizumoto, and W. Z. Qiao. Reasoning conditions on Kóczy's interpolative reasoning method in sparse fuzzy rule bases. *Fuzzy Sets Sys.* 75: 63–71. 1995.

395. Y. Yang and J. O. Pedersen. A comparative study on feature selection in text categorization. In *Proceedings of 14th International Conference on Machine Learning*. Margar Kaufmann, pp. 412–420. 1997.

396. J. Yang and V. Honavar. Feature subset selection using a genetic algorithm. *IEEE Intell. Sys.* 13(1): 44–49. 1998.

397. L. Yang, D. H. Widyantoro, T. Ioerger, and J. Yen. An entropy-based adaptive genetic algorithm for learning classification rules. In *Proceedings of Congress on Evolutionary Computation*, Vol. 2. IEEE Press, pp. 790–796. 2001.

398. J. Yao. Feature selection for fluoresence image classification. KDD Lab Proposal. Carregie Mellan University, 2001.

399. X. Yao. Average computation time of evolutionary algorithms for combinatorial optimisation problems. *Final IGR Report for EPSRC grant GR/R52541/01*. School of Computer Science, University of Birmingham,UK. 2003.

400. Y. Y. Yao. Combination of rough and fuzzy sets based on α-level sets. In T. Y. Lin, N. Cereone, eds., *Rough Sets and Data Mining: Analysis of Imprecise Data.* Dordrecht: Kluwer Academic, pp. 301–321. 1997.

401. Y. Y. Yao. A Comparative Study of Fuzzy Sets and Rough Sets. *Info. Sci.* 109(1–4): 21–47, 1998.

402. Y. Y. Yao. Decision-theoretic rough set models. In *Proceedings of International Conference on Rough Sets and Knowledge Technology*. Proc. of RSKT07. Springer, pp. 1–12, 2007.

403. D. S. Yeung, D. Chen, E. C. C. Tsang, J. W. T. Lee, and W. Xizhao. On the generalization of fuzzy rough sets. *IEEE Trans. Fuzzy Sys.* 13(3): 343–361, 2005.

404. F. W. Young and R. M. Hamer. *Theory and Applications of Multidimensional Scaling.* Hillsdale, NJ: Eribaum Associates. 1994.

405. S. Yu, S. De Backer, and P. Scheunders. Genetic feature selection combined with composite fuzzy nearest neighbor classifiers for hyperspectral satellite imagery. *Pattern Recog. Lett.* 23(1–3): 183–190. 2002.

406. Y. Yuan and M. J. Shaw. Induction of fuzzy decision trees. *Fuzzy Sets Sys.* 69(2): 125–139. 1995.

407. Y. F. Yuan and H. Zhuang. A genetic algorithm for generating fuzzy classification rules. *Fuzzy Sets Sys.* 84(1): 1–19. 1996.

408. L. A. Zadeh. Fuzzy sets. *Info. Control* 8: 338–353. 1965.

409. L. A. Zadeh. The concept of a linguistic variable and its application to approximate reasoning. *Info. Sci.* 8: pp. 199–249, 301–357; 9: 43–80. 1975.

410. L. A. Zadeh. A computational approach to fuzzy quantifiers in natural languages. *Comput. Math. Appl.* 9: 149–184. 1983.

411. L. A. Zadeh. Is probability theory sufficient for dealing with uncertainty in AI: A negative view. In *Proceedings of 1st Annual Conference on Uncertainty in Artificial Intelligence*. Elsevier, pp. 103–116. 1985.

412. L. A. Zadeh. Fuzzy logic. *IEEE Comput.* 21(4): 83–92. 1988.

413. J. Zhang. Fuzzy neural networks based on rough sets for process modeling. In *Proceedings of 5th International Symposium on Instrumentation and Control Technology*. SPIE, pp. 844–847, 2003.

414. L. Zhang and S. Malik. The quest for efficient boolean satisfiability solvers. In *Proceedings of 18th International Conference on Automated Deduction*. Springer, pp. 295–313. 2002.

415. M. Zhang and J. T. Yao. A rough sets based approach to feature selection. In *Proceedings of 23rd International Conference of NAFIPS*, Banff, Canada, June 27-30. IEEE Press, pp. 434–439. 2004.

416. Y. Zhao, X. Zhou, and G. Tang. A rough set-based fuzzy clustering. In *Proceedings of 2nd Asia Information Retrieval Symposium*. pp. 401–409, 2005.

417. N. Zhong, J. Dong, and S. Ohsuga. Using rough sets with heuristics for feature selection. *J. Intell. Info. Sys.* 16(3): 199–214. 2001.

418. L. Zhou, W. Li, and Y. Wu. Face recognition based on fuzzy rough set reduction. In *Proceedings of 2006 International Conference on Hybrid Information Technology*, Vol. 1. IEEE Press, pp. 642–646, 2006.

419. W. Ziarko. Variable precision rough set model. *J. Comput. Sys. Sci.* 46(1): 39–59. 1993.

420. W. Ziarko, R. Golan, and D. Edwards. An application of datalogic/R knowledge discovery tool to identify strong predictive rules in stock market data. In *Proceedings of AAAI Workshop on Knowledge Discovery in Databases*. ACM, pp. 89–101. 1993.

INDEX

www.ingramcontent.com/pod-product-compliance
Lightning Source LLC
LaVergne TN
LVHW011910200926
848484LV00002B/581